T0074036

Publications des Archives Henri Poincaré
Publications of the Henri Poincaré Archives

Textes et Travaux, Approches Philosophiques en Logique, Mathématiques et Physique autour de 1900
Texts, Studies and Philosophical Insights in Logic, Mathematics and Physics around 1900

Éditeur/Editor: Gerhard Heinzmann, Nancy, France

New Essays on Leibniz Reception

In Science and Philosophy of Science 1800–2000

Ralf Krömer
Yannick Chin-Drian
Editors

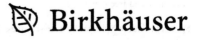

Editors
Ralf Krömer
Departement Mathematik
Universität Siegen
Siegen
Germany

Yannick Chin-Drian
Archives Poincaré
Université de Lorraine
Nancy
France

ISBN 978-3-0346-0503-8   ISBN 978-3-0346-0504-5 (eBook)
DOI 10.1007/978-3-0346-0504-5
Springer Basel Heidelberg New York Dordrecht London

Library of Congress Control Number: 2012934367

Printed on acid-free paper

Springer is part of Springer Science+Business Media (www.springer.com)

Contents

Introduction

Ralf Krömer and Yannick Chin-Drian

There is no doubt that ever since Leibniz's thinking became known, it influenced many research studies in the foundations and philosophy of logic, mathematics and exact sciences. Along with his general epistomological and metaphysical approach, there is a number of more particular concepts, problems and ideas which occupied Leibniz and later motivated much work in these fields: the concepts of continuum, space, identity, number, the infinite and the infinitely small, and the development of such constructs as a universal language, an analysis situs, a calculus of logic and so on. The Leibnizian influence has actually been so great and varied that a great part of it is still to be uncovered by historians of science and philosophy, and it has been so important for the development of modern science and philosophy that it merits being studied much closer than it has been so far. While the existing secondary literature on Leibniz's work could be said to be as vast as this work itself, only part of it concerns the Leibnizian influence, and relatively few works concern the Leibnizian influence in science and philosophy of science in the 19th and 20th centuries in particular.

It was for this reason that an international meeting entitled "Leibniz reception in the sciences and philosophy of science 1850–1950" was organized in April 2008 by the "Laboratoire d'Histoire des Sciences et de Philosophie (LHSP)-Archives Henri Poincaré" (UMR7117 CNRS) at Nancy/France. The basic idea of the meeting was to bring together two fields of research in the history of philosophy and of science: Leibniz research, and related research on history of science and its philosophy in the 19th and 20th centuries. Our aim was to characterize Leibniz's thought as it appears in the works of some authors from the latter period, and in this way to better understand Leibniz's influence on contemporary science and philosophy, but also to inspect this reception critically, in particular to confront it with the actual state of Leibniz research and publication of numerous editions of his work . The result of this effort is the present volume.

As the title suggests, ours is not the first collection of essays on Leibniz reception. Besides many isolated studies scattered around the literature,[1] another volume was devoted to such studies in 1986 [Heinekamp 1986]. We are convinced that 25 years after

---

[1] See the online-bibliography of the writings on Leibniz available at
http://www.leibniz-bibliographie.de for references.

this volume, it is a good time for a new collection of such essays to appear. In the remainder of this introduction, we shall point out the relations between the present volume and the 1986 volume whenever there is occasion to do so. A first observation is that in 1986 there were 27 contributions to cover the whole period 1716–1986, and there was no restriction concerning the topics covered, whereas our volume is restricted to the period 1800–2000 and to reception in the fields of logic, mathematics and philosophy of science – a set of restrictions under which only a handful of contributions to the 1986 volume would fall. (Our decision to concentrate precisely on this period and these fields actually was motivated by our feeling that they were underrepresented in the existing literature.) As a rule, in cases where the Leibniz reception by some author or some school has been studied before, the studies presented here cover new particular aspects and questions not in the scope of existing studies, or they use new material not known or evaluated before, or they are just more detailed than existing studies.

On the other hand, this is not (and actually, is not meant to be) a definitive study, nothing like a "handbook on Leibniz reception". A choice was made from the totality of questions relevant to Leibniz reception, even in the restricted sense explained above, and the major criterion for this choice was a very non-objective one, namely the willingness of the contacted authors to contribute, which in turn is at least partly related to the accident of knowing one of the editors. We are nevertheless convinced that the product of our collective effort constitutes an important step forward in the study of Leibniz's influence on science and its philosophy.

Let us now pass to a description of the contributions contained in this volume.[2] We organized the chapters roughly in chronological order with respect to the authors whose Leibniz reception is studied. By this structure, we hope to help the reader to read this book as a partial history of Leibniz's influence, and to follow up lines of development. Orderings of a more thematical kind would have been possible; we hope that the thoroughly prepared author and subject indexes will help the reader to access the work from particular thematical perspectives. Every chapter is followed by its own list of references (so that there is no general bibliography; see however the end of this introduction for some additional references).

In the first chapter, Philippe Séguin studies in a particular way to what degree Leibnizian viewpoints played a role in the development of the idea of number in the thinking of some 19th century German mathematicians: Gauss, Jacobi, Kummer and Dedekind. The history Séguin presents is rather a *cultural* than a *conceptual* history[3], a methodology he already employed in his thesis (supervised by Imre Toth) [Séguin 1996].

Séguin shows that German mathematics in the 19th century saw a phenomenon that today could seem surprising to us: the notion of number was ornamented with so much prestige that it became more than a simple concept, and certain German mathematicians allemands made of it what Séguin calls an "idea". Hidden behind this ennoblement is the multiple and diffuse presence of Leibniz's philosophical thinking, notably across a discipline which was the great German science from 1750 towards the end of the 19th

---

[2]The following presentation is based on abstracts provided in most cases by the authors.

[3]For the latter, see [Ferreiros 2007] and [Boniface 2007], for instance.

century: philology. This discipline considered itself as the science of the human mind. But the Leibnizian legacy has also been harvested and exceeded by the formidable vigor of German idealism which is not hostile to the conception of number as a true creation of the mind rather than a simple definition.

In Chapter 2, Volker Peckhaus studies the reception of Leibniz's logic in 19th century German philosophy. Although there is good evidence that modern mathematical logic was created in the second half of the 19th century independently of Leibnizian anticipations, it became quite early a commonplace for the proponents of the new logic that many of their ideas had been ingeniously anticipated by Leibniz. The sources of these insights were important publications in German philosophy. The central role of Johann Eduard Erdmann's edition of Leibniz's philosophical works (1839/40) and Adolf Trendelenburg's paper on Leibniz's sketch of a general characteristic (1856) must be recognized. These works conveyed Leibniz's considerations on logical calculi and his ideas about semiotics.

Chapter 3 by Françoise Willmann is on Kurd Lasswitz (1848–1910), the author of a history of atomism (*Geschichte der Atomistik vom Mittelalter bis Newton*). To this history published in 1890, critical thinking is central. Lasswitz tries to retrace, in the labyrinth of uncountable atomistic theories and their different motives, the long way towards conquest of the "thinking tools" that allowed access to modern science. A partisan of acinetic atomism, Lasswitz sets himself the task to show that the perfection of the corpuscular theory was achieved by Huygens. The passage to a dynamical theory of matter, as it can be found in Leibniz's writings, is a setback for Lasswitz. He sees it as subject to two major epistemological obstacles, anthropomorphic representations on the one hand, the metaphysical and theological interest on the other hand; according to him, these obstacles hindered Leibniz from making full use of the mathematical tools which were at his disposal.

Chapters 4 to 7 are connected by a greater common theme, namely by what could be called the "Leibniz renaissance" at the turn of the 19th to the 20th century. In Chapter 4, Erika Luciano studies the Leibniz reception by Giuseppe Peano and his school. She first analyzes the sources and the secondary literature consulted by Peano during his studies on Leibniz; reconstruction of this information has been possible due to the recent discovery of Peano's personal library. Next, she examines the second edition of the *Formulaire de mathématiques* (1898–1899), where for the first time excerpta of Leibniz's manuscripts on logic and arithmetic were published; she considers in particular Peano's marginalia and the additions and corrections due to his team of collaborators. Finally, she describes the reception in Italy of Leibniz studies by G. Vacca, G. Vailati, A. Natucci and L. Couturat, members of Peano's School, on the basis of their correspondence and of contemporary Italian periodicals.

This leads directly to the Leibniz reception by Couturat himself, studied by Anne Françoise Schmid in Chapter 5. The objective of Schmid's contribution is to show how Couturat constructed his interpretation of Leibniz, through his philosophical principles, his preference for the algebra of logic, and his conception of the history of philosophy. Among his writings, his book and his article on Leibniz as well as his edition of the unpublished manuscripts copied in Hanover remain key references in the history of philosophy. However, his interpretation of Leibniz more profoundly engages his own philosophy and allows us to understand how the former was a semi-failure for him. On the one hand, although his book on Leibniz was widely known and admired, its main thesis that logic is the heart of the system was not followed. Yet, Couturat hoped his book would carry weight in contemporary debates about logic and the philosophy of mathematics. On the other hand, his "rationalism" hampered him in his own studies on logic to the point that he progressively lost interest in keeping up with Russell's ongoing work on symbolism and refining the precision of the detail in demonstration. This situation, at the same time theoretical and biographical, had consequences in the history of this discipline because Couturat had completed two substantial works on these same questions, which are now lost. Here, Schmid reviews the criticisms made of Couturat's reading of Leibniz, and renders an account of his interpretation by hypothesising on the use he makes of logic's relations to other disciplines in Leibniz's work.

Chapter 6 by Nicholas Griffin is on the historical background and the contents of Russell's book on Leibniz. Griffin's main concern was to consider the ways in which Russell's study of Leibniz affected two positions Russell took up either immediately before or at the time he started his work on Leibniz, namely, his view that decompositional analysis was the proper method for philosophy and his rejection of the neo-Hegelian doctrine of internal relations. These views were central to Russell's rejection of the neo-Hegelian philosophy he had previously held and they paved the way for the major philosophical advances he made in the next few years. Even if Griffin stresses that Russell got neither of these views from Leibniz, he suggests that Russell's study of Leibniz was not only an event of major importance in Leibniz scholarship but of the greatest importance to the development of his own philosophy as well.

Griffin addresses two main themes: the impact of the containment principle on the possibility of contingent and/or synthetic propositions and the status of relations as internal or external. The former was of immense concern to Leibniz and is discussed in detail in Russell's book. The latter was not discussed by Leibniz, who seems to have simply accepted without dispute that all relations were what Russell would subsequently call internal, but it was of enormous importance to Russell.

Chapter 7 by Jean Seidengart is on Cassirer's reading, publishing and interpreting of Leibniz's philosophy.[4] Seidengart takes as his starting point Cassirer's presentation of Leibniz's philosophy which, according to Louis Couturat's criticism, made of Leibniz a sort of "pre-Kantian". While Couturat couldn't yet have a clear idea of Cassirer's philosophical intent when he wrote his review, Seidengart makes an effort to show that Cassirer rather aimed at correcting Kantianism with the help of Leibnizian philosophemes. Cassirer even took up so many basic elements of Leibniz's theory of knowledge in the building of his own philosophy, including the philosophy of symbolical forms, that one can, according to Seidengart, consider Cassirer's Neo-Kantianism as a comeback of Leibniz's thinking in the domain of philosophy of science, paradoxical as this might seem. Cassirer

---

[4]The importance of Leibniz to the Marburg school and to Cassirer in particular has been stressed and analyzed before, to the point that [Holzhey 1986] suggested that one could even speak about a "Neo-Leibnizianism" instead of a "Neo-Kantianism" in this case (p. 289). But while Holzhey discusses Cassirer only shortly, the more extensive study [Ranea 1986] is largely complementary to the one by Seidengart presented here.

is thus treated not as a simple historian but as a genuine philosopher in his reception of Leibniz's philosophy.

In Chapter 8, Vincenzo De Risi analyzes the debate between Hans Reichenbach and Dietrich Mahnke on Leibniz's Theory of Motion and Time. In the 1924 issue of *Kantstudien*, Hans Reichenbach published an article about Leibniz's theory of motion which was aimed at presenting Leibniz as a forerunner of Einstein's Theory of Relativity. In the following months, Reichenbach had an exchange with Husserl's disciple Dietrich Mahnke (who had published a number of books and articles on Leibniz in the preceding decades) about the views stated in his 1924 paper. This short Reichenbach-Mahnke correspondence is printed for the first time in the present volume as an appendix to De Risi's paper. The main issues treated in the correspondence are Leibniz's so-called causal theory of time and his quarrel with Newton on the relativity of motion and space. Reichenbach's reading of Leibniz's texts is strongly influenced by his own physicalistic and logicistic views, while Mahnke's verges towards pure phenomenology. The comparison and contrast between these two radically opposite interpretations, which intertwine with the first philosophical reactions to Einstein's theory of relativity, shed light both on Leibniz's peculiar way of addressing the problems of time and motion and on some aspects of the neo-empiricist and phenomenological ways of thinking. In this discussion of Leibniz's theory one can also appreciate the arguments that pushed Reichenbach away from the neo-Kantianism of his young years towards a perfected empiricist position, as well as those that eventually prompted Husserl to disown Mahnke's phenomenological enterprise.

In Chapter 9, David Rabouin analyzes some interpretations of Leibniz's *Mathesis universalis* at the beginning of the 20th Century. Rabouin starts with Dietrich Mahnke's 1922 thesis entitled "Leibnizens Synthese von Universalmathematik und Individualmetaphysik" and proposing a masterly panorama of the various interpretations of Leibniz of Mahnke's time. As the title suggests, Mahnke in these interpretations found a tension between two projects, that of a "universal mathematics" and that of a "metaphysics of individuation". Concerning the former project, Mahnke distinguished a central opposition between "logicist" interpretations (B. Russell, L. Couturat) and an anti-logicist tendency characterized by the attention payed to the role of the invention of differential calculus and thus to the connection of mathematics to natural sciences (E. Cassirer, L. Brunschvicg). Rabouin tries to reexamine these interpretations by deploying the challenges of these oppositions. But he also confronts these readings with Leibniz's original projects, better known today. Rabouin thus shows that these conflicts teach us more about the attempts to recover Leibniz, and more generally about the way in which the end of the 19th century positions itself with respect to the alleged legacy of "classical rationalism", than about Leibniz's own challenges.

In Chapter 10, Erhard Scholz is looking for Leibnizian traces in Hermann Weyl's *Philosophie der Mathematik und Naturwissenschaft*. Weyl's reading of Leibniz has been analyzed before [Breger 1986]; consequently, Scholz does not concentrate on the theme stressed by Breger, namely Leibniz's theory of the continuum. Neither does Scholz pretend to give a systematic evaluation of Weyl's way of presenting Leibniz. Coming from a background in the history of mathematics, Scholz just tries to present those aspects which apparently made Leibniz so important for Weyl in the middle of the 1920s. He basically stresses two such aspects, namely Weyl's use of Leibnizian ideas in his discussion of Hilbert's foundational program in the light of symbolic mathematics, and Weyl's use of Leibniz as a dialogue partner for understanding modern physics.

In Chapter 11, Gabriella Crocco presents an overview of what is known about Kurt Gödel's reading of Leibniz. The only published work in which Gödel explicitly mentions Leibniz's work is "Russell's mathematical logic" edited in 1944 by Schilpp for the Library of Living Philosophers. Crocco first presents briefly all other available evidence (including the published and unpublished material of the Gödel archives and the transcriptions of Gödel's conversations with Hao Wang in the 1970s) which prove the deep influence of Leibniz on Gödel's philosophical reflections and in his scientific work. Crocco then shows how, on the basis of the unpublished material, the intricate structure of Gödel's paper on Russell reveals a very clear and Leibnizian conception of the nature of logic and of the problems to be solved in order to achieve, in modern terms, Leibniz's program of the Characteristica Universalis.[5]

The book closes with Chapter 12 wherein Herbert Breger presents Gregory Chaitin's reading of Leibniz. Chaitin proved some interesting theorems in the mathematical theory of automata, developing ideas from Gödel (incompleteness theorem) and Turing (halting problem) in a different direction. Chaitin defined a real number Ω, the halting probability. The bits of Ω are logically irreducible; they cannot be deduced from axioms simpler than they are. Later Chaitin declared (evidently rhetorically) that he had been anticipated by Leibniz; Leibniz's distinction between "simple law" and "complicated law" (and not the distinction between "law" and "without any law") was the crucial idea. Chaitin took this as a starting point for his interest in Leibniz. In his general philosophical ideas Chaitin considers the computer as a paradigm for mathematics (just as 17th century mathematics took mechanism as a paradigm). Referring to the computer, Chaitin defines complexity as a mathematical notion. According to him, mathematics is a highway amidst a jungle; mathematicians should accept new axioms in order to be able to leave the highway. Chaitin's ideas, which can easily be combined with Lakatos's quasi-empiricism, have been integrated by others in a "digital philosophy" for mathematics and natural sciences.

It is clear that given this table of contents, our book leaves many questions untouched which would naturally belong to the overall theme of Leibniz reception in science and philosophy of science in the 19th and 20th centuries. We hope that others will say something on these questions. As to the literature already existing, we desisted from the task of producing an exhaustive bibliography of the topic, being convinced that such tools nowadays are better implemented electronically than printed. The reader is thus invited to browse the above-mentioned database, for example by looking up authors like Frege, Grassmann, Husserl, or Robinson.

---

[5]For a recent discussion of another aspect of Gödel's reading of Leibniz, see [van Atten 2009].

Acknowledgement

We wish to thank some people and institutions for their support during the production of the present volume. First of all, we thank the editor of the series "La science autour 1900", Gerhard Heinzmann, for his willingness to publish this work in the series, and the *Laboratoire d'Histoire des Sciences et de Philosophie (LHSP)-Archives Henri Poincaré* (UMR7117 CNRS) at Nancy, especially its director Roger Pouivet as well as Anny Bégard, Pierre-Edouard Bour and Lydie Mariani from its staff, for help with the organization of the 2008 meeting and for financial support of this volume. Other institutions having contributed financially to both the meeting and the volume are the *Institut Universitaire de France*, the *Communauté Urbaine du Grand Nancy*, the *Conseil Général de Meurthe-et-Moselle* and the *Université Nancy 2* (in particular its *Département de Philosophie*). Special thanks for their hospitality and creativity go to the staff at the *Goethe-Institut* at Nancy.

We wish to thank further the following colleagues for having contributed to the scientific quality of this volume by their critical remarks and helpful comments: Wolfgang Bonsiepen, Hélène Bouchilloux, Christophe Bouriau, Michel Fichant, Dominique Flament, Martine de Gaudemar, Philippe De Rouilhan, Didier Galmiche, Hartmut Hecht, Thomas Leinkauf, Eberhard Knobloch, Timothy McCarty, Alain Michel, Gregor Nickel, Helmut Pulte, Oliver Schlaudt, Francois Schmitz and Paul Ziche.

We also thank the staff at Birkhäuser, in particular Karin Neidhart, Thomas Hempfling, Anna Mätzener and Edwin Beschler, for their constant support during the production process.

Chapters 3 and 7 have been translated by Vivian Waltz, Chapter 5 by Garry White.

And finally, we thank our contributors for their collaboration and patience. We hope that our collective product justifies their efforts.

Siegen and Nancy, June 2011
Ralf Krömer and Yannick Chin-Drian

References

[Boniface 2007] Boniface, Jacqueline, "The concept of number from Gauss to Kronecker", in [Goldstein, Schappacher, Schwermer 2007] p. 315–342.

[Breger 1986] Breger, Herbert, "Leibniz, Weyl und das Kontinuum", in [Heinekamp 1986] p. 316–330.

[Cristin Sakai 2000] *Phänomenologie und Leibniz*, eds. R. Cristin and K. Sakai, Freiburg 2000.

[Ferreiros 2007] Ferreirós, José, "Ὁ Θεὸς Ἀριθμετίζει: The Rise of Pure Mathematics as Arithmetics with Gauss", in [Goldstein, Schappacher, Schwermer 2007] p. 235–268.

[Goldstein, Schappacher, Schwermer 2007] Goldstein, Catherine; Schappacher, Norbert; Schwermer, Joachim (eds.), *The shaping of arithmetic after C.F. Gauss's Disquisitiones Arithmeticæ*, Springer 2007.

[Heinekamp 1986] Heinekamp, Albert (ed.), *Beiträge zur Wirkungs- und Rezeptionsgeschichte von Gottfried Wilhelm Leibniz (= Studia Leibnitiana, Supplementa XXVI)*, Wiesbaden 1986.

[Holzhey 1986] Holzhey, Helmut, "Die Leibniz-Rezeption im 'Neukantianismus' der Marburger Schule", in [Heinekamp 1986], p. 289–300.

[Ranea 1986] Ranea, Alberto Guillermo, "La réception de Leibniz et les difficultés de la reconstruction idéale de l'histoire de la science d'après Ernst Cassirer", in [Heinekamp 1986], p. 301–315.

[Séguin 1996] Séguin, Philippe, *Vom Unendlichen zur Struktur: Modernität in Lyrik und Mathematik bei Edgar Allan Poe und Georg Cantor*, Peter Lang 1996.

[van Atten 2009] van Atten, Mark, "Monads and sets. On Gödel, Leibniz, and the reflection principle", in Rahman, Shahid, Primiero, Giuseppe (eds.), *Judgement and Knowledge. Papers in honour of B.G. Sundholm*, London: King's College Publications 2009, p. 3–33.

The Idea of Number from Gauss to Cantor. The Leibnizian Heritage and its Surpassing

Philippe Séguin

1. Introduction: number as an idea

In a letter to David Hilbert, datelined January 27, 1900, Georg Cantor takes a position against Dedekind's opinion that numbers are "*free* creations of the human mind" ("free" underlined by Cantor), and the editors comment as follows:

> In der Überzeugung, daß Zahlen auch transiente Realität besitzen [...], besteht denn auch die wesentliche Differenz zu Dedekind, dem die Zahlen "freie Schöpfungen des menschlichen Geistes" sind. Es ist der Gegensatz zwischen dem "Finden oder Entdecken" einerseits und dem "Erfinden oder Schöpfen" andererseits, der in den Auffassungen dieser beiden Mathematiker zum Ausdruck kommt. [Cantor 1991, 428–429]

They refer, as expected, to a very often quoted passage from the *Grundlagen einer allgemeinen Mannigfaltigkeitslehre* (1883), in which Cantor states that "the essence of mathematics just consists of its freedom" ("das *Wesen* der *Mathematik* liegt gerade in ihrer *Freiheit*") [Cantor 1932, 182]. I was very surprised, because I thought to have read in Cantor that he had created a new kind of numbers, which he called transfinite. A look at § 8 of the *Grundlagen* confirmed indeed that Cantor underscores in a rather exalted style the mathematician's freedom in introducing ("einführen") new numbers, since he is only bound by the non-contradiction of his concepts and by the correctness of his definitions [Cantor 1932, 182], but he doesn't mention at all a power of creation. Nevertheless, precisely in the strictly mathematical part, when he shows in §§ 11 and 12 how "you are led to definitions of new numbers", Cantor makes use of the following expressions: for the first principle of production (+1), "formation" ("Bildung"), for the second principle (omega as a limit), "the newly created number" ("die neu geschaffene Zahl") [Cantor 1932, 195]. Finally, after some neutral terms like "Einführung" and "Bildung" [Cantor 1932, 196], he comes at the end of § 12 to recapitulate his reasoning by stating clearly that he proceeds to "the creation of a new integer" ("die Schöpfung einer neuen ganzen Zahl") [Cantor 1932, 199].

    In this paper I would like to show that the creationist conception of the number notion proceeds from a tradition of German mathematical thinking, which considers the number as something higher than a concept, which I shall designate as an "idea". This conception is in my view related indirectly but by multiple links to Leibniz' philosophical thinking as well as to its surpassing, notably by the German new humanism which came to light in the middle of the 18th century, and by what we used to call German idealism. To this end I shall consider and analyse statements of mathematicians and fundamental texts, particularly by authors like Gauss, Jacobi, Kummer and Dedekind, whose names and works are quoted by Cantor.

2. Gauss (1777–1855), number theory and science for itself

Let us begin with two Gauss quotations, the one about life in general, the other about mathematics. In 1802, Gauss is just 25 years old, he writes to his former fellow student Wolfgang Bolyai:

> Möge der Traum den wir das Leben nennen, dir ein süßer seyn, ein Vorge-schmack des wahren Lebens in unsrer eigentlichen Heymath, wo den erwach-ten Geist nicht mehr die Ketten des trägen Leibes, die Schranken des Raums, die Geissel der irdischen Leiden und das Necken unserer kleinlichen Bedürf-nisse und Wünsche drückt. [Gauss 1899, 47]

    It is noteworthy that Gauss does not refer to the salvation of the soul, but to a problem of knowledge, since he mentions the deliverance of the mind, not of the soul, from the limits of space, not of the world. Of course we know that God did not create space ("Raum"), he created the world ("Gott schuf die Welt"). But for Gauss, space is more interesting, because it not only represents the world of divine creation, but it is the product of man's need for knowledge as well, therefore a product of the human mind, at least up to a certain point. Indeed, here it is what he wrote to Bessel on the 21st of January 1829 (this passage follows the well-known quotation about "das Geschrei der Böotier"):

> Nach meiner innigsten Überzeugung hat die Raumlehre in unserem Wissen a priori eine ganz andere Stellung, wie die reine Grössenlehre; es geht unserer Kenntnis von jener durchaus diejenige vollständige Überzeugung von ihrer Nothwendigkeit (also auch von ihrer absoluten Wahrheit) ab, die der letzteren eigen ist; wir müssen in Demuth zugeben, dass, wenn die Zahl bloss unseres Geistes Product ist, der Raum auch ausser unserem Geiste eine Realität hat, der wir a priori unsere Gesetze nicht vollständig vorschreiben können. [Gauss 1899, 497]

    Space is higher than number, which is "only" our mind's product, precisely because its reality is outside our mind. This declaration would not be surprising if it were not from Gauss. In fact, one year before writing the first letter, 1801, Gauss had published the *Disquisitiones arithmeticae*, with which number theory started again on a totally new basis, and Gauss himself became famous in the whole of mathematical Europe. In addition to this, we know many statements of Gauss himself: mathematics is the queen of science, number theory the queen of mathematics. To Lejeune-Dirichlet he confides that number theory is his favourite domain [Gauss 1863–1933 vol. 2, 515], and that he greatly prefers people who cultivate science for itself [Biermann 1975, 165]: from these assertions one can deduce that Gauss cultivates numbers and their theory for themselves. Therefore, what exclusively comes from the human mind is superior for Gauss, and it is number.

Of course we can connect this conception with the situation of mathematics in that time: particularly since the end of the 18th century, many problems have become more and more urgent: what was called "the parallel's problem", the foundations of analysis, the resolution of equations with radicals for instance. But Gauss's expression, "cultivate science [not "sciences", P. S.] for itself", goes beyond the scope of mathematics, and this attitude finds a confirmation in a famous quotation of a younger colleague of Gauss, which we will now analyse and comment, and which responds to a specific cultural background.

3. Carl Gustav Jacobi (1804–1851) and the number's honour

3.1. Philology and the unity of science

People interested in history of mathematics know this saying attributed to Jacobi: "Why are we doing mathematics? For the honour of the human mind." It belongs to the set of quotations which the public enjoys, particularly if this public consists of mathematicians, and which you can read for instance in Jeremy Gray's book about Hilbert [Gray 2000, 168]. But most of the time the quotation is wrong: Jacobi did not write "mathematics", but "science", and he wrote "Mr Fourier [...]should have known", and not "he should understand", as a renowned mathematician said one time. The exact quotation is to be found in a letter to the mathematician Legendre. Therefore it is in French:

> M. Fourier avait l'opinion que le but principal des mathématiques était l'uti-
> lité publique et l'explication des phénomènes naturels ; mais un philosophe
> comme lui aurait dû savoir que le but unique de la science, c'est l'honneur
> de l'esprit humain, et que sous ce titre, une question de nombres vaut autant
> qu'une question du système du monde. [Jacobi 1881–1891 vol. 1, 454]

Jacobi puts world and numbers on the same level. But why should Fourier have known this? After some research, the answer looks easily given: Jacobi was in possession of the *Histoire des Mathématiques depuis leur origine jusqu'à l'année* 1808 by Charles Bossut, a friend and collaborator of d'Alembert, who also wrote a successful handbook about calculus and was a colleague of Fourier at the Ecole polytechnique. And in his book about history of mathematics, Bossut asserts that calculus is part of the great achievements which contribute to the honour of the human mind [Bossut 1810 vol. 2, 3]. How could Fourier not have known this? As a matter of fact, it is not important whether Fourier knew this or not, because what Jacobi considers as evidence does not come from French thinking, it is grounded in a German tradition which was very strong in his time, the great German science in the second half of the 18th and during most of the 19th century, and which has now disappeared: it is between this science and mathematics that Gauss and Jacobi have been hesitating.

In Gray's book quoted above we can read that the first mathematics seminar at the university of Königsberg had been organized after a successful field in Germany in those times, linguistics [Gray 2000, 15]. It is only partly true, the notion of linguistics hides a much bigger reality. The science which had awakened Germany intellectually about 1750 was called philology. It had allowed Germany and its rising middle class, which was lacking any political power until the end of the 19th century, to revive the ideals of Greece and Rome. It had filled Gauss and Jacobi in such a manner that they had been hesitating about the choice of their profession. In the case of Gauss this hesitation was understandable: the very young university of Göttingen, which had been founded in 1734, had become rapidly a centre of attraction in the whole of Europe thanks to the work and the personality of such scholars as Christian Gottlieb Heyne, Johann Matthias Gesner and others, the grounding fathers of German philology, whereas Gauss had to attend the lectures of the professor of mathematics Abraham Kästner, who couldn't compete with the great mathematicians of his time, such as d'Alembert, Euler, Lagrange, etc.

If philology began with the study, editing and translation of old texts, most in Greek and Latin, it became in the beginning of the 19th century an encyclopaedia of all human knowledge, *the* science of the human mind, with August Boeckh (1785–1867). Jacobi was a brilliant student in his seminar, and Karl Weierstrass, who was his colleague in Berlin, called him in 1873, six years after his death, "our unforgettable Boeckh" [Weierstrass 1894–1927 vol. 3, 334].

Today, even educated germanists don't know Boeckh's name. In a lecture he has been giving during more than fifty years, Boeckh refers explicitly to Leibniz in order to define his subject and to demarcate it from philosophy:

> Leibniz, der unter allen Philosophen am meisten Philologe und Gelehrter war, verbindet mit dem Worte *Erudition* ungefähr den Sinn, welchen wir der Philologie beilegen; die Erudition hat es nach seiner Ansicht mit dem zu thun, was wir von den Menschen lernen, *quod est facti*, die Philosophie mit dem *quod est rationis sive juris.* [Boeckh 1877, 25]

In other words, "die eigentliche Aufgabe der Philologie [ist] das *Erkennen* des vom menschlichen Geist **Producirten**, d.h. des *Erkannten*" [Boeckh 1877, 10]. But Boeckh revivifies above all Leibniz's ideal which he mentions several times, unity in diversity, unity of all knowledge of the human mind. That it is, "the science", which is "the honour of the human mind". It is the heritage of the great Leibnizian project, which rules out nothing, where metaphysics and religion have their own place beside art, sciences etc. Jacobi is not frightened at using the word "revelation" ("Offenbarung"), which comes from the vocabulary of religion, when he speaks of science, for example in a letter of the year 1824 [Lejeune-Dirichlet 1889–1897 vol. 2, 248], but he is above all a mathematician, and one of the theses he defended at his graduation was: "Der Begriff der Mathematik ist der Begriff der Wissenschaft überhaupt. Alle Wissenschaften müssen daher streben, Mathematik zu werden." [Jacobi 1881–1891 vol. 3, 44]. Kurt Biermann quotes this aphorism [Biermann 1975, 378], but in doing so he forgets an important detail: it is not by Jacobi, but by Novalis, the romantic poet to whom we owe the "blue flower". Jacobi hints expressly at Novalis ("Egregie asserit Novalis poëta" [Jacobi 1881–1891 vol. 3, 44]), hence it doesn't seem to have shocked the jury. Nevertheless, the reference to Novalis could be surprising, since this fragment means nothing more than a very famous declaration by Kant in the *Metaphysische Anfangsgründe der Naturwissenschaft* ("[...] so wird Naturlehre nur so viel eigentliche Wissenschaft enthalten, als Mathematik in ihr angewandt werden kann" *Vorrede, IX*). But in a letter to his brother, Jacobi has recourse to Novalis in the same way in order to justify the happiness he feels to have chosen mathematics as a lifetask ("Das Leben der Götter ist Mathematik, sagt Novalis mit Recht, denn mein Leben jetzt ist das Leben der Götter" [Koenigsberger 1904, 118]). What does this enthusiasm mean for the romantic poet?

3.2. Novalis and mathematics as a religion

A partial answer is given to us by a poem quoted in Tobias Dantzig's book *Number, the Language of Science* [Dantzig 1930, 179], as being by Jacobi, but which is in reality a parody of a well-known poem by Schiller, who was one of the founders of German idealism with the philosopher Fichte: both had tried, in their own way, to surpass the Kantian criticism in order to go back to a unique, unified knowledge. Leibniz's mark is to be found as well in Schiller as in Fichte, even if it is not obvious in the first's work, and dissimulated in the second's. Jacobi takes over Schiller's idea that art is divine, but that it had been so before the world existed, because "What in the cosmos thou seest is but the reflection of God, / The God that reigns in Olympus is Number Eternal." [Dantzig 1930, 179]. Then for Jacobi there is undoubtedly a unity between the numbers (mathematics) and the world, but the numbers have priority, and they are at the same time divine and human as products of the human mind. Actually, Jacobi is a perfect idealist: of course Kant was convinced of the mind's superiority over matter, but for Schiller, Fichte, Novalis, Hegel and for Jacobi, the mind's superiority and "honour" have a divine dimension which Kant, the great separator of science and metaphysics, did not recognize.

It is not certain whether Jacobi knew the origins of Novalis's nowadays strange ideas about mathematics. People knew of course that Novalis, like Alexander von Humboldt, had attended the prestigious Bergakademie at Freiberg in Saxony, where the geologist Abraham Werner was teaching. But the editors of his work had set him up as a romantic dreamer and deleted the scientific and philosophical aspects of his activity. His readers could only recognize the influence of Frans Hemsterhuis's thinking on him, particularly the relation between mathematics and divinity. Actually, Novalis was really initiated into the mathematics of his time, calculus, which he learned with Bossut's handbook, and in 1798 already, in the year of the translation into German of Lagrange's *Théorie des fonctions analytiques*, he had the book in his possession. He was impressed in such a way by the theory of the development in power series that the notion of "Potenzenreihe" (nowadays "Potenzreihe") is at the very center of the founding fragment of his romanticism theory [Novalis 1960 vol. 2, 545]. But in addition to this he was fascinated by the theory of the combinatorial school of Carl Friedrich Hindenburg (1741–1808), whose goal was nothing other than to found analysis, and then the whole of mathematics, on one formula, the formula of the multinome. And the fact is that Hindenburg constantly refers to Leibniz, so that he tries to pass off his theory, combinatorial analysis, for the realization of Leibniz's *ars combinatoria*. Here is the rather overbearing result of his research at the end of his founding treatise, *Der polynomische Lehrsatz, das wichtigste Theorem der ganzen Analysis* (1796):

> Das mag genug seyn, den Nutzen der Einführung einer *allgemeinen Charak-teristik* von fest gesetzter unabänderlicher Form und Bedeutung zu bewähren; [...] Irre ich mich nicht, so habe ich das, was *Leibniz* von einer *wahren* und *ächten* Verbindungskunst fordert [...] nach Möglichkeit erreicht. [Hindenburg 1796, 302]

He did not doubt being capable of giving a universal method in order to solve all mathematical problems, and what is more, the problems of all sciences. The conclusion of *Der polynomische Lehrsatz* is worth being quoted:

> Die *combinatorische Analysis* hat endlich den Schleyer aufgedeckt, und es bleibt hinfort nicht mehr dem blinden Ungefähr überlassen, *ob* und *wenn* es der Legem naturae herbeyführen will. Die Spur, auf welcher die Göttin wandelt, ist hier überall deutlich vorgezeichnet, und kann man sie nunmehr festen und sichern Fußes verfolgen. [Hindenburg 1796, 304]

The middle-aged Hindenburg writes like the young romantics Novalis and Schlegel and the idealists Schiller and Fichte. Novalis was so much under the influence of Hindenburg's combinatorial ideal, that he set about compiling an encyclopeadia, although not along the line of d'Alembert's descriptive kind, but after a combinatorial scheme, all branches of knowledge answering each other. Novalis's goal was that all fields fructify the others in order to produce new ideas, the combinatorial art being a new *ars inveniendi*.

Of this attempt to an absolutization of human knowledge which rested particularly upon mathematics, physics, chemistry etc., we can see a first draft in a thick collection of fragments called *Brouillon général*, but it was itself soon surpassed and embedded in a far more universal project. This project, in which mathematics and religion had a chief part, was called magic idealism by Novalis, and we have to understand certain enigmatic fragments, which were called after Novalis's death *Hymns to mathematics*, in relation to Novalis's design ("Die reine Mathematik ist die Anschauung des Verstandes, als Universum." [Novalis 1960 vol. 3, 593] "Reine Mathematik ist Religion." [ibid., 594]). Part of these *Hymns* are the two aphorisms quoted above by Jacobi, and other fragments were quoted later on by Pringsheim, Minkowski and other mathematicians. Unlike Novalis, Jacobi does not seem to have been very interested in religion, but it is not the case of all German mathematicians of his time. Ernst Eduard Kummer for instance, the successor in Berlin of Jacobi, whom he very much revered, was like him intellectually with a wide range of interests, but in addition to that he put religion at the top of man's activities.

4. Kummer (1810–1893), religion and the ideal of freedom

Kummer is particularly interesting for us because he was not only a first rate mathematician who dedicated most of his research to number theory, but he also was like Novalis a child of Protestantism and idealistic philosophy, and he did not disavow them. It is very surprising to read in a letter of the year 1842 to Kronecker (Kronecker was nineteen, Kummer thirtytwo), that he considered Schelling, who at the age of sixtysix had just been appointed at the university of Berlin, as the greatest philosophical personality still living

[Kummer 1975 vol. 1, 78]. But let us first observe that Schelling, in his attempt to surpass Kant, had been looking for philosophical inspiration in Leibniz. Here is what he wrote in one of his major works, *Ideen zu einer Philosophie der Natur* (1797):

> Die Zeit ist gekommen, da man seine Philosophie wieder herstellen kann. Sein Geist verschmähte die Fesseln der Schule, kein Wunder, daß er unter uns nur in wenigen verwandten Geistern fortgelebt hat und unter den übrigen längst ein Fremdling geworden ist [...]. Er hatte in sich den allgemeinen *Geist der Welt*, der in den mannigfaltigsten Formen sich selbst offenbart und wo er hinkommt, Leben verbreitet. [Schelling 1976 vol. 5, 77]

Can we imagine Kummer's letter to Kronecker in the hands of German physicists, chemists, physiologists of that time, who looked upon Schelling's and Hegel's from Pre-romantism proceeding "Naturphilosophie" as the worst obscurantism? Besides, wasn't the title of Schelling's last great work, *Philosophie der Offenbarung*, (Offenbarung = revelation) the best confirmation for their opinion? But like Jacobi, Kummer wasn't afraid of the word "revelation", and a proof of it is a passage of a speech made in 1848 about academic freedom, in which he speaks of the divine as the origin of the human mind's freedom and of the "revelations of the divine" as the guides of its aspirations [Kummer 1975 vol. 2, 710]. For Kummer, religion, philosophy, sciences, and especially his domain, mathematics, are a whole, they are "revelations of the divine". In that what he writes to his mother when he is only eighteen, what he declares constantly until the presentation of Jacobi's works in 1881, what he gives as a justification in his introduction of the complex divisors, it is freedom which guides his mind, freedom with which god inspired him, the whole of mankind, that is the human mind. The human mind, he writes on the same page, is "the infinite divine contents of its freedom", therefore it makes man capable of making "justice, art, religion and sciences". But referring like Jacobi to a poem by Schiller, Kummer does not only mean that "freedom is the most inward essence of the mind itself" ("die Freiheit [ist] das innerste Wesen des Geistes selbst"), he adds to this a historical perspective inherited from Hegel, whom he revered, saying for instance that the final end is "absolute freedom, therefore the divine itself", because "universal history" should be recognized as "a work of the divine mind as well as of the human mind" [Kummer 1975 vol. 2, 710]. Nevertheless we have to underscore that the expression "human mind", which Kummer often makes use of, does not come from Hegel, who does not use it.

Kummer wrote these lines in 1848, the year of the last uprising of the German middle class. The idea of a liberation applied to mathematics is also to be found in 1866 in an address to the king, therefore without technicity, where he hints at "the idea of the liberation of the quantity concept of the limits [...] which are not essential to it and then disturbing". [Kummer 1975 vol. 2, 786] As an example, and in order to remain understandable, he mentions calculation with letters (algebra), which he qualifies as a "first grand creation". [ibid.] In the same way he points to "the creative power of the mind" in his introductory note about Jacobi [Kummer 1975 vol. 2, 696], and later his biographer, K. Hensel [Kummer 1975 vol. 1, 39], as well as Dedekind [Dedekind 1930–1932 vol. 3, 490], did not hesitate to speak of "the creation of ideal numbers" by Kummer. However, it is remarkable that Kummer himself, although he gave a name to his complex divisors and surely had the feeling of having "created" new numbers, since he first called them "ideal complex numbers", then more simply "ideal numbers", did not go so far as to name himself a creator. Here is the quotation:

> Es ist mir gelungen, die Theorie derjenigen complexen Zahlen [...] zu vervollständigen und zu vereinfachen; und zwar durch Einführung einer eigenthümlichen Art imaginärer Divisoren, welche ich ideale complexe Zahlen nenne; [Kummer 1932 vol. 1, 203]

Four times on this first page of *Zur Theorie der complexen Zahlen* (1847), Kummer makes use of the term "introduction" ("Einführung"), even when he connects his new theory with the theory of Gauss, the *princeps mathematicorum*, who was still living. Then, why "ideal"? Isn't it a reminiscence of Schiller, Hegel and Boeckh, all representatives of this "higher" current of thinking, inherited from Leibniz but surpassed by German idealism, to which he belonged as well, and the goal of which was the liberation of the mind? Indeed, ideal numbers allow one to extend a well-known property of the integers, the unique divisibility in prime factors, to other numbers, the complex integers, therefore allowing the mind an ever bigger endeavour of generalization to free itself of ever more non-"essential" impediments. But one could also be reminded of these lines of Leibniz about imaginary numbers:

> Verum enim vero tenacior est varietatis suae pulcherrimae Natura rerum, aeternarum varietatum parens, vel potius divina Mens, quam ut omnia sub unum genus compingi putiatur. Itaque elegans et mirabile effugium reperit in illo Analyseos miraculo, idealis mundi monstro, pene inter Ens et non-Ens Amphibio, quod radicum imaginarium appellamus. [GM 5, 357]

Leibniz's ideal world, this manifestation of the divine mind, is consistent with Kummer's religious conception, as the thinking movement, in its historicity, goes through mathematics with the invention (the creation?) of algebra, then of Newton's and Leibniz's calculus etc. in order to celebrate the human mind's ascension to an ever greater, and particularly to an ever higher freedom. We know that Dedekind himself made Kummer's ideas his own and laid the foundation to a new theory which he designed as "theory of ideals". But instead we are going to consider his work with real numbers.

5. Dedekind (1831–1916) "en état de Créateur absolu"[1]

In an often quoted letter of Dedekind to Heinrich Weber of 24th January 1888 we can read (recalling that Gauss was an inventor, or creator, with Wilhelm Weber, of the telegraph):

> Wir sind göttlichen Geschlechtes und besitzen ohne jeden Zweifel schöpferische Kraft nicht bloß in materiellen Dingen (Eisenbahnen, Telegraphen), sondern ganz besonders in geistigen Dingen. [Dedekind 1930–1932 vol. 3, 489]

---

[1][Novalis 1960 vol. 3, 415].

If Dedekind had been content with expressing his opinions about mathematics confidentially, it would have been of no consequence, but it was the contrary, and this makes Dedekind all the more interesting. Unlike Jacobi and Kummer, Dedekind almost never deals with philosophy, but rather shows no interest in it, and even exhibits a full misunderstanding. Once he mentions Fichte in a letter to his sister, just to turn to ridicule the latter's supposed pretension to deduce the world from man's mind alone [Scharlau 1981, 33]. Dedekind loved music, romantic music, not Wagner, and he frequented the house of Dirichlet, who was a musician too. Neither Dirichlet nor Dedekind seem to have understood Jacobi's philosophical interests, but it is precisely Dedekind, who clearly puts man into God's place; he makes of him a creator.

To the question "what are numbers and what is their meaning?" (*Was sind und was sollen die Zahlen?* 1888) he answers in a calm and collected manner in the preface: "Numbers are the free creation of the human mind." [Dedekind 1930–1932 vol. 3, 335]. In doing this, he confirms the motto, "αει ο ανθροπος αριθμετιζει" [ibid.], a parody of the Greek aphorism: "God always geometrizes." The paradigm of the truth is no longer geometry, but numbers, which are creations of the human mind. Not only did Dedekind make this statement, but he put it into practice already in 1872, since he names the fourth chapter of *Continuity and irrational numbers* (*Stetigkeit und irrationale Zahlen*) "Creation of the irrational numbers" [Dedekind 1930–1932 vol. 3, 323], and in this place he formulates his famous definition of the cut:

> Jedesmal nun, wenn ein Schnitt (A_1, A_2) vorliegt, welcher durch keine rationale Zahl hervorgebracht wird, so **erschaffen** wir eine neue, eine irrationale Zahl a, welche wir als durch diesen Schnitt (A_1, A_2) vollständig definirt ansehen [Dedekind 1930–1932 vol. 3, 323].

In doing so and in conformity with what he had declared without pathos in the first chapter entitled "Properties of the rational numbers" [Dedekind 1930–1932 vol. 3, 317], he placed himself at the end of an evolution of man's creativity: according to Dedekind, the human mind creates by counting first the succession of the positive integers, then, in going over their own limits, negative integers, then rational numbers, thereupon coming to the notion of a number field, which was created by Dedekind himself, and eventually to the irrational numbers. On this level, Dedekind stands at the end of an evolution in possession of absolute knowledge, in a position not unlike Hegel's. In that time Cantor does not write about creation. It will be the case much later, as we have seen it, without Dedekind's spontaneity.

6. Conclusion: Leibniz and no end

We have arrived at the provisory end of our exploration. I hope to have shown that the idea of creation in mathematics is an indirect heritage, and the surpassing of this heritage, of the Leibnizian conception's capacity of the human mind to encompass the whole of knowledge, this knowledge having been produced, created by the human mind itself. The notion of number, as well as all creations of the human mind (Boeckh) which lead man to an ever greater freedom (Hegel), to ever higher harmonies, according to a poem by

Schiller, *The artists* (*Die Künstler*), which was in the 19th century known by heart by all members of the educated German middle class (Bildungsbürgertum), has become for some German mathematicians something higher, an idea, like justice, like freedom. To bring, provisionally, I repeat, to an end, these reflections about the rising of the idea of number, and since I mentioned Hegel in connection to Dedekind, let us quote some lines out of the *Vorlesungen über die Philosophie der Geschichte*:

> Daß die Weltgeschichte dieser Entwicklungsgang und das wirkliche Werden des Geistes ist, unter dem wechselnden Schauspiele ihrer Geschichten – dies ist die wahre *Theodizee*, die Rechtfertigung Gottes in der Geschichte. Nur *die* Einsicht kann den Geist mit der Weltgeschichte und der Wirklichkeit versöhnen, daß das, was geschehen ist und alle Tage geschieht, nicht nur nicht ohne Gott, sondern wesentlich das Werk seiner selbst ist. [Hegel 1969–1971 vol. 12, 540]

François Châtelet, in his little book about Hegel, remarks that this "théodicée" is rather a "noodicée", that is a mind's justification [Châtelet 1968, 162]. For our problematic discussion it seems indeed hardly doubtful that the likes of Gauss, Jacobi, Kummer, Dedekind, maybe even Cantor, who were all children of the mind's divination arising out of Fichte's and Schiller's idealism, perceived numbers as the offspring of their interiority, and considered their activity as mathematicians as a calling for mankind, the whole of mankind, like the artists in Schiller's poem. But in 1872, the real number's year of birth, this way of thinking was already outdated. Boeckh had died in 1867, having been fought and eventually side-lined by the philologist Theodor Mommsen, who was a specialist on Rome and the law, not like Boeckh on Greece and knowledge. He was a convinced nationalist and fervent backer of German unity. In 1870, Bismarck did not only realize Germany's unity. He also gave the German liberal middle class to understand that nobleness of mind was of little value compared to the nobility of the great Prussian landowners. The human mind (der menschliche Geist) had to admit its defeat before the triumphant march (der Siegeszug!) of the German spirit (der deutsche Geist). So, when Weierstrass exclaimed in 1873 "our unforgettable Boeckh", things looked bad for the latter's ideal, but German mathematicians had been keeping his heritage alive, in the first place by going through the "cantorian paradise". And on that occasion, Leibniz's thinking was honoured again; but this will be the subject of another contribution. So we can openly conclude, with reference to Goethe's "Shakespeare und kein Ende": in mathematics, Leibniz and no end.

References

[Biermann 1975] Biermann, Kurt: *Carl Gustav Jacob Jacobi.* In: *Biographien bedeutender Mathematiker.* Ed. by Hans Wussing & Wolfgang Arnold. Berlin: 1975.

[Boeckh 1877] Boeckh, August: *Encyclopädie und Methodologie der philologischen Wissenschaften.* Leipzig: 1877.

[Bossut 1810] Bossut, Charles: *Histoire des mathématiques depuis leur origine jusqu'à l'année 1808,* 2 vol. Paris: 1810.

[Cantor 1932] Cantor, Georg: *Gesammelte Abhandlungen mathematischen und philosophischen Inhalts*. Berlin: Springer, 1932.

[Cantor 1991] Cantor, Georg: *Briefe*. Ed. by H. Meschkowski & W. Nilson. Berlin Heidelberg New York: Springer-Verlag, 1991.

[Châtelet 1968] Châtelet, François: *Hegel*. Paris: Seuil, 1968.

[Dantzig 1930] Dantzig, Tobias: *Number, the Language of Science*. New York: 1930.

[Dedekind 1930–1932] Dedekind, Richard: *Gesammelte mathematische Werke*, 3 vol. Braunschweig, 1930–1932.

[Gauß 1863–1933] Gauß, Carl Friedrich: *Werke*, 12 vol. Göttingen: Königliche Gesellschaft der Wissenschaften, 1863–1933.

[Gauß 1880] Gauß, Carl Friedrich: *Briefwechsel Carl Friedrich Gauß – Friedrich Wilhelm Bessel*. Leipzig: Engelmann, 1880.

[Gauß 1899] Gauß, Carl Friedrich: *Briefwechsel zwischen Carl Friedrich Gauß und Wolfgang Bolyai*. Ed. by F. Schmidt & P. Stäckel. Leipzig: Teubner, 1899.

[Gauß 1990] Gauß, Carl Friedrich: *Carl Friedrich Gauß – Der "Fürst der Mathematiker" in Briefen und Gesprächen*. Ed. by Kurt-R. Biermann. München: Beck, 1990.

[Gray 2000] Gray, Jeremy J.: *The Hilbert Challenge*. Oxford New York: OUP, 2000.

[Hegel 1969–1971] Hegel, G.W.Fr.: *Werke in 20 Bänden*. Frankfurt am Main: Suhrkamp, 1969–1971.

[Hindenburg 1796] Hindenburg, Carl Friedrich: *Der polynomische Lehrsatz, das wichtigste Theorem der ganzen Analysis*. Leipzig, 1796.

[Jacobi 1881–1891] Jacobi, Carl Gustav: *Gesammelte Werke*, 7 vol. Berlin: Reimer, 1881–1891.

[Koenigsberger 1904] Koenigsberger, Leo: *Carl Gustav Jacob Jacobi Festschrift zur Feier der 100ten Wiederkehr seines Geburtstages*. Leipzig: B.G. Teubner, 1904.

[Kummer 1975] Kummer, Ernst Eduard: *Collected papers*, 2 vol. Berlin New York: Springer-Verlag, 1975.

[Dirichlet 1889–1897] Dirichlet, Peter Gustav Lejeune: *Mathematische Werke*, 4 vol. Berlin: Reimer, 1889–1897.

[Novalis 1960] Novalis: *Schriften,* 6 vol. Ed. by Paul Kluckhohn, R. Samuel etc. Stuttgart: Kohlhammer, 1960 etc.

[Scharlau 1981] Scharlau, Winfried (ed.): *Richard Dedekind 1831–1916*. Braunschweig Wiesbaden, 1981.

[Schelling 1970] Schelling, Fr.W.J.: *Historisch-kritische Ausgabe*, 13 vol. Ed. by H.M. Baumgartner, W.G. Jacobs & H. Krings. Stuttgart: Frommann-Holzboog, 1976 etc.

[Weierstraß 1894–1927] Weierstraß, Karl: *Mathematische Werke*, 7 vol. Berlin: 1894–1927.

Philippe Séguin
44 rue du grand verger
F-54000 Nancy, France
philippe.seguin11@wanadoo.fr

The Reception of Leibniz's Logic in 19th Century German Philosophy

Volker Peckhaus

1. The problem

Leibniz's impact on the emergence of mathematical (algebraic, algorithmic or symbolic) logic is an important topic for understanding the emergence and development of the current views on logic.[1] However, the question whether Leibniz had any influence at all, or whether his ideas were not more than ingenious anticipations of later developments, is still disputed. The significance of this problem can be shown by referring to Louis Couturat who claimed that in respect to the logical calculus Leibniz had all the principles of much later logical systems of the algebra of logic (George Boole, Ernst Schröder) and in some points he was even more advanced than they (Couturat 1901, 386). One important step in dealing with the problem can be seen in an answer to the question whether early "modern" logicians like Boole, Schröder, or Frege had had any knowledge of Leibnizian logic, i.e., could Leibniz have had any influence on these pioneers of modern logic?

1.1. Diverging claims

In dealing with these questions we are faced with different theses:

> Thesis 1: Leibniz had no impact on modern logic because his contributions were not known.

Wolfgang Lenzen, e.g., wrote that Leibniz was the most significant logician between Aristotle and Frege, but despite the enormous significance of his logic it played hardly any role in the history of logic.[2] According to Lentzen, Leibniz's mature logical theory was present in his *Generales Inquisitiones de Analysi Notionum et Veritatum* which was only published in Louis Couturat's edition of Leibniz's small writings and fragments (C 356–399). Couturat, however, referred to it already in his book on Leibniz's logic which appeared two years earlier (Couturat 1901).

---

[1] For a full scale study on the topic see Peckhaus 1997.
[2] Lenzen 2004a, 15. See also Lenzen 2004b.

If this thesis is accepted, it is possible to connect the discovery of the logician Leibniz to the Leibniz renaissance in early 20th century. Besides Couturat's book *La logique de Leibniz d'après des documents inédits* (1901), with a presentation of Leibniz's logic in the spirit of the new logic, the following landmark publications have to be mentioned: Bertrand Russell's *A Critical Exposition of the Philosophy of Leibniz* (1900), providing an axiomatic deductive reconstruction of Leibnizian metaphysics, and Ernst Cassirer's *Leibniz' System in seinen wissenschaftlichen Grundlagen* (1902), focusing on a Neo-Kantian interpretation of Leibniz's philosophy. Undoubtedly, Louis Couturat's edition of Leibniz's *Opuscules et fragments inédits de Leibniz*, taken from the manuscripts in the Royal Library in Hanover and published in 1903, gave access to the wealth of Leibniz's different approaches to logic for the first time.

Thesis 2: Leibniz had an impact on the emergence of modern logic.

Among the proponents of this thesis, Eric J. Aiton is to be mentioned; he wrote that the Leibnizian project of a universal characteristic and the logical calculi resulting from it "played a significant role in the history of logic" (1985, ix). Franz Schupp assumed, starting from Couturat's evaluation quoted earlier "that the Leibnizian logic might be relevant for the further development of modern logic, beyond the historically interesting aspect of an 'ingenious anticipation' " (Schupp 1988, 42). Every step in the development led to new insights into the Leibnizian logic, but sometimes dealing with Leibniz influenced the development itself.

1.2. Referring to the Leibnizian heritage

The second thesis has a big advantage over the first one, as it helps to explain why the pioneers of modern logic themselves referred to Leibniz. Mary Everest Boole, i.e., George Boole's widow, wrote that her husband felt "as if Leibnitz had come and shaken hands with him across the centuries," after having been informed of Leibniz's anticipations of his own logic.[3] William Stanley Jevons, who was responsible for the great public success of modern logic in Great Britain, claimed that "Leibnitz' logical tracts are [...] evidence of his wonderful sagacity" (Jevons 1883, xix). Ernst Schröder, the German pioneer of the algebra of logic, thought that Leibniz's ideal of a logical calculus had been brought to perfection by George Boole (Schröder 1877, III). The particular controversy between Ernst Schröder and Gottlob Frege became dicisive for the later distinction between the two kinds of modern logic: the algebra of logic and the Frege style mathematical logic. Both circled around the question how far the Leibnizian heritage was present in the respective variations of logic. In his *Begriffsschrift*, Frege wrote that the idea of a general characteristic of a *calculus philosophicus* or *ratiocinator* was too mammoth to be achieved by Leibniz alone. With his own *Begriffsschrift*, Frege wanted to supplement the first steps towards this goal of a general characteristic which can be found in the formula languages of arithmetic and chemistry (Frege 1879, VI). In his review of Frege's *Begriffsschrift*, Schröder criticized the title "Begriffsschrift" as promising too much (Schröder 1880, 82). Frege's system does not bend towards a 'General Characteristic' but towards the Leibnizian *calculus ratiocinator* which could be called commendable, if a significant part of

---

[3] Mary Everest Boole 1905, quoted in Laita 1976, 243.

it had not already been achieved by others (esp. by Boole). Frege replied that he had attempted to express content, contrary to Boole (Frege 1883, 1). Therefore the *Begriffs-schrift* is not a mere *calculus ratiocinator*, but a *lingua characteristica* in the Leibnizian sense, although he accepted that deductive calculation [*schlussfolgernde Rechnung*] was a necessary constituent of the *Begriffsschrift*.

2. The first editions

Given these examples it seems to be clear that referring to Leibniz was a common place in the initial period of the development of modern mathematical logic. In order to determine Leibniz's influence on this development, answers to the following questions may be helpful: (a) "Was Leibniz's philosophy of logic available in the 19th century?", (b) "Were Leibniz's attempts to create logical calculi available in the 19th century?"and (c) "Who read what?" A big problem is, of course, that Leibniz did not publish much during his lifetime. Thus answers have to be found by analyzing the early editions of Leibniz's works.

The edition of Leibniz's philosophical works in Latin and French, published by Rudolph Erich Raspe (Leibniz 1765), contains some up to then unpublished letters and six pieces from the unpublished papers, among them "Difficultates quadam logicae" and "Historia et commendatio linguae charactericae". The most important feature of Raspe's edition was the first publication of the "Nouveaux Essais sur l'entendement humain" which were missing for 60 years. This publication caused a stir, maybe the reason for Johann Gottfried von Herder to call Raspe "the man who found Leibniz".[4] In 1768 Louis Dutens published the *Opera omnia nunc primum collecta in Classes distributa praefation-ibus & indicibus exornata* (Dutens), a rather complete collection of Leibniz's published works. It contained some hitherto unpublished correspondences.

It may be sufficient to proceed in an exemplary way in order to determine how far contemporaries could go into the details of Leibniz's philosophy of logic. The *Nouveaux Essais* may serve as an example, although they are usually not regarded as a core text of logic.

3. Logic in the *Nouveaux Essais*

The *Nouveaux Essais* count as Leibniz's main work in epistemology. They were written between 1703 and 1705 containing criticism of John Locke's *Essay in Human Under-standing* of 1690. Locke died in 1704 when Leibniz was still working on the essays. They are composed as a dispute between Philalethes, the *alter ego* of Locke, and Theophilus who represents Leibniz's position. Logical considerations can be found in the fourth book "De la connaissance". There Leibniz distinguishes primitive ideas, i.e., simple and original truths which can be found by intuition, into two groups: necessary truths of reason and contingent truths of matters of fact (Ch. II, § 1). The truths of reason are logically

---

[4] Herder in a letter to Raspe of May 1774, quoted according to Hallo 1934, 175.

relevant, among them in predominantly identical truths. In its affirmative form an identical truth claims that everything is as it is (principle of identity). In its negative form it follows the principle of contradiction. In a very general version this principle says that a sentence is either true or false. The principle of contradiction, thus, constitutes bivalence. This principle includes the principle of excluded contradiction according to which a sentence cannot be true and false at the same time. It also implies the decidability of the system, i.e., the impossibility that a sentence is neither true nor false (ibid.).

Leibniz rejects the view that identical sentences are superfluous because of being not informative, and therefore do not serve any purpose. He argues that all inferences in logic are proved by identical sentences. Furthermore, all indirect proofs in geometry are done with the help of the principle of contradiction. Finally, the second and the first figure of the syllogism are justified with the help of the principle of contradiction. Leibniz, hence, uses pragmatic arguments by hinting at the benefit these sentences offer. He consequently stresses their character as tools (ibid.).

Leibniz does not discuss logical calculi which were in the focus of his interest in the 1690s, but syllogistics, the traditional theory of inferences. He discusses analytical and synthetic aspects. Analysis in syllogistics means, as already in Aristotle's work, the art of discovering the idea mediating between the two premises in a syllogistic inference which makes the inference possible (*terminus medius*). The art of analysis serves for evaluating given sentences in respect to the question, if they can be derived from premises recognized as true with the help of syllogistic inferences (Ch. II, § 7). This kind of analysis is therefore regressive analysis used to solve problems (heuristics).[5] In this context Leibniz stresses his preference for synthetic procedures. It is more important, he says, to find truth by oneself than to find proofs for truths found by other persons. It is very difficult, however, to find the tools for discovering what one is looking for exactly when one is looking for it. The combinatorial method does not help, although Leibniz had already called it the germ of a *logica inventiva*, logic of invention, in the subtitle of his *Dissertatio de arte combinatoria* of 1666. Frequently, it seems to be rather easy to drink up an ocean than to set up all required combinations. Therefore it is necessary to find the Ariadne thread through this labyrinth (ibid.).

The tool of choice is syllogistics which is part of some sort of universal mathematics, an art of infallibility. This art, however, is not restricted to syllogisms. It concerns all kinds of formal proofs, i.e., all reasoning in which inferences are done by virtue of their form (Ch. XVII, § 4). According to Leibniz there are some problems with algebra. It is still far from being an art of invention. It has to be supplemented by a general art of signs or an art of characteristic (Ch. XVII, § 9).

In sum one can say that Leibniz's rationalistic philosophy of logic with its characteristic demands to use logic as a tool box for finding new truths and to determine the validity of given hypotheses was "on the market". So it is not astonishing that there were several authors writing in the Leibnizian spirit, to name only the most important Christian Wolff (1679–1754) and his school, Johann Heinrich Lambert (1728–1777), and Gottfried Ploucquet (1716–1790). This rationalistic movement dominating German philosophy in

---

[5]For regressive analysis cf. Peckhaus 2002. For varieties of the notion of analysis cf. Beaney 2008.

the late 18th century was stopped by Immanuel Kant's Transcendental Idealism and by German Idealism (Fichte, Hegel, Schelling). A combination of Kant's Critical Philosophy and elements of Leibnizian Rationalism can be found in the Critical Realism of Johann Friedrich Herbart (1776–1841) and his school.

4. Second wave of reception

When access to Leibniz's papers stored in Hanover became possible in the 1830s, a new interest in Leibniz arose almost immediately. It can be said that the opening initiated German research on Leibniz (Glockner 1932, 60). The pioneers in this period of research were the first editors of the papers. Although the philological interest stood in the center, an emerging interest in Leibniz's logic could be observed. The following works have to be named: Gottschalk Eduard Guhrauer (1809–1854) edited the *Deutsche Schriften* (Leibniz 1838/40). Georg Heinrich Pertz directed the edition of the collected works of which a first series with the mathematical writings was edited by Carl Immanuel Gerhardt (GM). Pertz also edited Leibniz's *Annales imperii occidentis Brunsvicenses* (Leibniz 1843–1846).

4.1. Johann Eduard Erdmann

The edition of Leibniz's philosophical works *God. Guil. Leibnitii opera philosophica quae extant Latina Gallica Germanica omnia* (Leibniz 1839/40) was the most important among the editorial projects. It was prepared in two volumes by Johann Eduard Erdmann (1805–1892) in which, for the first time, fragments were published containing elaborations of Leibniz's ideas concerning logical calculi. Among the papers edited, Leibniz's letter to Gabriel Wagner, written in 1696, can be found which contains the famous definition of logic or the art of reasoning as the art of using the intellect ["Verstand"], i.e., not only to evaluate what is imagined, but also to discover (invent) what is hidden. The edition also contains the seminal fragments "Specimen demonstrandi in abstractis" and "Non inelegans specimen demonstrandi in abstractis" (E 94–97), the last with the algebraic plus minus calculus, i.e., a central specimen of Leibniz's various attempts to formulate logical calculi.

Johann Eduard Erdmann studied theology and philosophy at Tartu and Berlin. Friedrich Schleiermacher and Georg Friedrich Wilhelm Hegel were among his teachers. He later became a member of the right wing Hegelian school. In 1839, he was appointed full professor of philosophy at the University of Halle. Erdmann became well-known for his comprehensive history of modern philosophy entitled *Versuch einer wissenschaftlichen Darstellung der Geschichte der Neueren Philosophie*, published in seven volumes (Erdmann 1834–1853). This history of philosophy covers the period between Descartes and Hegel. Shortly after having published the edition of Leibniz's philosophical works, he presented a discussion of Leibniz and the development of idealism before Kant in pt. 2 of vol. 2 of his history published in 1842. There he stresses the connection between mathematics and philosophy. Erdmann deals with Leibniz's logic in the section on the philosophical method. He mentions Leibniz's definition of "method" as the way of deriving all knowledge with the help of "principles of knowledge" (*Erkenntnisprinzipien)* (Erdmann 1842, 109). These principles are the law of contradiction and the law of sufficient reason. Given the definition that logic is the art of using the intellect, it is the key to all sciences and arts. According to Erdmann, Leibniz identifies the logical method with the mathematical method being the true philosophical method. Erdmann, furthermore, deals at length with Leibniz's "mathematical treatment of philosophy" not only because it was important for Christian Wolff and his school, but also "because just this point is usually ignored in presentations of Leibniz's philosophy" (ibid., 114). He has good reasons for this evaluation because most of the relevant writings became only accessible by his own edition (E). Erdmann deals with Leibniz's calculi as "methodic operations" with data in the "way of calculating". He discusses Leibniz's idea of a character script for the calculus which allows using signs without always remembering their meaning. Such "pasigraphy" would erase the differences between the languages; however, according to Erdmann's evaluation, the idea of a universal language was not in the center of Leibniz's interests. Leibniz's main point was that "all mistakes in reasoning will at once show up in a wrong combination of characters, and therefore the application of the characteristic script provides a means to discover the mistake in a disputed point like in every other calculation" (ibid., 122–123).

Erdmann's discussion of Leibniz can be evaluated as follows. He opens the way for Leibniz's conception of logic into the actual philosophical debates on logic. This is the more astonishing as Erdmann was a Hegelian. Hegel was known and heavily criticized for his depreciation of formal logic. On the other hand, stressing the close connection between philosophy and mathematics fits into a time when many philosophers tried to bring philosophy back into contact with sciences.

Erdmann reported that, while preparing his history, he became unsatisfied with the available editions of Leibniz's works. He therefore intended to unite Raspe's edition with the philosophical parts of Dutens' edition and some pieces from the unpublished papers. He started editorial work at the archive in Hanover in 1836.

4.2. The impact of Erdmann's edition

Erdmann's edition immediately stimulated further research on Leibniz's logic. Gottschalk Eduard Guhrauer criticized extensively Leibniz's universal characteristic in the first volume of his biography of Leibniz (Guhrauer 1846). He stressed its absurd and utopian character: one can hardly avoid putting the general characteristic and the philosophical calculus in one box with the philosopher's stone and the secrets of producing gold.

In a paper entitled "Über Leibnitz'ens Universal-Wissenschaft" (1843), the Austrian philosopher Franz Exner referred explicitly to Erdmann's edition. For Exner the edition throws a brighter light on Leibniz's conception of a universal science. It has its weaknesses, but Exner prognoses a healthy impact on philosophy. Exner wrote (Exner 1843, 39):

> For him, the universal science is the true logic; both, universal science and logic, are the arts of judgment and invention; writing mathematically means for him writing *in forma*, what he believes to be possible outside mathematics; for him, the logical form of reasoning is a calculus; formulas, relations and operations of his universal science correlate with the concepts, judgments and inferences of his logic; finally, the second part of the universal science, the art of invention, is an epitome of relatively general methods. We cannot

accuse him of having overestimated logic. It was not his opinion that simple knowledge of logical rules would do great things, but its application. There, however, men who had the knowledge to a great extent had shown weaknesses.

In 1857, the Herbartian philosopher from Bohemia, František Bolemír Květ (1825–1864), published a booklet entitled *Leibniz'ens Logik*. Květ reconstructed the elements of Leibniz's *scientia generalis* stressing that although the elements might not be original, their combination is. He discussed the "extremely meager" fragments concerning the philosophical calculus. They showed, Květ wrote, how far their author stood behind his aims. He dismissed Leibniz's *ars inveniendi*, calling it embarrassing because of its infirmity, defects and impossibility.

4.3. Friedrich Adolf Trendelenburg on general characteristic

The most important figure in this second period of reception was Friedrich Adolf Trendelenburg (1802–1872). He had studied philology, history and philosophy at the Universities of Kiel, Leipzig and Berlin. Among his teachers were Karl Leonhard Reinhold and Johann Erich von Berger. In 1833, Trendelenburg became professor, followed by a professorship of practical philosophy and pedagogic at the Friedrich Wilhelms University, Berlin, 1837, where he grew into one of the main leaders of Prussian education and German philosophy. Being an ordinary member of the Royal Prussian Academy of Science at Berlin since 1846, he became 1847 secretary of the Philosophical-Historical Section of this Academy. Trendelenburg was an anti-Hegelian who started from Hegelian philosophy. His fame as a neo-Aristotelian goes back to his *Elementa logices Aristotelicae*, first published in 1836 with five further editions. In his systematic work on logic he pleaded for a unity of logic and metaphysics as being present in the Aristotelian organon. This systematic attitude is developed in a comprehensive work containing heavy criticism on logical systems of his time: his *Logische Untersuchungen*, published in two volumes 1840.

As a secretary of the Academy, Trendelenburg had to care for the memory of Leibniz who had been the first president of the "Societät der Wissenschaften" at Berlin, founded on his initiative in 1700 and preceding as an institution the Royal Prussian Academy of Science. In 1856, he delivered a seminal lecture "Über Leibnizens Entwurf einer allgemeinen Charakteristik" at the Leibniz ceremony of the Royal Academy of Science at Berlin (Trendelenburg 1857). This paper was reprinted in the third volume of his *Historische Beiträge zur Philosophie* (1867). In this discussion of Leibniz, Trendelenburg stresses the essential role of signs in communication and reasoning. There is no logical relation between sign and intuition. Science has given us the opportunity to "bring the composition of the signs into immediate contact with contents of the concept" (Trendelenburg 1857, 3). The composition of the sign presents the characteristic distinguished and comprehended in the concept in a distinguishing and comprehending way (ibid.). The beginnings of such a "Begriffsschrift" (Trendelenburg's term) are made, e.g., in the decadic number system. Trendelenburg sees the objectives of the Leibniz program in widening such an approach to the whole domain of objects aiming at a "characteristic language of concepts" and a "general language of matter". He mentions the different names used by Leibniz: *lingua characterica universalis* (in fact Trendelenburg's term), alphabet of human thoughts, *calculus philosophicus*, *calculus ratiocinator*, *spécieuse générale*. These names prove the significance of Leibniz's philosophy for this program. According to Trendelenburg, Leibniz aimed at "an adequate and therefore general signification of the essence, namely such an analysis into the elements of concepts, that it becomes possible to treat them by calculation" (ibid., 6). He mentions historical precursors, Raymundus Lullus' *ars magna*, and other concepts of universal languages. Because of its generality, Leibniz's *characteristica universalis* stands out compared with competing conceptions by George Dalgarno (1661) and John Wilkins (1668) which were mixed from "choice, nature and chance", and leaned upon existing languages (ibid., 14–15).

Trendelenburg, however, heavily criticizes the practical side of the program, in particular calculation in logic. The connection of properties in a concept is much more complicated than it can be expressed with Leibniz's operations (ibid., 24). He gives the advice to abstain from calculation (ibid., 25):

> If the side of calculus, invention and discovery is excluded from general characteristic, still an attractive logical task remains: [to find] the sign distinguishing the element and therefore being clear, avoiding contradictions; to reduce blind intuition to a sharply thought content; to reduce the intricate to the simple contained in it. It remains the task to find a sign that is, like our number script, determined by the concept of the matter itself.

For the last task it is necessary to analyze the concepts completely. However, this is not always possible given the actual state of science. Therefore arbitrary assumptions would have to be accepted until they can be replaced by better knowledge.

4.4. Evaluation

Erdmann's edition induced a second wave of reception. This reception is characterized by an interest in Leibniz's ideas on logic. Its context was the reorganization of the philosophical scene after Hegel's death (1831). This process was connected with a discussion on the so-called "Logical Question", a term presented by Adolf Trendelenburg (Trendelenburg 1842). The discussion concerned the role of formal logic in the system of philosophy overcoming Hegel's identification of logic and metaphysics. The philosophical dominance of metaphysics was subsequently replaced by that of epistemology.

Trendelenburg's results were typical for a line of reception stressing the metaphysical character of Leibniz's philosophy: He was interested in the *characteristica universalis* as a tool for knowledge representation, although he stressed its utopian character. He had no interest in the logical calculus due to a philosophical scepticism towards mechanical tools. They cannot explain creativity in the emphatic sense and have no relations to the contemporary philosophy's interest in dynamical (temporal) logics which should help to model the movement of thought ("Denkbewegung").

Given Trendelenburg's special emphasis in presenting the Leibnizian system, his significance for the mathematical reception of Leibnizian ideas in the context of the emergence of formal mathematics and mathematical logic in the second half of the 19th century is astonishing. Trendelenburg's paper on Leibniz's program of a general characteristic became, e.g., a point of reference of the logical pioneers such as Gottlob Frege and Ernst Schröder.

5. Discovery of Leibniz in mathematical logic

The discovery of Leibniz in mathematical logic can be shown exemplarily in the case of George Boole, the founder of the algebra of logic. In his first writing on logic, the booklet *The Mathematical Analysis of Logic* of 1847, he gave an algebraic interpretation of traditional logic. His fame of being one of the founders of modern logic goes back to his *An Investigation of the Laws of Thought* of 1854. According to Boole's own evaluation, his main innovation was the Index Law (1847), later revised to the Law of Duality, also called "Boole's Law", expressing the idempotency

$$A = AA.$$

What are the connections to Leibniz's logic? Are there anticipations of the Boolean calculus in the work of Leibniz? One of those authors looking for anticipations was Robert Leslie Ellis (1817–1859) who edited Francis Bacon's *Novum Organon* in *The Works of Francis Bacon* (1858–1874; vol. 1: 1858). During his editorial works he found a parallel to Boole's Law (p. 281, footnote 1): "Mr. Boole's *Laws of Thought* contain the first development of ideas of which the germ is to be found in Bacon and Leibnitz; to the latter of whom the fundamental principle in logic $a^2 = a$ was known." As reference he gave Erdmann's edition (E p. 130). Robert Harley (1828–1910), Boole's first biographer, discussed this information in a paper entitled "Remarks on Boole's Mathematical Analysis of Logic" (1867). He did not find the proper quote at the place indicated by Ellis, but he found other relevant texts. About the significance of Ellis' remark he wrote: "Boole did not become aware of these anticipations by Leibnitz until more than twelve months after the publication of the 'Laws of Thought,' when they were pointed out to him by R. Leslie Ellis" (p. 5).

Harley's research was taken up by the Manchester economist and philosopher William Stanley Jevons (1825–1882). He posited his philosophy of science as being present in the *Principles of Science* (1877, 2nd ed. 1883) against John Stuart Mill's predominant inductive logic. His alternative was the Principle of Substitution, the "substitution of similars": "So far as there exist sameness, identity or likeness, what is true of one thing will be true of the other" (Jevons 1883, 9). He included a section "Anticipations of the Principle of Substitution", which was enlarged in the later edition with a long discussion of Leibniz's anticipations. There he expressed his thanks to Robert Adamson for the information that the Principle of Substitution can be traced back to Leibniz. Jevons asked for the reasons for the long ignorance of Leibniz's anticipations. Only Dutens' edition was available in Owens College Library, Manchester. He regrets having overlooked Erdmann's edition, but this was also done by other "most learned logicians".

Finally John Venn (1834–1923) has to be mentioned. His *Symbolic Logic* (1881) is important for the historical contextualization of the new logic. He criticized Jevons' statement on the Law of Duality according to which "the late Professor Boole is the only logician in modern times who has drawn attention to this remarkable property of logical terms" as being simply false. Besides Leibniz, Lambert, Ploucquet and Segner had anticipated the law "perfectly explicitly" and Venn had no doubts "that any one better acquainted than myself with the Leibnitzian and Wolfian logicians could add many more such notices" (Venn 1881, xxxi, footnote 1).

6. Conclusions

No doubt, the new logic emerging in the second half of the 19th century was created in Leibnizian spirit. The essentials of Leibniz's logical and metaphysical program and of his idea concerning a logical calculus were available at least since the 1840s. Erdmann's edition of the philosophical works and Trendelenburg's presentation of Leibniz's semiotics were the most important steps towards the further reception of Leibnizian ideas among mathematical logicians at the end of the 19th century. As soon as these logicians were aware of the Leibnizian ideas they recognized Leibniz's congeniality and accepted his priority. However, the logical systems had already been established. Therefore there was no initializing influence of Leibniz on the emergence of modern logic in the 2nd half of the 19th century.

References

[Aiton 1985] Aiton, Eric J, *Leibniz. A Biography*, Adam Hilger: Bristol and Boston 1985.

[Bacon 1858] Bacon, Francis, "Novum Organum sive indicia vera de interpretatione naturae", in Francis Bacon, *The Works of Francis Bacon*, edited by James Spedding, Robert Leslie Ellis and Douglas Denon Heath, 14 vols., Longman & Co. etc.: London 1858–1874, reprinted Friedrich Frommann Verlag Günther Holzboog: Stuttgart-Bad Cannstatt 1963, vol. 1 [1858], 149–365.

[Beaney 2008] Beaney, Michael, "Analysis", *The Stanford Encyclopedia of Philosophy (Winter 2008 Edition)*, Edward N. Zalta (ed.), URL=<http://plato.stanford.edu/archives/win2008/entries/analysis/>.

[Boole 1951] Boole, George, *The Mathematical Analysis of Logic. Being an Essay towards a Calculus of Deductive Reasoning*, Macmillan, Barclay, and Macmillan: Cambridge and George Bell: London 1847; reprinted Basil Blackwell: Oxford 1951.

[Boole 1958] Boole, George, *An Investigation of the Laws of Thought, on which are Founded the Mathematical Theories of Logic and Probabilities*, Walton & Maberly: London 1854; reprinted Dover: New York nd. [1958].

[Boole 1931] Boole, Mary Everest, "Letters to a Reformer's Children" [1905], in Mary Everest Boole, *Collected Works*, 4 vols., eds. E.M. Cobham, C.W. Daniel: London 1931, vol. 3, 1138–1163.

[Cassirer 1902] Cassirer, Ernst, *Leibniz' System in seinen wissenschaftlichen Grundlagen*, Elwert: Marburg 1902.

[Couturat 1901] Couturat, Louis, *La logique de Leibniz d'après des documents inédits*, Alcan: Paris 1901.

[Erdmann 1853] Erdmann, Johann Eduard, *Versuch einer wissenschaftlichen Darstellung der Geschichte der Neueren Philosophie*, 7 vols., Vogel: Berlin 1834–1853.

[Erdmann 1842] Erdmann, Johann Eduard, *Versuch einer wissenschaftlichen Darstellung der Geschichte der neueren Philosophie*, vol. 2, pt. 2: *Leibniz und die Entwicklung des Idealismus vor Kant*, Vogel: Leipzig 1842.

[Exner 1843] Exner, Franz, "Über Leibnitz'ens Universal-Wissenschaft," *Abhandlungen der Königlichen Böhmischen Gesellschaft der Wissenschaften*, 5th series, vol. 3 (1843–44), Calve: Prag 1845, 163–200; separat in Commission at Borrosch & André: Prag 1843.

[Frege 1977] Frege, Gottlob, *Begriffsschrift, eine der arithmetischen nachgebildete Formelsprache des reinen Denkens*, Louis Nebert: Halle 1879; reprinted in Gottlob Frege, *Begriffsschrift und andere Aufsätze*, 3rd ed., with E. Husserl's and H. Scholz's comments edited by Ignacio Angelelli, Wissenschaftliche Buchgesellschaft: Darmstadt 1977.

[Frege 1883] Frege, Gottlob, "Ueber den Zweck der Begriffsschrift," *Jenaische Zeitschrift für Naturwissenschaft* 15 (1883), supplement: *Sitzungsberichte der Jenaischen Gesellschaft für Medicin und Naturwissenschaft für das Jahr 1882*, 1–10; again in Gottlob Frege, *Begriffsschrift und andere Aufsätze*, 3rd ed., with E. Husserl's and H. Scholz's comments edited by Ignacio Angelelli, Wissenschaftliche Buchgesellschaft: Darmstadt 1977, 97–106.

[Glockner 1932] Glockner, Hermann, *Johann Eduard Erdmann*, Fr. Frommanns Verlag: Stuttgart 1932 (*Frommanns Klassiker der Philosophie*; 30).

[Guhrauer 1966] Guhrauer, Gottschalk Eduard, *Gottfried Wilhelm Freiherr v. Leibnitz. Eine Biographie*, 2 vols., Hirt: Breslau 1842; new edition 1846; reprinted Olms: Hildesheim 1966.

[Hallo 1934] Hallo, Rudolf, *Rudolf Erich Raspe. Ein Wegbereiter deutscher Art und Kunst*, Kohlhammer: Stuttgart and Berlin 1934 (*Göttinger Forschungen*; 5).

[Harley 1867] Harley, Robert, "Remarks on *Boole's* Mathematical Analysis of Logic," *Report of the Thirty-sixth Meeting of the British Association for the Advancement of Science; Held at Nottingham in August 1866*, John Murray: London 1867.

[Jevons 1883] Jevons, William Stanley, *The Principles of Science. A Treatise on Logic and Scientific Method*, 2 Bde., Macmillan and Co.: London 1874 [New York 1875]; 2nd ed. Macmillan and Co: London and New York; [3]1879; "stereotyped edition" 1883.

[Květ 1857] Květ, František Bolemír, *Leibnitz'ens Logik. Nach den Quellen dargestellt*, F. Tempsky: Prag 1857.

[Laita 1976] Laita, Luis María, *A Study of the Genesis of Boolean Logic*, Ph.D. Notre Dame 1976.

[Leibniz 1990] Leibniz, Gottfried Wilhelm, *Dissertatio de arte combinatoria*, Fick und Seubold: Leipzig 1666; Academy edition vol. 6.1, [2]1990.

[Leibniz 1765] Leibniz, Gottfried Wilhelm, *Œuvres philosophiques latines et françaises de feu Mr de Leibnitz, tirées des ses Manuscrits qui se conservant dans la Bibliothèque royale à Hanovre et publiées par M. Rud. Eric Raspe*, Jean Schreuder: Amsterdam/Leipzig 1765.

[Leibniz 1838] Leibniz, Gottfried Wilhelm, *Deutsche Schriften*, ed. G.E. Guhrauer, Berlin 1838/40.

[Leibniz 1846] Leibniz, Gottfried Wilhelm, *Annales imperii occidentis Brunsvicenses*, ed. G.H. Pertz, Hannover 1843–1846.

[Leibniz 2000] Leibniz, Gottfried Wilhelm, *Die Grundlagen des logischen Kalküls*, ed., transl., with a commentary by Franz Schupp with collaboration of Stephanie Weber, Meiner: Hamburg 2000.

[Lenzen 2004a] Lenzen, Wolfgang, "Leibniz und die (Entwicklung der) moderne(n) Logik", in Wolfgang Lenzen, *Calculus Universalis. Studien zur Logik von G.W. Leibniz*, Mentis: Paderborn 2004, 15–22.

[Lenzen 2004b] Lenzen, Wolfgang, "Leibniz's Logic", in Dov M. Gabbay and John Woods (eds.), *Handbook of the History of Logic*, vol. 3: *The Rise of Modern Logic. From Leibniz to Frege*, Elsevier North Holland: Amsterdam et al. 2004, 1–83.

[Locke 1690] John Locke, *An Essay Concerning Human Understanding*, Thomas Ballet: London 1690.

[Peckhaus 1997] Peckhaus, Volker, *Logik, Mathesis universalis und allgemeine Wissenschaft. Leibniz und die Wiederentdeckung der formalen Logik im 19. Jahrhundert*, Akademie-Verlag: Berlin 1997 (*Logica Nova*).

[Peckhaus 2002] Peckhaus, Volker, "Regressive Analysis," in Uwe Meixner and Albert Newen (eds.), Philosophiegeschichte und logische Analyse. Logical Analysis and History of Philosophy, vol. 5, Mentis: Paderborn 2002, 97–110.

[Russell 1900] Russell, Bertrand, *A Critical Exposition of the Philosophy of Leibniz*, The University Press: Cambridge 1900.

[Schröder 1966] Schröder, Ernst, *Der Operationskreis des Logikkalkuls*, Teubner: Leipzig 1877; reprinted as special edition Wissenschaftliche Buchgesellschaft: Darmstadt 1966.

[Schröder 1880] Schröder, Ernst, Review of Frege, *Begriffsschrift, Zeitschrift für Mathematik und Physik, Hist.-literarische Abt.* **25** (1880), 81–94.

[Schupp 1988] Schupp, Franz, "Einleitung. Zu II. Logik," in Albert Heinekamp and Franz Schupp (eds.), *Leibniz' Logik und Metaphysik*, Wissenschaftliche Buchgesellschaft: Darmstadt 1988 (*Wege der Forschung*; 328), 41–52.

[Schupp 2000] Schupp, Franz, "Einleitung," in Leibniz 2000, VII–LXXXVI.

[Trendelenburg 1862] Trendelenburg, Friedrich Adolf, *Elementa logices Aristotelicae. In usum scholarum ex Aristotele excerpsit, convertit, illustravit*, Bethge: Berlin 1836, ⁵1862.

[Trendelenburg 1862] Trendelenburg, Friedrich Adolf, *Logische Untersuchungen*, 2 vols., Bethge: Berlin 1840, 2nd ed. Hirzel: Leipzig 1862.

[Trendelenburg 1843] Trendelenburg, Friedrich Adolf, "Zur Geschichte von Hegel's Logik und dialektischer Methode. Die logische Frage in Hegels Systeme. Eine Auffoderung [sic!] zu ihrer wissenschaftlichen Erledigung," *Neue Jenaische Allgemeine Literatur-Zeitung* 1, no. 97, 23.4.1842, 405–408; no. 98, 25.4.1842, 409–412; no. 99, 26.4.1842, 413–414; separately published as Friedrich Adolf Trendelenburg, *Die logische Frage in Hegel's System. Zwei Streitschriften*, Brockhaus: Leipzig 18

[Trendelenburg 1867] Trendelenburg, Friedrich Adolf, "Über Leibnizens Entwurf einer allgemeinen Charakteristik," *Philosophische Abhandlungen der Königlichen Akademie der Wissenschaften zu Berlin. Aus dem Jahr 1856*, Commission Dümmler: Berlin 1857, 36–69; separatly published in Friedrich Adolf Trendelenburg, *Historische Beiträge zur Philosophie*, vol. 3: *Vermischte Abhandlungen*, Bethge: Berlin 1867, 1–47.

[Venn 1881] Venn, John, *Symbolic Logic*, Macmillan & Co.: London 1881.

Volker Peckhaus
Universität Paderborn
Institut für Humanwissenschaften: Philosophie
Warburger Str. 100
D-33098 Paderborn, Germany
volker.peckhaus@upb.de

Leibniz's Metaphysics as an Epistemological Obstacle to the Mathematization of Nature: The View of a Late 19th Century Neo-Kantian, Kurd Lasswitz

Françoise Willmann

Among the neo-Kantians of the late 19th century, we find a few figures who are less well known than the renowned Cohen, Natorp, Windelband, Rickert, and Cassirer. One of these is Kurd Lasswitz (1848–1910), physicist by education and author of *Geschichte der Atomistik vom Mittelalter bis Newton*, a history of atomism published for the first time in 1890, reprinted in 1984 and again very recently in 2010. In the last part of this history, Lasswitz devotes about forty pages to the complexity of the position held by Leibniz, who, in his mind was deterred, by his metaphysical bent, from the path his mathematical genius should have led him to take. So rather than pursuing the work of Huygens – which Lasswitz sees as the culmination point of his history – Leibniz (like Newton for that matter) is thought to have contributed to a genuine epistemological regression. We shall see that, as an overt partisan of a kinetic atomism, Lasswitz looks essentially at the phase of Leibniz's work during which he moved away from the atomism to which he had adhered in his youth. In regards to this phase, Lasswitz strives to uncover the reasons why Leibniz had such difficulty conceptualizing bodies and motion, relying mainly on the *Hypothesis Physica Nova*, but also on his correspondence and more generally his theoretical dialogues with other thinkers. In addition to understanding the Leibnizian stances, Lasswitz's goal was also to defend his own theoretical position by bringing its legitimacy to the fore, precisely by examining the difficulties encountered by Leibniz.

Kurd Lasswitz: Some biographical landmarks

In truth, Kurd Lasswitz has not been totally forgotten, but his protean work was received in what one might call a piecemeal way. At the present time, he is seen above all as a precursor of science fiction in Germany; discovered as such in the second half of the 20th century, his name was thereafter associated with one of the two great German science fiction prizes awarded every year. By 1871, he had published his first short story *Bis zum*

Nullpunkt des Seins, which takes place in 2371. Six years later, he produced the sequel entitled *Gegen das Weltgesetz*, set in 3877. The two short stories were published together in 1878 under the title *Bilder aus der Zukunft*, which has been reprinted time after time since then. It is his novel *Auf zwei Planeten,* published in 1897, that is generally regarded as his masterpiece. A book of nearly a thousand pages, it appeared in numerous shortened editions before being republished in its entirety in 1979. Lasswitz was to publish novels and "scientific" short stories throughout his life. One of the latter, *Auf der Seifenblase*, which came out in 1887, is clearly of Leibnizian inspiration.

But Lasswitz was more than just a writer, or at least more than just a writer of fiction. He obtained his Doctor's degree on the behavior of water droplets under the effects of weightiness in 1873, and he also studied mathematics and philosophy in Berlin and Breslau (Wroclaw), where his professors included Galle the astronomer, Kummer and Weierstrass the mathematicians, and Dilthey the philosopher. However, after obtaining his Doctor's degree in physics, Lasswitz went on to get a secondary-school teaching certificate and pursued a career as a teacher of mathematics, physics, philosophy, and geography at the Ernestinum High School in Gotha.

Alongside this, Lasswitz spent a great deal of time on what one might call the popularization of philosophy. He co-founded an intellectual exchange group in Gotha named *Mittwochsgesellschaft* aimed at promoting communication and debate about a wide variety of subjects, with a focus on philosophy and science, and he published numerous reviews, often in the "sciences" column of general or literary journals. The purpose of this facet of his work can be summarized as follows: in the face of progress and success in the sciences and their impact on modern life, Lasswitz attempts to harmoniously reconcile ways of thinking inherited from the past, with those of modern times. In this endeavor, he does not claim to be philosophically innovative, but openly bases his proposals on Kant, from whom he borrows a number of basic principles: the critical approach and thought of the limit. The history of 19th century philosophy thus granted him a small place among the neo-Kantians surrounding Marburg, particularly Hermann Cohen. His *Geschichte der Atomistik* owes much to this epistemological anchor point, and can be considered as a truly Kantian work.

Geschichte der Atomistik

As an extension of his Doctor's degree in physics, Lasswitz worked on a project for about fifteen years which culminated in 1889 with two volumes of his *Geschichte der Atomistik vom Mittelalter bis Newton* on the history of atomism between the Scholastic Period and Newton. His idea was to take a historical and systematic approach to tell the story of the theory of matter, that is to say, the emergence of modern physics. According to the author, the history of the theory of matter provided the material *par excellence* from which it had to be possible to inquire into the possibility conditions of knowledge. He envisages the history of atomism as a history of the arising of an interest in scientific knowledge about nature detaching itself from speculation, or as a history of the emergence of an interest in physics detaching itself from an interest in metaphysics.

Elaboration and reception

For about ten years – since its creation in fact – Lasswitz contributed regularly to *Viertel-jahrsschrift für Wissenschaftliche Philosophie*, a journal (headed at first by R. Avenarius, with the collaboration of C. Göring, M. Heinze, and W. Wundt) that moved back and forth between positivism and neo-Kantism. Lasswitz's publications therein correspond for the most part to the main chapters of his *Geschichte der Atomistik*, which was to be published in 1890 as a synthesis covering nearly a thousand pages. The first edition must not have been received as well as the author had hoped. It could very well be that his critical Kantian approach was held against him. That, in any case, is what was insinuated in a review of the second edition, published exactly as is thirty years later, Lasswitz having died in 1910. He was – or so the author of the recension implies – reproached outright for having tampered with the history of atomism to give it a Kantian reading. The second edition of 1920, on the other hand, met with largely favorable reverberations and it seems that, little by little, the idea that this was indeed a remarkable piece of work took over; or, in any event, one was thereafter to hear of it being referred to as a "classic". Among many others, Ernst Cassirer refers to it in his *Philosophie der Symbolischen Formen* [Cassirer, 1994, p. 531], and Gaston Bachelard, who in 1933 presented his endeavor to classify "atomistic intuitions", abandoned his plan for a "développement historique des doctrines atomistiques", which he described as a "tâche vraiment inutile après l'admirable ouvrage de Lasswitz" [Bachelard, 1933, p. 9].

History of atomism and the problem of understanding nature

For Lasswitz, the question of knowledge is raised in Kantian terms. For him, criticism is the decisive conquest in the order of thought, being of a nature that can clarify the relationship between philosophy and science. It is the task of a theory of matter to go beyond physics, for its purpose is to progress toward the elaboration of unitary principles of science. From there, Lasswitz makes the distinction between useful theories and epistemologically satisfactory theories, and in doing so, he argues resolutely for kinetic atomism and against dynamic theories, thereby opposing not only the Kant of *Metaphysische Anfangsgründe*, but also Hermann Cohen, who, in *Das Prinzip der Infinitesimalmethode und seine Geschichte*, spoke in 1883 of a "systematische Überschätzung der Atomhypothese" [Cohen, 1968, p. 197] and who challenged that hypothesis. Lasswitz's article, *"Zur Rechtfertigung der Kinetischen Atomistik"*, published two years later in *Vierteljahrsschrift für Wissenschaftliche Philosophie*, undertakes a critiquing dialogue with this work by Cohen and seizes the opportunity to clarify his own position. To complete his conception in Kantian terms, he recapitulates therein:

> Die Anwendung der Kategorien auf die Sinnesanschauungen liefert die Grundsätze des reinen Verstandes, und die Anwendung dieser auf die empirischen Daten der Empfindung bestimmt die Axiome der mathematischen Naturwissenschaft. Diese Daten werden mathematisch darstellbar, insofern sie als bewegte Materie auftreten. Als Eigenschaften derselben ergeben sich extensive

Grössen der Raumtheile infolge der Kategorie der Quantität, intensive Raum-
erfüllung infolge der Kategorie der Realität, mechanische Wechselwirkung
beharrlicher individueller Raumelemente infolge des Zusammenwirkens der
Kategorien der Relation mit den erstgenannten. Die beiden ersten Kategorien
bedingen noch nicht den Atombegriff, sondern derselbe tritt erst in der Mechanik unter dem Einflusse der Relation auf. Die Relation bedingt nämlich,
dass die Eigenschaften der Materie von einem Subjecte prädicirt werden, und
dass diese so bestimmten materiellen Subjecte unter einander in causale Verbindung treten. Die actuelle Bewegung jedes einzelnen materiellen Subjects
muss daher in gesetzlichem Zusammenhange mit der Bewegung der übrigen
stehen, dadurch aber wird es eben Individuum, das sich von den anderen un-
terscheidet und abgrenzt und seine Selbständigkeit zugleich gewinnt durch die
gemeinsame Beziehung zu den übrigen als Subject der Bewegung. Erst durch
das Zusammen der Kategorien der Quantität, Qualität und Relation entsteht
der Atombegriff, erst damit ist die Realisierung der Materie vollendet, damit
aber auch die Wechselwirkung zugleich mit der Beweglichkeit und Raumer-
füllung der individualisierten Raumtheile gewährleistet. Die Atome sind durch
ihre Bewegungsgesetze, die Bewegungsgesetze durch beharrliche Subjecte der
Bewegung gegenseitig bestimmt. [Lasswitz, 1885, p. 159–160]

This epistemological summary of a conception of matter founded on kinetic atomism ends with the assertion of the uselessness of the dynamic hypothesis, which adds
the hypothesis of acting forces claimed by Lasswitz to be as useless as it is confusing.
It would surely be reductionist to see this recapitulative presentation as a mere frame of
reference based on which Lasswitz manages to unravel the complexity of the conceptions
encountered in the history of atomism; a better view is that it is as much a methodological
matrix for the work, as it is something output by it.

Indeed, Lasswitz's history of atomism tells us how scholarly thinking about the
question "What is matter?" evolved over time, starting from the renewal of the corpuscular theory with the Fathers of the Church, up through to the peak of atomist thinking
incarnated (according to Lasswitz) by Huygens, and then on to its decline. The history of
science is most certainly not to be understood as a regular and gradual accumulation of
knowledge, nor as the linear acquisition of conceptual tools. Even if Lasswitz is guided
by the Kantian conception evoked here, his account strives to be more than simply teleological. He relies notably on the model of the spiral, as he formulates it precisely by
attempting to situate Leibniz and Newton with respect to Huygens. What Lasswitz hopes
to bring out is how lengthy and difficult it was for these thinkers and scholars to conquer
what he calls the necessary "means for thinking" ("Denkmittel"), a notion that falls vis-
ibly in line with the Kantian approach to the theory of knowledge. In the history of the
theories of matter unraveled and examined by Lasswitz, we discover a proliferation of
questions and solutions, which become entangled as the centuries go by. To cut a path
through the jungle of theories he interrogates, Lasswitz begins by positing the need to find
a connecting thread that can do justice to the multiple positions uncovered. To this end,
he distinguishes three interests or pursuits, which, in his mind, dictate the elaboration of

theories and hypotheses: metaphysics, knowledge, and physics *per se*. In the course of the period studied, the accent is placed on the fact that the physical interest cannot really come to completion until the metaphysical one ceases to be preponderant. In the Scholastic Period in particular, metaphysics is the prevailing interest, in such a way that science cannot – or rather has no reason to – develop.

In going back over these centuries of inquiry into the essence of matter, all revolving around the question of its ultimate makeup, one is especially struck by the great complexity of the theories built. The history that Lasswitz relates is a history of exchanges, controversies, borrowings, laborious attempts to cross existing boundaries, misunderstandings, steps forward and steps backward. He reconstructs a collective intellectual work, while emphasizing the unbounded investment of intelligence and ingeniousness underlying this strong will to understand, and while also striving to bring to the fore its key moments, major obstacles, and decisive victories. In the end, two main steps are needed to exit from the labyrinth. The first involves going beyond an anthropomorphic approach consisting of ascribing sense qualities to the atom, in effect, of seeing it as a body in miniature (we have traces of this view in today's French expression *"atomes crochus"* or "hooked atoms", used to refer to two people with a great affinity for each other); this step permits distinguishing between corpuscular theories and atomist theories. The second step involves going beyond the metaphysical approach, which tries, for example, to find the origin of motion. These two breakthroughs imply an approach to phenomena that is simultaneously mathematical *and* mechanical, one that manages to conceive of matter in a way that permits finding in it a legality, one that masters it from the quantification point of view. This of course means doing without the hypothesis of God, but also without all other opaque interventions such as remote force, a trap into which Newton himself is thought to have fallen again. It even and above all means understanding exactly what is at stake in the atomist hypothesis, i.e., just how far-reaching it is, epistemologically speaking. Now Leibniz, thanks to infinitesimal calculus – made possible by mechanics which in turn enabled the conception of intensive magnitudes – and thanks to the very thought process required by his approach, could have bypassed the question of the motion of atoms in its anthropomorphic form. And yet, he still ended up placing priority on the metaphysical account, and this is what prevented him from drawing full benefit from his mathematical thinking.

Huygens, the height of the theory of matter

If Lasswitz sees Huygens's work as marking the apogee of the theory of matter, it is precisely because he was able to ground kinetic atomism even without formulating the last stage (due to his premature death). Exchanges with Leibniz provide the material to Lasswitz for pinpointing the decisive phase of his thinking. Indeed, their correspondence reveals diverging points of view about the question of the solidity of atoms, and in Huygens we find the following remark, "Ich habe eine Art, die Kohäsion der zusammengesetzten Körper zu erklären, welche von dem Druck von außen und noch von einer andern Sache abhängt." [Lasswitz, 1890, vol. II, p. 363] Yet it is precisely that "other thing" that

Huygens will not have the time to put down in words. When Lasswitz is writing this, he has no accurate information about what Huygens proposes to "publish soon" in order to explain how atoms can be absolutely solid and still not lose motion when colliding. Let us cite Lasswitz:

> Und es scheint uns kein Zweifel, dass Huygens die Grundlegung der kineti-schen Atomistik dadurch vollendet hat, dass er zuerst die Bewegung der Ato-me nicht, wie es Leibniz für nötig hielt, von den Stoßgesetzen abhängig mach-te, sondern dieselben begründete auf Prinzipien der Mechanik. Die Erhaltung der algebräischen Summe der Bewegungsgrößen und die Erhaltung der Ener-gie, das sind die beiden Prinzipien der Mechanik, welche als Grundsätze der physikalischen Erfahrung an der Voraussetzung individuell unveränderlicher, substanzieller, als ganzes bewegter, diskontinuierlicher Raumteile – der Ato-me – hinzutreten, um die kinetische Atomistik zu begründen. [Lasswitz, 1890, vol. II, p. 368]

Leibniz obstinately opposes the idea (supported by Gassendi and Borelli, however) that the ultimate components of matter must be unalterable and solid, i.e., the idea that they must be elastic unless the motion imprinted on them by collisions ends up being re-sorbed. Now this reasoning, which rests on a representation of atoms as bodies that truly exist and act as such, reveals a poor understanding of their true epistemological dimen-sion. Huygens, on the other hand, has freed himself of anthropomorphic representations. To interpret motion, he no longer has to worry about what happens when atoms collide, nor about whether they have consistency; he no longer needs to call upon the senses to imagine the movements and reactions of bodies, elastic or not. It is "das rationale Bedürf-nis gesicherter Naturerkenntnis" [Lasswitz, 1890, vol. II, p. 370], as Lasswitz puts it, that leads him to start, on the contrary, from the laws of mechanics. He replaces sensible representations of atoms with rational concepts, i.e., mathematical tools (even though the latter were still largely insufficient), or, to take up on another formulation by Lasswitz: "Er hat [...] der Korpuskulartheorie ihren Abschluß gegeben, indem er die sinnliche Tatsache der Wechselwirkung der Körper in einem mechanischen Prinzip *objektivierte*." [underlined by Lasswitz, 1890, vol. II, p. 376] This is something that Leibniz was unable – or did not want – to understand or do.

Leibniz's place

The study of Huygens, which is the purpose of the fourth book of *Geschichte der Atom-istik*, is followed by a last book whose main protagonists are Leibniz and Newton, to both of whom Lasswitz also devotes the conclusion, where he deplores the metaphysical impli-cations of remote force ("das mathematische Gesetz wird unter dem Namen der Fernkraft versinnbildlicht und anthropomorphisiert" [Lasswitz, 1890, vol. II, p. 580]) and thus sees a regressive tendency in both Leibniz and Newton: "Und so sehen wir den großen Physiker und Mathematiker Newton bei demselben Resultate anlangen, wie den großen Philosophen und Mathematiker Leibniz." [Lasswitz, 1890, vol. II, p. 580] One can add here that it was not until Kant that science was founded at last on a clear separation of different sources of knowledge.

Lasswitz focuses his study of Leibniz's role in the history of atomism on his efforts to tackle motion. He shows him as starting from Gassendi's atomism, moving rather quickly on to a corpuscular theory, and seeking the solution to motion in ether. At this point, Leibniz manages to grasp intensive magnitude, thanks to his understanding of the infinitesimal, but, staying too close to Aristotle and Hobbes, he falls back on an anthropomorphic approach to bodies and their relationships, looking for their cohesion by way of a logic that again takes him away from the possibility of a quantitative approach. Finally, instead of "realizing motion", which he could do if he were to accept the atom, he substantializes force as the cause of motion and thereby adopts a dynamic metaphysical theory.

Concrete motion

Under the influence of Bacon and especially Gassendi, young Leibniz was nevertheless a fervent advocate of atomism, which in 1665 he saw as the only route for knowing nature. However, his initial writings, notably *Ars Combinatoria*, reduce the original coherence of atoms, as well as their interaction, to an act of creation by a divine being endowed with the attributes of unity, wisdom, and power. Granted, this recourse to a theological foundation does not interfere, according to Lasswitz, with his mechanical, atomist conception of nature. However, while Leibniz settles for this *initial* divine intervention, it seems as if what warrants it leads him closer and closer to a way of thinking that separates matter from motion; this slides him over in the direction of dynamism. To make this claim, Lasswitz relies on Leibniz's 1668 and 1669 attempts to reconcile mechanical theory with Aristotle, attempts by means of which he expresses the idea that the cause of motion, if it cannot be *in* bodies, can only come from the mind, and in the end, from God. Later, moreover, we will see that Cotes reasons in the same way, starting from Newton's remote forces. Leibniz nonetheless strives to limit divine intervention to an initial act of creation.

Hypothesis Physica Nova of 1670 – which Lasswitz uses as a basis for marking off the major phases of Leibniz's reflection about motion – has two parts, the first in which Leibniz attempts to relate all natural phenomena to his ether hypothesis and to the formation of bubbles, the second aimed at providing rational grounds for the concepts of the continuousness of matter and motion, assumed in the first part.

Lasswitz carefully traces back over this "new physical hypothesis" developed by Leibniz, which consists of reducing all changes in the universe to a single principle, the general circulation of ether. Everything from the world of planets filled with liquid ether serving as a medium for transmitting motion (Leibniz is especially interested in the sun and the earth), to growth processes observed by the first users of microscopes (Kircher and Hooke), can be explained in terms of the circulation of ether and the formation, actions, and reactions of bubbles produced when ether encounters harder bodies. A body is solid if its parts move in the same way, if they are coherent; it is liquid if the parts move in different ways and the ether that penetrates into them is retained by a bubble. Such bubbles are the basis of bodies; they give bodies their specific shape and determine their diversity. From there, everything can be explained: the elasticity of bodies, the propagation of light and sound, chemical or even organic reactions. Water is an accumulation of these bubbles, and air and ether differ only by the fact that air is heavy; circulating ether causes weightiness; fire is nothing more than the union of spurting ether and air, and so on. Due to the variety of their envelopes and contents, bubbles are of infinite diversity, and the set of all qualities of bodies can be related to this diversity. In speaking of some of these explanations, Lasswitz notes in passing that they are rather forced or even unclear, sometimes merely suggested. This, as we have seen, is but a hypothesis. However, if Leibniz makes it plain that he wishes to force no one, he asserts nonetheless that his hypothesis has a certain degree of necessity, that it is more than just a simple arbitrary conjecture. On the other hand, traditional representations of atoms, he says, with their hooks, their little balls, and their swirls always seemed to him to be fabricated in an artificial fashion and to be no more than intellectual pastimes.

But what Lasswitz stresses above all is Leibniz's desire to reconcile his theory with the diverse and sometimes very far-removed representations of a myriad of scholars such as Anglus, Digby (whom he calls a remarkable man while nonetheless being surprised by the "remains of metaphysical concepts" that make him accept the idea of an initial push), Fabri, and even Aristotle, Boyle, and others. Lasswitz insists on these efforts to accommodate, but also on the great open-mindedness of Leibniz, who strives to rely on other theories too, including physiological or chemical ones. It is thus all the more puzzling to find that Leibniz makes no mention of his great proximity to Hobbes, that he dwells, on the contrary, on what he considers to be the latter's weaknesses, criticizing his mathematical concepts and his refusal to believe in incorporeal souls and indivisibles. Yet, his points in common with Hobbes are true points of convergence, and they are fundamental ones: the two scholars share the idea that the universe is filled with matter, that there is a liquid ether, that motion cannot be immanent in matter, which is divisible ad infinitum, etc.

The importance one can grant to this only-partially-acknowledged Hobbesian descendence is that it will not fail to put Leibniz in a rut, which, as we shall see, is revealed later by the difficulties he encounters in getting a handle on bodies. Lasswitz settles for stating, without further comment, that he is surprised to see Leibniz's apparent unawareness of the real similarity between Hobbes's thinking and his own. The reader is forced to discern therein that Leibniz's lack of lucidity on this particular point is a key to his erring – it not only prevents him from realizing exactly what he retains from Hobbes but also leads him, on the other side, to retain elements that are incompatible with his own view. By bringing this forth from the study of texts, Lasswitz puts critical analysis to work in the fullest sense of the term: he strives to clarify the thought process by attempting to locate the boundaries that escaped the thinkers each step of the way.

Lasswitz dwells somewhat on this first part of *Hypothesis Physica Nova*, perhaps in order to stress not only the ingeniousness of the solutions, but also their coherence and the multiple efforts made to explain the greatest possible number of phenomena with the smallest possible means. From this approach, Lasswitz notes, the philosopher was to retain above all the effect of ether and his concept of continuity. But Leibniz himself recognizes that he still doesn't have the necessary mathematical tools: he doesn't yet understand that something else is needed to explain motion, namely force. In regards to the various hypotheses about ether, Lasswitz asserts that none have much merit in comparison to other corpuscular theories. The famous bubbles can only support a sense-based representation of phenomena. From this hypothesis, neither mathematical laws nor mechanical principles can be derived. However, for the second part of his writings, Leibniz states that he intends, precisely, to proceed in a purely rational manner, without recourse to sense experience.

The theory of abstract motion

Lasswitz accentuates two aspects. The first concerns Leibniz's greatest accomplishment, differential calculus; the second explores the dead-end road taken by Leibniz, precisely due to his poor grasp of his proximity to Hobbes.

Seeking to base his theory of motion on a purely rational foundation, Leibniz's reflections on continuity bring him to some essential thinking. To define what he calls *inextensum*, he starts from a definition of "a point" that departs from both Euclid and Hobbes:

> Ein Punkt ist nicht dasjenige, was keine Teile besitzt, noch dasjenige, dessen Teil nicht in Betracht kommt [...] sondern dasjenige, dem keine Ausdehnung zukommt oder dessen Teile keinen Abstand haben, dessen Größe sich nicht in Betracht ziehen oder angeben läßt, weil sie kleiner sind als das, was durch ein Verhältnis zu einer andern sinnlichen Grösse (es sei denn ein unendliches) darstellbar ist, und kleiner, als man angeben kann. [p. 464]

He draws from the mathematics of his time, Cavalieri's *Geometria Indivisibilium* published in 1635, and claims to ground it himself with his own reasoning; in doing so, Lasswitz tells us, he clarifies the concept of infinitely small and paves the way to differential calculus, an achievement whose epistemological importance cannot be overestimated. What he achieves is "die Ableitung der sinnlichen Ausdehnung aus dem rationalen Begriff des Kontinuums." [p. 464] He can then start from indivisible elements (indivisible because they are non-extensive) and construct the extension in a continuous fashion by applying Cavalieri's method, i.e., in the same way as a line produces a plane when moved in a regular manner. Extensiveness has an objective reality in the concept of *tending toward* the extension – the conceptual tool of variability – no longer as a fact but from the angle of the law governing its constitution. And for objectification to be possible, an element of measurement must exist; the extension must become a magnitude, which can happen as soon as time is brought into the picture.

Note the importance of the notion of construction in this presentation: Lasswitz preferentially uses the term "erzeugen" (to generate), more concrete than "konstruieren" (to construct) and closer to production, stressing even more the creative-activity dimension of the thought process, which cannot be a mere reading of reality but, on the contrary, is only fruitful if it manages to free itself of the senses.

But – and this is what will prevent Leibniz from following through with the "construction" process implemented in his attempt to understand *tending toward* the extension – his insufficiently clarified adherence to Hobbes takes him off track. His reflections on motion make use of the Hobbesian concept of "conatus", which allows him to grasp motion up to the point where, interrupted by an obstacle, it is still possible to apprehend it in the infinitely small of a temporal moment, in terms of *tending toward* motion. Lasswitz captures the historical importance of this meeting of minds between Leibniz and Hobbes:

> Der Begriff des galileischen Moments ist durch Hobbes' Vermittlung an den Mann gelangt, welcher bald darauf das arithmetische Zeichen für denselben fand und nunmehr die Realisierung der Bewegung im Differenzial vollzog. [p. 467]

But in the immediate present, Leibniz falls back onto a metaphysical conception of bodies that is apparent in the way he again raises the question of body cohesion in a letter to Hobbes on the 13th/23rd of July 1670. In this letter, he proposes an answer based on a representation of pressures that has them penetrating each other.

For Hobbes, the body's cohesion comes from the mechanical equilibrium of the different conatus, but, Lasswitz reminds us, Hobbes's concept of conatus objectifies pressure as virtual motion, which Leibniz does not do. For Hobbes, it is possible to distinguish and quantitatively define unequal forces and their differing effects. Leibniz, on the other hand, confounds geometric continuity and mechanical cohesion by defining the body in terms of mathematical continuousness, and he then loses sight of the physical meaning of cohesion, which requires being able to compare and measure. Lasswitz quotes this "geistreiche[s] Aperçu" [p. 469] found in the letter from Leibniz to Hobbes, where we clearly see how far Leibniz is from a truly physical conception when he thinks he can answer the question of body cohesion by reasoning about the effects of pressures on the disappearance of their boundaries. He thereby moves away from any possible way of drawing a mathematical advantage from the notion of conatus in question here.

The problem, according to Lasswitz, lies in the fact that although Leibniz did indeed attempt to distinguish the geometric point (the spatial element) from the mechanical point (the conatus, the element of a trajectory), he failed to follow through with this approach, borrowing from Hobbes the concept of conatus without sharing his conception of the substantiality of the extension of bodies. His study of the continuum had led him to understand it as being made up of non-extensive elements that one could represent as lines, granted, but that should not have been conceived of as lines in the definition of the concept itself. Leibniz made this mistake in taking up on Hobbes's idea – irreconcilable with his own view – of the presence of a point at different places in space during motion. In fact, it wasn't long before he himself realized that his body-cohesion hypothesis was unsatisfactory, and made it known to Fabri that the only conceivable explanation would have to concern the concordance of the movements of the elements of bodies.

The contribution – and the limitations – of infinitesimal calculus

In the meantime, Leibniz's theory of motion had many gaps, and the failure of his search for how to explain the cohesion of bodies brings out a problem we might state as follows: what is missing is a concept capable of giving substantiality to bodies, or reality to motion.

We can go back and start over from the common ground shared by Leibniz and Hobbes, whose notions of instant, point, and conatus reveal the same goal: to found the continuum on the concept of the infinite. But there is a fundamental difference between Leibniz and Hobbes – Hobbes bases the non-extensive on a logical decision to disregard the extension, whereas Leibniz cannot accept that the extension might exhaustively explain the nature of bodies. It is his mathematical research from the late 1670's onward that allows him, via the algorithm of differential calculus, to define ratios in the infinitely small, transpose them onto extensive magnitudes, and perform operations on them. Indeed, explains Lasswitz, the transition from the non-extensive (the infinitely small) to the extensive (the magnitude) was possible for Leibniz because, unlike Hobbes, the magnitude was not given by the extension but by the number. He notes, however:

> Die stetige Auffassung der Zahl und ihre Anwendung auf die Elemente von Raum und Zeit gaben reiche Entdeckungen in Arithmetik, Geometrie und Phoronomie, aber sie gaben keine Dynamik, keine Darstellung der Kraftwirkungen der Körpermassen. [p. 472]

It seemed indeed that, in bodies, there had to be something else, something other than the simple extension. Or in Lasswitz's words, which paint a picture of the problem as Leibniz must have posed it himself, "Zwar ließen sich die Conatus mathematisch zusammensetzen, aber die Körper wollten ihnen nicht folgen!" [p. 474], the interaction of bodies did not follow from this. And as noted by Leibniz, quoted at length by Lasswitz in regards to these difficulties:

> [...] so sah ich doch, daß dies aus jenen Regeln der Bewegung, wie man sie auch kombiniere, allein nicht erhalten werden könne, sondern daß es dazu eines gewissen höheren Prinzips bedürfe, um Regeln einer systematischen Bewegung zu erhalten, da die Körper selbst, wenn sie nur in ihrem mathematischen Begriff bestehen, die zukünftigen Wirkungen und daher die Gesetze ihrer Bewegungen nicht enthalten können. [p. 476]

From this, he draws the conclusion that the principles of mechanics and the laws of motion do in fact come from a "necessity of matter", while nonetheless reusing the notion of "higher principle":

> Die mechanischen Prinzipien selbst und die Gesetze der Bewegung stammen, glaube ich, aus der Notwendigkeit der Materie, aber aus einem gewissen höheren Prinzip, das von der Anschaulichkeit und dem Mathematischen unabhängig ist. [p. 476]

It was Huygen's work that allowed him to come back to physics. But paradoxically, it supported him in his search for the origin of motion outside of the corporeal world.

From the substantialization of force to the monad

Indeed, from Huygens's experiments, Leibniz was to draw out the idea that since the conservation of force only holds true for collisions of elastic bodies, the ultimate parts of bodies have to behave as if they were elastic. But, unlike Huygens, he did not see matter as finding its substantiality in solid atoms, or motion as finding its reality in the principles of mechanics. For him, matter had no substantiality; it was motion that had to supply it. Lasswitz quotes Leibniz:

> Der Ausdehnung ist die Aktion hinzuzufügen. Der Körper ist also ein ausge-
> dehntes Agens. Man könnte sagen, es ist ausgedehnte Substanz, wenn man
> nur festhält, dass alle Substanz agiert und jedes Agens Substanz genannt wird.
> [p. 479]

Lasswitz refers to the "Substanzialisierung der Kraft" [p. 470] as being Leibniz's basic error, but for him, what characterizes his thinking the best is that the "ambiguous" concept of action reduces action to a spiritual power that is put into things by the Creator. In order for the interaction of bodies to be guaranteed by the principles of mechanics, explains Lasswitz, two things are required: changes in space and time must acquire both objectivity (i.e., identity) and continuity. Identity, or substantiality, creates units of space; continuity, which Lasswitz also calls reality or variability, creates units of time. Through its substantiality, an element of a body is seen as identical to itself; through its variability, it is seen as an element of the continuous change, including the law governing that change. Taken together, these two thinking tools define the concept of atoms endowed with energy. But in a dynamic way of thinking like that of Leibniz, the concept of body is made up solely of the law of continuous change; substantiality is eliminated and moved over to the concept of force. This is what becomes apparent when we look at how Leibniz switches from a mechanics-based theory to a dynamic theory by searching for substance, not in the extension but in something that is "behind" it, thereby substantializing force and making it into a metaphysical entity.

It is as such that Leibniz arrives at a conception of formal atoms, which, to make them more understandable, he calls "original forces" rather than first entelechies, as Aristotle did, and he makes monads out of them. In fact, they are "metaphysical points" (as for Lasswitz), which have something alive about them, and it is only their union that makes them into physical points. But the monad cannot and must not be used to *explain* physical facts; it can only *give reality to* those facts. Given the purpose of Lasswitz's book, his study of Leibniz stops here, because with the monad, he insists, Leibniz infringes upon the domain of the laws of mechanics by "grounding" ("begründen", "fundieren") physics in metaphysics.

Leibniz the "philosopher and mathematician" and epistemological obstacles

Lastly, we see that Leibniz, while experiencing a profound attraction for metaphysics, starts from atomism in his attempt to understand the physical world, and then confronts his own ideas with those of Hobbes, with whom he shares a keen interest in the questions of continuity and the infinitely small. Unlike Hobbes, he manages to devise a mathematical instrument that will allow Newton to formulate the laws of gravitation. Leibniz himself, however, will not arrive at a concept of matter that is free from metaphysical assumptions.

Firstly, despite his fundamental interest in mathematics, Leibniz remains very close to an author to whom Lasswitz devotes but a few pages, Spinoza. Not that he takes him lightly: on the contrary, he recognizes the decisive role that Spinoza played in the advancement of scientific knowledge when he unyieldingly defended the autonomy and liberty of science. Moreover, he mentions the influence Spinoza had on Leibniz, which, in the end, proved to be detrimental to kinetic atomism. It was indeed his notion of substance (whose impact on the progression of Leibniz's work was noted by Lasswitz) that Lasswitz saw as responsible for reviving in Leibniz the necessity of a metaphysical foundation for thought.

The second reason is related to the first: Leibniz has infinitesimal calculus playing a metaphysical role, since he starts from his reflection about continuity and the infinitely small, which he resolves mathematically, to arrive at the theory of the monad: one is not thinkable without the other. We can thus legitimately say that in Leibniz, the epistemological obstacle is not just philosophical, it is also mathematical.

Conclusion

Saying that Leibniz stumbled on a metaphysical obstacle does not mean that his conception of the world was hindered by his faith or by a dogmatic corpus that might have blinded him. Granted, Lasswitz gives the example of Leibniz's trans-substantiation problems, which he acknowledges as having possibly been a genuine concern for the philosopher. But he does not make this a critical factor. While religious and dogmatic beliefs may have oriented Leibniz's thinking, they are not thought to be the basis of his positions. Instead, the latter are seen as firmly rooted in the metaphysical dimension of his understanding of bodies, especially force, which in his mind had to spring, as we have seen, from a "higher principle". In no case does Lasswitz reproach Leibniz for his metaphysics – and his own religious faith probably accounts for a certain wavering around the concept of metaphysics – but he does reject the mixing of genres.[1] What this study of the missed meeting between Leibniz and atomism truly has to offer, then, is a consequence of the neo-Kantian – or criticist – reading that is made of it. The main merit Lasswitz attributes to the criticist method he claims as his own lies in the idea that Kant knew how to definitively separate science and metaphysics, so that the latter discipline could no longer claim to ground the former. The problem then becomes one of the relationship between thought and reality. In relying on Kant, in relying on his inquiry into the relationship between the subject and the object of knowledge, between knowledge and experience, Lasswitz will be then able to define what is supposed to happen in a successful process of knowledge generation about matter:

> Die Aufgabe der erkenntnistheoretischen Untersuchung der Materie ist, die-
> jenigen Bedingungen aufzufinden, welchen der beschriebene Bewußtseinsin-

---

[1] See Françoise Willmann, "De la compatibilité entre science et religion: l'argumentation de Kurd Lasswitz au service de l'une et de l'autre", *Recherches Germaniques*, No. 39, 2009, p. 27–45.

halt genügen muss, damit er Gegenstand mathematischer Behandlung werden kann. [Lasswitz, 1885, p. 138]

For Lasswitz, atoms are not real and/or transcendental objects; they are the product of the cognizing we do to make sense of the world; they are rational formations, or in Kantian terms, they constitute a transcendental base for experience. "Atome sollen also nichts anderes bedeuten als diejenigen Teile des Raumes, welche für sich als einzelne Ganze bewegt sind, d.h. die durch den Substanzbegriff fixierten Raumelemente." [Lasswitz, 1885, p. 151] The difficulties with which Leibniz battles are linked to the fact that his mathematical approach causes him to lose "the real", i.e., the substance, a loss of which he is well aware. This is why he reintroduces substance in a place where it doesn't belong – in force – a notion that is only freed of its metaphysical character when defined as the product of mass and velocity, that is, in mechanical terms.

There is no question here of reducing this study of Leibniz – seen from the angle of its significance in the history of atomism, that is to say, in the history of the emergence of modern physics – to an attempt to assess its importance in a teleological narrative. Lasswitz's goal is not to enumerate the geniuses involved in this history, nor to evaluate or outline the limitations of each of their contributions. The table of contents, which basically lists the names of the protagonists, is reminiscent of the hagiographical histories of the heros of science like those readily attributed to 19th-century writings. Notwithstanding, Lasswitz's work is aimed instead at understanding the movement in the thinking itself, through a meticulous study of texts that takes into account various aspects of the sociological or psychological context (without however going further into them for their own sake). As an example, we often find Leibniz's conciliating temperament being mentioned along with its epistemological implications: Leibniz's thought process is shown as being hindered by sometimes contradictory connections, and is inserted in the broader issue of whether it is fruitful to stand apart from reasoning by analogy. As a general rule, physics, i.e., understanding of nature, appears indeed as a conquest, but as a collective conquest, one in which the very example of Leibniz demonstrates just how complex and difficult that conquest really was, and above all, how laborious it was to develop the necessary tools of thought, while striking a delicate balance between perception and understanding, and engaging in a lengthy, ongoing process of adaptation of one's perspective.

Errings, however, should not be thrown out of history. Following his doctorate in physics Lasswitz defended four theses, as was customary. One of them was that science contains an abundance of poetic elements. In this respect, the history of atomism is particularly rich. Lasswitz makes good use of the Leibnizian representations of continuity in the universe. In the passage studied above, for example, he reminds us of how Leibniz hypothesized that if our eyesight were good enough, we could no doubt see in miniature what we see in full scale, to infinity; and of how this has to be possible because whatever is continuous is infinitely divisible, each atom necessarily representing a sort of universe of infinite forms, with universes inside universes, ad infinitum.

Lasswitz ends this paraphrasing of Leibniz with the following words: "Wer dies tiefer betrachtet, wird nicht umhinkönnen von Begeisterung bei der Bewunderung des Schöpfers der Dinge hingerissen zu werden." [Lasswitz, 1890, vol. II, p. 457] The contemporary reader tempted to detect a hint of irony in this conclusion – mentioning such an emotion in this context is somewhat surprising – would surely be mistaken. Lasswitz himself defends the right to religious faith, for which he advocates a legitimacy equal to that of science – the bliss of Leibniz in the face of the works of the Creator is also his own. From the scientific point of view, Lasswitz's first contribution to *Vierteljahrsschrift für Wissenschaftliche Philosophie* was a reply to an article by Wilhelm Wundt entitled "Ein Beitrag zum Kosmologischen Problem und zur Feststellung des Unendlichkeitsbegriffes". Discussing the questions raised by Wundt on the issue of knowing whether the universe should be represented as finite or infinite, Lasswitz arrives at the following hypothesis, via a line of reasoning about nature and the meaning of infinite magnitudes in mathematics and physics: The universe can be conceived of as an infinite succession of finite systems. If this representation echoes the Leibnizian one, all the while being based on other grounds, the short story *Auf der Seifenblase* drawn from it makes this idea into a humorous and satirical fiction in which soap bubbles turn out to be miniature globes of the world: "Welten in Welten ins Unendliche" [p. 457].

References

[Bachelard 1933] Bachelard, Gaston, *Les intuitions atomistiques, (Essai de classification)*, Boivin, Paris, 1933.

[Cassirer 1994] Cassirer, Ernst, *Philosophie der symbolischen Formen*, Teil 3, Wissenschaftliche Buchgesellschaft Darmstadt, 1994.

[Cohen 1968] Cohen, Hermann, *Das Prinzip der Infinitesimalmethode und seine Geschichte*, Suhrkamp, Frankfurt/Main, 1968.

[Lasswitz 1885] Lasswitz, Kurd, "Zur Rechtfertigung der kinetischen Atomistik", *Vierteljahrsschrift für wissenschaftliche Philosophie*, 1885.

[Lasswitz 1890] Lasswitz, Kurd, *Geschichte der Atomistik vom Mittelalter bis Newton*, Georg Olms, Hamburg/Leipzig, 1890.

[Lasswitz 1890] Lasswitz, Kurd, *Seifenblasen, Moderne Märchen*, Hamburg, 1890.

[Lasswitz 1897] Lasswitz, Kurd, *Auf zwei Planeten*, Weimar, 1897.

[Kollektion Lasswitz 2010] Kollektion Lasswitz, *Neuausgaben der Schriften von Kurd Lasswitz in der Fassung der Texte letzter Hand*, Dieter von Reeken, Lüneburg, 2010.

[Willmann 1997] Willmann, Françoise, "La lampe de la science et les bulles de la fiction: Kurd Lasswitz, 1848–1910", *Alliage*, n°32, automne 1997, p. 49–54.

[Willmann 2009] Willmann, Françoise, "De la compatibilité entre science et religion: l'argumentation de Kurd Lasswitz au service de l'une et de l'autre", Recherches Germaniques n°39, 2009, p. 27–45.

Françoise Willmann
Université Nancy 2 – Campus Lettres Sciences Humaines
B.P. 13397
F-54015 Nancy Cedex
francoise.willmann@univ-nancy2.fr

Peano and His School
Between Leibniz and Couturat:
The Influence in Mathematics and
in International Language

Erika Luciano

1. The influence of Leibniz on Giuseppe Peano

There are many significant testimonies to the legacy of Leibniz's thinking on Peano: at least two hundred textual references and quotations from Leibnizian writings, which can be identified in his output.[1] The influence of some of these suggestions in various mathematical disciplines can be recognised, above all, where Peano and his students consider and present their work as being designed to complete or to extend some of the lines of research indicated by Leibniz. Thus certain *Leitmotive* emerge: the desire to situate the developments of ideography in a historical context, finding in Leibniz a precursor of their work; the sharing of Leibniz's philanthropic ideals and, in particular, of his conviction of the usefulness of an international language to facilitate collaboration and communication among peoples; the inheritance of an interest in the history of mathematics, with a view to establishing the paternity of concepts, methods and symbols and lastly, the desire to take from Leibniz's work cues for the development of new results of pure and applied mathematics.

Nonetheless, if until 1894 the references are fragmentary and sporadic, from that date they become frequent in the works of Peano and of his collaborators Giovanni Vacca and Giovanni Vailati. The reason for this can be traced to the 'discovery' of algorithmic logic, to its use in mathematics and to the beginning of the plan of the *Formulaire*.

In fact, one feature of Peano's mathematical and linguistic studies is his resorting to the original sources, numerous excerpts from which he carefully transcribed. Attention to the philological aspect and preference for the faithful critical editions are aspects that are

---

[1] The *corpus* of quotations is organized around a single conceptual axis, which can be recognized in the semantic link between sign, algorithm and language.

continually repeated by Peano. It is important, then, to ask what were the sources of Leibnizian literature that he consulted. Peano uses three different editions of Leibniz's works. Specifically, the results of Logic, Geometry and Arithmetic are taken from the *Opera philosophica* edited by Erdmann (1840) and, after 1898, are gathered from unpublished manuscripts examined by Vacca in Hanover in summer 1899. For the studies on Binary Arithmetic Peano makes reference to the oldest edition of Leibniz, the one edited by Louis Dutens (1778). Lastly, he makes systematic use of the volumes of the *Mathematische* and of the *Philosophische Schriften* published by Karl Immanuel Gerhardt. Peano does not carry out a philosophical and philological critique of Leibniz's works, but constantly updates his knowledge of the contemporary literature about the German mathematician and philosopher. For example, he follows Couturat's studies with interest, appreciating the two volumes *La Logique de Leibniz* and *Opuscules et fragments inédits de Leibniz*.[2] A further source of information must finally be identified in the writings of Ernst Schröder, whom Peano met at the International Congress of Mathematicians in Zurich in 1897 and who, on that occasion, defined Leibniz as a precursor of pasigraphy.[3] As regards the bibliography of Leibniz, in Peano's Library[4] the edition of the works by Dutens, the volumes by Couturat and a number of little-known extracts, among them the *Esquisse d'une grammaire de langue conforme aux idées de Leibniz* by V. Hely, have come to light.[5]

2. The *Formulaire* and the 'Leibnizian dream'

In 1894 Peano and a team of colleagues, collaborators and students launched an ambitious project, the *Formulaire de mathématiques*. At first it was simply a *Collection of Formulas* (*Raccolta di Formule*), but it soon became the engine of Peano's entire research activity, and of that of his School. Conceived in the threefold nature of encyclopedia of the elementary mathematics of earlier centuries, translated into logical language, as a teaching handbook, and as a repertoire of research programs, the *Formulaire* soon became an object of fierce debate, at national and international level.

It is in the *Formulaire* above all that Peano's debt to Leibniz's ideas about the *Encyclopedia generale* and the *Characteristica universalis* can be perceived. From 1669, Leibniz describes in numerous notes his plan for a universal dictionary of all knowledge, the compilation of which is made possible by the use of the *characteristica*:

---

[2] L. COUTURAT 1901, 1903.

[3] E. SCHRÖDER 1898, *Über Pasigraphie, ihren gegenwärtigen Stand und die pasigraphische Bewegung in Italien*, *Verhandlungen des Ersten Internationalen Mathematiker-Kongresses in Zürich vom 9 bis 11 August 1897*, Leipzig, Teubner, p. 147–162, in particular p. 147–148.

[4] It is possible today to have very exact knowledge of the sources consulted by Peano thanks to the discovery, in February 2007, of his personal Library, which had been believed lost. Cf. website http:// www.peano2008.unito.it/ *Catalogo della Biblioteca di Peano* and E. LUCIANO 2007, *La biblioteca "ritrovata" di Giuseppe Peano*, in L. BONO, S. CHIAVERO, D. DAMIANO (eds.), *Rendiconti Cuneo*, Cuneo, Nerosubianco, p. 184–188. This is an important heritage, comprising approximately 1300 volumes and more than 2500 documents. Many of the volumes are peppered with notes handwritten by Peano with observations, corrections, notes on mathematics, history and bibliographic cross-references.

[5] V. HELY, *Esquisse d'une grammaire de la langue internationale*, Langre, M. Berret, 1905.

"Consilium de Encyclopaedia condenda, velut Inventario cognitionis humanae condendo in quod referantur utiliora, certiora, universaliora et magis sufficientia pro reliquis omnibus determinandis; additis semper rationibus eorum quae fiunt originibusque inventionibus. Quod opus non nimis erit prolixum [...]. Hujus operas usus erit ut occurratur confusioni librorum eadem repetentium, paucaque interdum utilia sub magna farragine obruentium, si sit Basis aliqua ad quam omnia imposterum nova per modum supplementorum referri possint."[6]

The distinctive features of such an encyclopedia are taken up by Peano, who liked to present the *Formulaire* as the fulfillment of Leibniz's dream of the construction of the universal encyclopedia, saying that:

«après avoir tombé depuis longtemps dans l'oubli, est maintenant réalisé, grâce a la logique mathématique, la nouvelle science qu'a pour objet les propriétés des opérations et des relations et donc les résultats sont merveilleux, et bien dignes des éloges de Leibniz à la science qu'il avait deviné».[7]

The *Formulaire* is, in any case, a work that created a rift in the publishing panorama of the time, so it is not surprising that Peano and his collaborators should look for its natural 'archetype', with a view to situating this treatise in a correct historical, mathematical and philosophical perspective.

In his prefaces to the works on logic, Peano repeatedly states that Leibniz is the mathematician who has understood the most general problem ever posed and faced in the course of the centuries, namely the problem of developing

«pendant toute sa vie, depuis son premier travail jusqu'à ses dernières lettres, une spécieuse générale ou une manière de langue ou d'écriture universelle, où toutes les vérités de raison seraient réduites à une façon de calcul.»[8]

The *Spécieuse générale* is conceived by Leibniz as a support for the mind and the memory and it made possible the discovery of the gaps and inaccuracies in the deductive procedures, since «sophismes et paralogismes ne sont rien d'autre que solécismes et barbarismes».[9]

Peano's ideography, analogously, is as expressive and economical as possible and serves, so to speak, as a 'filo d'Arianna' for thinking. Algorithmic logic is a set of procedures which transform reasoning into calculus and, at the same time, it is a tool that brings reflection closer to writing. Symbols, as Leibniz had already recommended, are not just abbreviations but represent ideas, hence «ideography is not tachigraphy».[10] Peano rediscovered Leibniz's reflections on Chinese ideography and made them once more of

[6]G.W. LEIBNIZ 1679, *Initia et Specimina Scientiae Generalis de instauratione et augmentis scientiarum*, GP, p. 58.
[7]G. PEANO 1896b, *Introduction au tome II du "Formulaire...*, RdM, 6, p. 2. Cf. also G. PEANO 1891c, *Principii di logica matematica*, RdM, 1, p. 9 ; 1894g*, *Notations de Logique Mathématique*, p. 3, 52 ; 1894e, *Un precursore della logica matematica*, RdM, 4, p. 120 ; 1896b, *Introduction au tome II du "Formulaire de mathématiques"*, RdM, 6, p. 1, 3 ; 1901b, *Formulaire de Mathématiques*, p. IX.
[8]G. PEANO 1896i, *Réponse n. 719. (Lausbrachter)*, L'intermédiaire des mathématiciens, 3, p. 169.
[9]G.W. LEIBNIZ, A I, 2, p. 240.
[10]G. PEANO 1896j, *Studii di logica matematica*, Atti della Reale Accademia delle Scienze di Torino, 32, p. 566.

current interest, for this ideography allows the reciprocity between characters and ideas to be preserved at best. Moreover, ideography not only enables the statement of theorems and definitions in a clear form, but is in general the indispensable tool for analyzing the principles of a theory and identifying its primitive and derived ideas, the axioms and the theorems. For Peano and for Leibniz, the *Characteristica* has consequently a sort of amphibious nature, for it records and at the same time it organizes and produces knowledge. This last aspect is the most problematic. In fact Peano several times states, under the influence of Leibniz, that symbols make the invention of new, elegant theories easy[11] and that:

> «l'utilità principale dei simboli di logica si è che essi facilitano il ragionamento. [...] Perciò il simbolismo è più chiaro; permette di costruire serie di ragionamenti quando l'immaginazione sarebbe interamente inabile a sostenere se stessa senza aiuto simbolico.»[12]

Nevertheless, in Italy and in France, many mathematicians were to continue to maintain that original results cannot be concretely obtained only by means of symbolism and without recourse to intuition and other forms of synthetic or *a priori* reasoning.

In the light of these quotations, the conviction emerges, largely shared by C. Burali-Forti, A. Padoa, G. Vacca, G. Vailati, M. Pieri and U. Cassina, that the new researches on pasigraphy fit harmoniously into a tradition which, after Leibniz, continued with the studies on the algebra of logic, to culminate in the birth of a new discipline.[13] In fact, this conviction would fade markedly with the passing of time. The tumultuous development of mathematics at the end of the 19th century had aroused the wish to found and contextualize in a specific historical perspective a vast mass of results of recent acquisition and for Peano "suddenly, what Leibniz had said about the *Characteristica Universalis* assumed the value of a prophecy".[14] His attempt to present the modern directions of logico-foundational researches as a repercussion of Leibniz's intuitions can, and must, today be broadly revaluated, but it seems to be the result of a carefully thought-out choice.

3. Leibniz's manuscripts in the *Formulaire des Mathématiques*

The *Formulaire* was from the very beginning conceived as a work of collaboration, carried out by a team of mathematicians, historians and secondary-school teachers.

---

[11]G.W. LEIBNIZ, *Linguae philosophicae Specimen in geometria edendum*, 1680, A VI, 4 (1677–1690), Teil A, Band 1, Berlin, p. 384–385.

[12]G. PEANO 1915j, *Importanza dei simboli in matematica*, Scientia, 18, p. 170, 172.

[13]Cf. for example C. BURALI-FORTI 1897, *Introduction à la Géométrie différentielle suivant la méthode de H. Grassmann*, Paris, Gauthier-Villars, p. VI; C. BURALI-FORTI 1919, *Logica matematica*, Milano, Hoepli, p. XVII-XVIII; U. CASSINA 1933, *L'oeuvre philosophique de G. Peano*, Revue de Métaphysique et de Morale, 40, p. 489–491; A. PADOA 1933, *Il contributo di G. Peano all'ideografia logica*, Periodico di Matematiche, 4, 13, p. 15–18; M. PIERI 1906, *Uno sguardo al nuovo indirizzo logico-matematico delle scienze deduttive*, Annuario della Università di Catania, p. 394–396; G. VACCA 1946, *Origini della Scienza*, Roma, Partenia, p. 31.

[14]M. MUGNAI 1973, *Leibniz e la logica simbolica*, Firenze, Sansoni, p. 3.

The first edition was printed in 1894–1895, but work for a new edition began immediately, as Peano wrote to F. Klein: «every day is a new part that is translated into symbols».[15]

However, the first edition lacks virtually all the historical indications, which are essential for readers to perceive the origins, developments and links between propositions and theories.

At this time Peano happened to read an article by Vacca, a young student of mathematics at the University of Genoa.[16] In this text, by means of the correspondence between Leibniz and Johann Bernoulli, and thanks to the analysis of the *Nova Algebrae promotio* published by Gerhardt, Vacca shows that the German mathematician was the first to have established the formula which gives the coefficient of any term in the development of the power of a polynomial. Appreciating the norms of the historical research carried out by Vacca – distinguished as it was by the literal transcription of excerpts and precise references to the sources consulted – Peano did not hesitate to write to him, proposing that he collaborate on the *Formulaire*.[17] Thus, in 1894 a relationship was formed between Peano and Vacca, a very solid and long-lasting working and human alliance which is testified to by an intense correspondence. Vacca answered Peano a few days later, accepting his invitation, and commenting on his article:

> «Giunsi a trovare la dimostrazione di Leibniz leggendo le opere matematiche di Leibniz, delle quali è sperabile si faccia una edizione più facilmente consultabile che non quella del Gerhardt, la quale oltre ad essere priva di ogni indice analitico, necessario allorchè si vuole iniziare una qualche ricerca, è poco ordinata e confusa. Forse molte altre utili cognizioni si possono trovare in Leibniz, oggi trascurato e poco letto. Alle note storiche alla parte II § 10 si potrebbe aggiungere in relazione alla P.4 Leibniz 1695 lettere X e XII Joh. Bernoulli, ove si trova l'espressione form[ale] di una potenza qualunque di un polinomio».[18]

In fact, from this time on, the *Formulaire* is enriched by a myriad of references to the works of Leibniz on Logic, Geometry, Arithmetic, Algebra and Analysis.[19] Vacca

---

[15] G. Peano to F. Klein, 29.8.1894, M. Segre 1997, *Le lettere di Giuseppe Peano a Felix Klein*, Nuncius. Annali di Storia della Scienza, 12, p. 119–121.

[16] G. Vacca 1894, *Intorno alla prima dimostrazione di un teorema di Fermat,* Bibliotheca Mathematica, 2, p. 46–48.

[17] G. Peano to G. Vacca, 15.5.1894, c. 1r-v, E. Luciano, C.S. Roero (eds.) to appear.

[18] [«I came upon Leibniz's proof when reading his mathematical works, of which it is to be hoped there will be made an edition easier to consult than Gerhardt's, which in addition to being quite without an analytical index, necessary when one wants to undertake any research, is rather disorganized and confused. Perhaps much more useful knowledge can be found in Leibniz, today neglected and not much read. To the historical notes on part II § 10 there might be added in relation to P.4 Leibniz 1695 letters X and XII Joh. Bernoulli, where the form[al] expression of any power of a polynomial can be found».] G. Vacca to G. Peano, 31.5.1894, c. 1r-v, E. Luciano, C.S. Roero (eds.) to appear. Cf. also G. Vacca to G. Loria, [may 1894] and G. Loria to G. Vacca, 18.6.1894, P. Nastasi, A. Scimone (eds.) 1995, p. 90–91.

[19] Cf. G. Peano 1895aa, *Formulaire de mathématiques*, p. 1, 2, 3, 4, 25, 127, 128, 129, 132; G. Peano 1899b, *Formulaire des mathématiques*, p. 5, 6, 8, 9, 10, 11, 12, 14, 15, 16, 18, 20, 40, 42, 55, 66, 71, 72, 76, 80, 83, 85, 86, 99, 118, 119, 126, 144, 149, 165, 167, 169, 195; G. Peano 1903f, *Formulaire mathématique*, p. 3, 5, 8–10, 18, 21, 27, 50, 65, 71, 72, 91, 93, 142, 155, 156, 159, 169, 170, 174, 179, 189, 191, 206, 207, 234, 239, 264,

makes a close examination of both Gerhardt's collections and Leibniz's correspondence, progressively annotating the results of his studies.[20] The items gathered were first inserted in the *Additions et corrections* published in the *Revue de mathématiques* of which Peano was director[21]; they are recorded by Peano and by Vacca themselves in their autograph notes made on their personal copies of the *Formulaire*[22] and on the proofs of the treatise and, finally, they were brought together in subsequent editions. Vacca's researches are meticulous and very careful, but they lack orderliness. His collaboration on the edition of the historical notes to the *Formulaire* was greatly appreciated by Peano and further intensified when Vacca was staying in Turin, in the years 1897–1902, becoming Peano's assistant at the University. He was to remain in Turin until 1905, editing most of the historical and bibliographical apparatus of the *Formulaire*.

At first, in collaboration with Vailati, Vacca was occupied with historical notes to the chapter on Logic; subsequently, in 1898, he began to write sections on Arithmetic and, as a result, he studied Leibniz's contributions on the theory of numbers and on binary arithmetic, contributions which at the time were entirely forgotten and neglected by historiography.[23] Both Vacca and Peano gave particular importance to the article *Explication de l'arithmétique binaire*, in which Leibniz provides an interpretation of Fohy's system of hexagrams, as well as illustrating the laws of periodicity in many numerical dyadic progressions. Vacca's interest in this field continued until 1903, when he presented at the Second Congress of Historical Sciences in Rome a talk on the history of binary arithmetic, elaborated thanks to the information gathered in view of the historical notes to the *Formulaire*. In this note he goes back over the history of the dyadic system, starting from Fohy's hexagrams and arriving at the most recent developments of E. Lucas, also mentioning the applications to the technique of the calculating engines.[24] Once again it was to Leibniz that Vacca attributed the most brilliant and fruitful intuitions on the binary system.

265, 291, 306, 318, 358, 375; G. Peano 1908a, *Formulario Mathematico*, p. VII, 3, 4, 16, 17, 56, 61, 62, 92, 94, 129, 146, 223, 224, 256, 258, 263, 264, 277, 336, 342, 343, 395, 431.

[20]Cf. Vacca's manuscripts in the Library of the Dep. of Mathematics of the Turin University, *Fondo Peano-Vacca*, envelopes nn. 19, 25, 32, 33, 34, 35, 36.

[21]G. Peano et alii, 1898d, *Additions et corrections à F_2*, RdM, 6, p. 69, 70, 71, 73. Cf. also G. Vailati 1903, *Aggiunte alle note storiche del Formulario*, RdM, 8, p. 60–63.

[22]Cf. G. Peano 1894g*, *marginalia* p. 3; 1895aa*, *marginalia* p. 1, 2, 3, 88, 127, 128, 129, 141; 1897b*, *marginalia* p. 5, 18, 27, 32, 35, 39; 1899b*, *marginalia* p. 9, 10, 14, 15, 17, 18, 71, 135, 195; 1900a*, *marginalia* p. 19; 1901b*, *marginalia* p. 13, 14; 1903f*, *marginalia* p. 5, 7, 21; 1906g*, *marginalia* p. 32, 52, 124, 223; 1908a*, *marginalia* p. 276, 277, 401, 438.

[23]Under the influence of suggestions from Leibniz, in this period Peano planned and made a shorthand machine which worked on the basis of the binary system. Cf. G. Peano 1898m, *La numerazione binaria applicata alla stenografia*, Atti della Reale Accademia delle Scienze di Torino, 34, p. 47–49. Cf. also E. Luciano, C.S. Roero 2004, *La macchina stenografica di Giuseppe Peano*, Le Culture della Tecnica, AMMA, 15, p. 5–28 and E. Luciano, C.S. Roero 2004, *Dagli esagrammi di Fo-hy all'aritmetica binaria: Leibniz e Peano*, in E. Gallo, L. Giacardi, O. Robutti (eds.), *Conferenze e Seminari 2003–2004*, Torino, Ass. Sub. Mathesis, p. 49–69. The prototype has unfortunately been lost, but three postcards survive, written by Peano to Vacca in binary shorthand. Cf. G. Vacca to G. Peano, 2.11.1898, 28.10.1899 and 20.5.1903, E. Luciano, C.S. Roero, *La macchina stenografica di Giuseppe Peano*, 2004 cit., p. 20–22 and E. Luciano, C.S. Roero 2008, *Giuseppe Peano Matematico e Maestro*, Turin, Dep. of Mathematics, 2008, p. 167.

[24]G. Vacca 1904, *Sulla storia della numerazione binaria*, Atti del II Congresso Internazionale di Scienze Storiche, Roma 1–9.4.1903, v. 12, Roma, Tip. della R. Accademia dei Lincei, p. 63–67.

The second edition of the *Formulaire de Mathématiques*, published in 1898–99, is a key moment for the study of Leibniz's writings in the School of Peano because, when he came to write its historical notes, Vacca realized that there were some gaps in Gerhardt's *Schriften*. The ruling for the compilation of the apparatus to the *Formulaire* was very strict[25]. Peano had established that it was not sufficient to simply state that a certain result is found in a particular author or work. The exact place had to be specified and, where possible, the statement of every proposition was to be transcribed. It is not surprising, then, that in order to fulfil the task with which he had been entrusted, Vacca decided to go to Hanover to study Leibniz's manuscripts, which had been 'buried'[26] in the city library there for almost 200 years.

In 1903 Peano recalled that, after two editions of Leibniz's works, it was the general opinion at the time that the manuscripts no longer offered great novelties, but rather «suo importantia magis pate».[27] Vacca stayed in Hanover for only a matter of days in the month of August and immediately sent Peano the transcriptions of some unpublished papers so that they could be inserted in the *Formulaire*. Leibniz's were to be the only manuscripts extensively quoted in the historical notes to this treatise. Vacca's attention was directed above all to some *brouillons* that had been neglected till that moment: the ms. *Philosophie* VII B which comprises reflections on mathematical logic and the international language and the ms. *Mathematik* III A and III B, which are interesting for the results of the theory of numbers and of combinatorial mathematics.

In anticipation of the new edition of the *Formulaire*, Vacca also gave a brief account of his discoveries in the *Bollettino di Bibliografia e Storia delle Scienze Matematiche*,[28] a specialist journal on the history of mathematics, which had been founded in Turin in 1898 and was edited by the Genovese Gino Loria. Vacca here expressed his gratitude to E. Bodemann, who had made available to him the catalogues of Leibniz's unpublished works, allowing him to consult the manuscripts on the Fermat-Wilson theorem and thus to reconstruct the route followed by Leibniz in order to arrive at the proof of this theorem. At the end of his article Vacca wrote:

> «Un preciso esame delle scoperte di Leibniz relative alla logica matematica, che uscirebbe dall'indole di questa nota, sarà tra breve pubblicato nel *Formulaire de mathématiques* N° 3».[29]

In fact the second edition of this treatise is very different from the first, from the point of view of the history of mathematics, above all in its publishing Leibniz's fragments. These quotations were taken up several times by Peano and his students in the

---

[25]Cf. G. PEANO 1898e, *Sul § 2 del Formulario, t. II: Aritmetica*, RdM, 6, p. 83, 89.

[26]G. PEANO 1903c, *De latino sine flexione*, Cavoretto (Torino), Tip. Cooperativa, p. 8. Cf. also G. PEANO, 1914f, *Prof. Louis Couturat*, Revista universale (U. Basso), a. 4, 40, october 1914, p. 79.

[27]G. PEANO 1903c, *De latino sine flexione*, Cavoretto, Tip. Cooperativa, p. 8.

[28]G. VACCA 1899, *Sui manoscritti inediti di Leibniz*, Bollettino di Bibliografia e Storia delle Scienze Matematiche (G. Loria), 2, p. 113–116.

[29][«An exact examination of Leibniz's discoveries regarding mathematical logic, which would go beyond the nature of this note, will shortly be published in the *Formulaire de mathématiques* N° 3».] G. VACCA 1899 cit., p. 115.

years that followed, to the point that they became a sort of 'shared patrimony of knowl-edge', a *topos* of their works on the history of mathematics and of socialization of math-ematical culture.

The expression «an exact examination of Leibniz's discoveries regarding mathe-matical logic will be published in the *Formulaire de mathématiques*» might be perplexing since, on closer inspection, what is included in the *Formulaire* are simply a number of transcriptions of passages from Leibniz's manuscripts, without any kind of critical reflec-tion *a latere*. Nevertheless, this is a typical aspect of the historiographic methodology adopted in the School of Peano, as Vacca himself made clear to Couturat:

> «Je veux vous faire encore une petite description relative au F[ormulaire], pour les indications historiques. Lorsque j'ai commencé à ajouter des notes, je l'ai fait presque au hasard. En avançant dans le travail j'ai vu qu'il y avait là une nouvelle méthode historiques. Qu'est que c'est l'histoire d'une science ? On peut penser que ce soit l'exposé impartial des idées scientifiques de ceux qui nous ont précédé. Mais on ne peut pas les exposer toutes, si l'on veut les expo-ser toutes avec impartialité il faut les reproduire presque en entier. Ce travail prépare l'histoire, ce n'est pas encore l'histoire. La seule conception qui per-mettre de choisir dans les travaux des anciens c'est de se mettre à notre point de vue. Faire l'histoire des vérités d'une science, c'est chercher et exposer dans le passé tous les essais qui ont produit successivement les vérités que nous connaissons. [...] L'histoire d'une science est alors l'exposition ordon-née des vérités de cette science suivie d'un nome ou d'un date.»[30]

4. The collaboration with Louis Couturat

In 1900, in Paris, two important international congresses were held, one on philosophy and the other on mathematics. The philosophy congress in particular proved to be a fertile opportunity for cultural exchanges for the School of Peano and, for Vacca, there were very stimulating conversations with the French philosopher Louis Couturat.[31] This is what he said on the subject, in a lecture dedicated to the memory of Peano, given in Rome in 1946:

> «In quei due congressi, ai quali partecipavo, ebbi occasione di conoscere Bertrand Russell, il quale aveva pubblicato allora un volume su Leibniz (tra-dotto più tardi in francese nel 1908). Io stesso già da più di un decennio avevo studiato gli scritti di Leibniz, e da questo studio era sorta la mia amicizia per Giuseppe Peano. Feci quindi da intermediario tra Peano e Russell, il quale era allora in relazione con Louis Couturat [...]· Meravigliai Couturat quando

---

[30]G. Vacca to L. Couturat, december 1901, P. Nastasi, A. Scimone (eds.) 1995, p. 51–52.
[31]Couturat was one of the most active promoters in France of Peano's logic and of his *Formulaire*, to which he devoted a number of articles in the *Bulletin des Sciences Mathématiques* and in the *Revue de Métaphysique et de Morale*. Cf. L. Couturat 1901, *Peano G., professeur à l'Université de Turin. Formulaire de Mathématiques ... Turin, Bocca et Clausen*, Bulletin des Sciences Mathématiques, 2, 25, p. 159 et E. Luciano, C.S. Roero (eds.) 2005, p. IX–LX.

gli descrissi rapidamente la massa dei manoscritti inediti di logica matematica di Leibniz esistenti ad Hannover, che io avevo studiato colà nel 1899 e di
cui avevo già dato alcuni saggi nella *Rivista di Matematica*, nel *Formulario* di
Peano e nel *Bollettino di Storia della Matematica* del prof. Gino Loria.»[32]

Vacca then urged Couturat to take up systematic studies of Leibniz's manuscripts,
and Couturat – who had no wish to return to teaching after the sabbatical year he had been
granted to collaborate on the organization of the philosophy congress – asked Louis Liard
to authorize him to go to Hanover, with a ministerial mission, in order to carry out this
research.

In this way the contacts between Couturat and the School of Peano were intensified,
and they further consolidated on the occasion of his journey to Italy in 1902.[33] In addition
to Peano, Vacca e Vailati,[34] Couturat exchanged letters with R. Bettazzi, C. Burali- Forti
and M. Pieri, and all these mathematicians collaborated in various ways on the spread of
his works in Italy.[35]

When Couturat met Vacca in Paris, he was on the point of publishing the volume
La Logique de Leibniz, which appeared in 1901. Vacca's suggestions regarding Leibniz's
unpublished works were therefore of the greatest importance for him, and Couturat immediately asked Peano for some clarifications regarding Vacca's articles,[36] telling him of
his intention to further his assistant's by editing a collection of leaflets and fragments by
Leibniz. At the same time, Couturat collaborated with the School of Turin on writing the
historical notes for the third edition of the *Formulaire*, adding further references to Leibniz's manuscripts.[37] Meanwhile, the preface to the *Formulaire* announced the impending
publication of his *Logique*,[38] the progress of which was attentively followed by Vacca,
Vailati and Peano.

---

[32][«At these two congresses, in which I took part, I had the opportunity to meet Bertrand Russell, who had
then published a volume on Leibniz (later translated into French in 1908). I myself had already been studying
the writings of Leibniz for more than ten years, and out of this study had arisen my friendship with Giuseppe
Peano. I thus acted as intermediary between Peano and Russell, who at the time was in close contact with
Louis Couturat [. . .]. I astonished Couturat when I quickly described him the mass of unpublished manuscripts
on mathematical logic by Leibniz existing in Hanover, which I had studied there in 1899 and of which I had
already given samples in the *Rivista di Matematica*, in Peano's *Formulario* and in Prof. Gino Loria's *Bollettino
di Storia della Matematica*.»] G. Vacca 1946, *Origini della Scienza*, Roma, Partenia, p. 31.

[33]L. Couturat to G. Peano, 15.10.1902, E. Luciano, C.S. Roero (eds.) 2005, p. 48–49; L. Couturat to G. Vacca,
13.8.1902 and 15.10.1902, P. Nastasi, A. Scimone (eds.) 1995, p. 54–55, 56; G. Vailati to G. Vacca, 29.9.1902;
M. Calderoni to G. Vailati, 6.10.1902 and 22.10.1902, G. Lanaro (ed.) 1971, p. 210, 644, 645.

[34]Cf. E. Luciano, C.S. Roero (eds.) 2005, in particular p. VII, XIV–XVI, XXI–XXIII, LXI, LXV, 12–13, 15,
18, 21, 24, 26, 39, 46, 47, 49, 50, 51, 52, 54, 55, 58; P. Nastasi, A. Scimone (eds.) 1995, p. 48–57. Unfortunately,
the correspondence between Couturat and Vailati has been lost and only indirect traces remain, thanks to the
letters of Peano and Vacca.

[35]Cf. L. Couturat to R. Bettazzi, 5.3.1899, L. Couturat to C. Burali-Forti, 12.1.1906, 22.1.1906, E. Luciano,
C.S. Roero (eds.) 2005, p. 189–191, 221–224, 226; L. Couturat to M. Pieri, 15.6.1899, 26.7.1899, 24.4.1900,
30.5.1900, 8.2.1901, 7.7.1901, 29.3.1905, 2.3.1906, G. Arrighi (ed.) 1997, p. 42–50.

[36]L. Couturat to G. Peano, 15.1.1901, E. Luciano, C.S. Roero (eds.) 2005, p. 38–39.

[37]Cf. L. Couturat to G. Peano, 12.5.1899 and L. Couturat to G. Peano, 18.3.1902, E. Luciano, C.S. Roero (eds.)
2005, p. 15, 47.

[38]G. Peano 1901b, *Formulaire de Mathématiques*, t. III, Turin, Bocca-Clausen, p. IV.

The French philosopher constantly kept Peano and his collaborators up to date on his studies,[39] several times showing them his conviction that the new research on logic could profit from the reading of these fragments. For example, he wrote to Pieri in February 1901:

> «... ma *Logique de Leibnitz*, en ce moment sous presse, contient tout un chap. consacré à la Characteristica geometrica (par suite beaucoup plus étendu et détaillé que les 4 ou 5 pages de M. Cantor). Vous savez peut-être que, suivant l'exemple de M. Vacca je suis allé à Hanovre fouiller les mss. de Leibnitz et que j'en ai rapporté des inédits fort importants, notamment pour son Calcul logique. J'espère que leur publication contribuera à faire valoir les travaux modernes sur le même sujet, d'autant plus que presque toutes les formules fondamentales de la Logique algorithmique se trouvent déjà dans Leibniz»[40]

and also to Peano in 1903

> «Comme je le dis à Vacca, je suis heureux de voir que ma publication des *Inédits* de Leibniz profite aux travaux de Logique mathématique; c'est d'ailleurs, je puis l'avouer, dans cette intention que je l'ai entreprise.»[41]

There was a very intensive exchange of opinions between Vacca, Vailati and Couturat in 1901–1902.[42] For example, while in Hanover, Couturat asked Vacca his opinion of the most important of Leibniz's unpublished works regarding algorithmic logic and mathematics.[43] Vacca urged him to examine the correspondence with the Jesuit missionaries in China, in which he had been particularly struck by the reflections on the universal characteristic and on binary arithmetic.[44]

Couturat sent Peano,[45] Vacca and Vailati the proofs of his *Opuscules* and, in 1902, when the editing was at an advanced stage, he asked Vacca to compare the first printed pages with the notes he had taken in 1899:

> «Si par hasard vous pensiez que dans vos notes de Hanovre se trouvent quelques passages que je n'aie pas, vous pourriez me les envoyer, et je publierais sous votre nom ceux qui je n'aurais pas et que je vous emprunterai. De toute façon, d'ailleurs, votre nom figurera dans la *Préface*, pour la raison que vous

---

[39]L. Couturat to G. Peano, 15.1.1901, 22.12.1901, 2.1.1903, 13.9.1903, 6.12.1903, E. LUCIANO, C.S. ROERO (eds.) 2005, p. 38–39, 46, 49, 51–52, 54.

[40]L. Couturat to M. Pieri, 8.2.1901, G. ARRIGHI (ed.) 1997, p. 45. Cf. also L. Couturat to G. Peano, 15.1.1901, E. LUCIANO, C.S. ROERO (eds.) 2005, p. 39.

[41]L. Couturat to G. Peano, 6.12.1903, E. LUCIANO, C.S. ROERO (eds.) 2005, p. 54.

[42]L. Couturat to G. Vacca, 18.1.1901, G. Vacca to L. Couturat, 23.8.1901, L. Couturat to G. Vacca, 23.10.1901, G. Vacca to L. Couturat, december 1901, L. Couturat to G. Vacca, 16.1.1902, L. Couturat to G. Vacca, 20.7.1902, L. Couturat to G. Vacca, 13.8.1902, G. Vacca to L. Couturat, autumn 1902, P. NASTASI, A. SCIMONE (eds.) 1995, p. 48–56; G. Vailati to G. Vacca, 7.2.1901; G. Vailati to G. Vacca, spring 1901; G. Vailati to G. Vacca, 21.10.1901, G. Vailati to G. Vacca, 28.12.1901, G. Vailati to G. Vacca, 31.1.1902.

[43]L. Couturat to G. Vacca, 16.1.1902, P. NASTASI, A. SCIMONE (eds.) 1995, p. 52–53.

[44]G. Vacca to L. Couturat, 23.8.1901, P. NASTASI, A. SCIMONE (eds.) 1995, p. 48.

[45]Cf. L. Couturat to G. Peano, 22.12.1901, E. LUCIANO, C.S. ROERO (eds.) 2005, p. 46.

savez. Oui, je crois qu'il y a encore aujourd'hui du profit à tirer de Leibniz pour s'orienter dans les doctrines philosophiques modernes.»[46]

Vacca and Vailati sent some comments on this second book by Couturat:

«J'avais aussi remarqué le système de lecture des nombres décimaux en syllabes. Mais ce procédé n'est pas du à Leibniz, mais à Ariabatta (voir le *Form*. T. 2). C'est pourquoi je ne l'avait cité dans le *Formulaire*.»[47]

Finally, thanks to their suggestions, Couturat revised *ex novo* a number of sections of this work. His gratitude to Vacca, who had opened up a new line of research for him, making him aware of the importance of Leibniz's manuscripts, is testified to by the very cordial acknowledgment at the beginning of the *Opuscules*:

«Notre ouvrage sur *La logique de Leibniz* était presque terminé (nous le croyions du moins) lorsque nous eûmes le plaisir de faire la connaissance de M. Giovanni Vacca qui avait compulsé, un an auparavant, les manuscrits de Leibniz conservés à Hanovre, et en avait extrait quelques formules de Logique insérées dans le *Formulaire de Mathématiques* de M. Peano. C'est lui qui nous révéla l'importance des œuvres inédites de Leibniz, et nous inspira le désir de les consulter à notre tour. [...]. C'est à ce concours de bonnes volontés, de conseils et de protections que notre ouvrage doit le jour ; nous nous faisons un plaisir et un devoir de le déclarer, et d'exprimer à MM. Liard, Bodemann et Vacca toute notre reconnaissance.»[48]

Throughout Europe, Couturat's books stirred interest and controversy with their theses and, especially, with the author's hypothesis according to which Leibniz's metaphysical system was mainly constructed starting from his logical system. Some critics considered this interpretation risky and debatable.[49] In contrast, Peano's disciples expressed enthusiastic approval. In spring 1902, for example, Vacca wrote to Couturat, thanking him for sending his volume:

«Votre splendide œuvre sur la *Logique de Leibniz* ! J'y avais pensé autrefois, vous avez réalisé mon songe. J'ai parcouru d'un seul trait tout le volume en y trouvant une clarté et une vision nette des idées de Leibniz que j'avais entrevues mais que je ne savais pas écrire. J'aime parfois à chercher des indices qui mesurent le mérite des ouvrages contemporaines et il me semble qu'on peut juger de la valeur d'un livre en s'imaginant l'intérêt qu'il pourrait présenter pour les savants célèbres des siècles passés. S'il pouvaient un moment revivre qu'il seraient contents de savoir que ... Après avoir vu votre livre j'ai tout de suite pensé : quels remerciements vous auriez de Leibniz s'il pouvait encore vivre pour vous connaitre !»[50]

---

[46]L. Couturat to G. Vacca, 16.1.1902, P. Nastasi, A. Scimone (eds.) 1995, p. 53.

[47]G. Vacca to L. Couturat, autumn 1902, P. Nastasi, A. Scimone (eds.) 1995, p. 56. Cf. also G. Vailati to G. Vacca, 21.10.1901, G. Lanaro (ed.) 1971, p. 193.

[48]C, p. I, II.

[49]Cf. for example M. Ferrari 2006, p. 173, 181–183.

[50]G. Vacca to L. Couturat, autumn 1902, P. Nastasi, A. Scimone (eds.) 1995, p. 55.

Two reviews of Couturat's books, which Peano had been eager to have,[51] appeared in the *Revue de Mathématiques*.[52] Vailati also published an analysis of the *Logique de Leibniz* in Loria's *Bollettino*.[53] Vacca's and Vailati's reviews benefited from a continuous 'three-way dialog' with Couturat,[54] who gratefully wrote to Vacca:

> «Je viens de recevoir la Rev. de Math. et de lire votre article sur la Logique de Leibniz. Je vous remercie de toutes les choses flatteuses que vous dites de mon ouvrage. Mais ce qui me fait encore plus de plaisir, c'est de voir la partie que vous (et les autres collaborateurs de M. Peano) tirez de ma publication d'inédits [...]. Je souhaite qu'elle donne une nouvelle impulsion aux travaux de Logique mathématique et qu'elle attire sur eux l'attention des philosophes.»[55]

Concerning the volume *La Logique de Leibniz*, Vailati remarked that only a scholar who was master equally of mathematical and of philosophical culture could attempt to interpret Leibniz's writings, in which these two components are inextricably intertwined. The greatest merit of Couturat's work was therefore that it showed this interaction, in the light of which it was possible to fully appreciate the importance of the project of construction of the *characteristica*:

> «Ciò – said Vailati – rende anche, nello stesso tempo, ragione di un altro fatto notevole, messo chiaramente in luce dal presente volume del Couturat, che, cioè, perfino negli scritti del Leibniz, già da tempo pubblicati, le parti che toccano più davvicino gli argomenti a cui abbiamo sopra alluso, cioè in particolare i vari metodi di rappresentazione simbolica dei ragionamenti deduttivi e il concetto generale di un algoritmo operatorio (*calculus ratiocinator*), sembrano non aver quasi richiamato sopra di sé alcuna attenzione ed essere giaciute non meno neglette o ignorate di quelle altre parti, ad esse affini, che gli editori delle opere di Leibniz non avevano finora neppur stimate degne della pubblicazione.»[56]

In any case – concluded Vailati – it was precisely to Leibniz's reflections that reference should be made if the origins of and the links between his most original philosophical and mathematical ideas were to be understood.

---

[51]G. Peano to G. Vacca, 9.1.1903, c.p., E. LUCIANO, C.S. ROERO (eds.) to appear. Couturat had urged the publication of a review of his work in the RdM the previous week. Cf. L. Couturat to G. Peano, 2.1.1903, E. LUCIANO, C.S. ROERO (eds.) 2005, p. 49.

[52]G. VAILATI 1901a, p. 148–159; G. VACCA 1903, p. 64–74.

[53]G. VAILATI 1901b, p. 103–110. The two reviews are very different one from each other.

[54]Cf. G. Vailati to G. Vacca, 7.2.1901, [february-march 1901], [may 1901], 21.10.1901, 28.12.1901, 31.1.1901, 19.2.1902, 10.6.1902, 7.7.1902, 7.11.1902, 8.11.1902, 12.2.1903, 20.4.1903; 7.12.1903, G. Vacca to G. Vailati, 15.2.1903, cc. 1r-2v, M. DE ZAN, E. LUCIANO, C.S. ROERO, to appear.

[55]L. Couturat to G. Vacca, 6.12.1903, P. NASTASI, A. SCIMONE (eds.) 1995, p. 57.

[56][«This also, at the same time, confirms another important fact, clearly highlighted in this volume of Couturat, namely, that even in Leibniz's writings, already published some time ago, the parts that come closest to the subjects we have alluded to above, i.e., the various methods of symbolic representation of deductive reasoning and the general concept of an operating algorithm (*calculus ratiocinator*), seem hardly to have attracted any attention to themselves and to have lain no less neglected or ignored than those other parts, similar to them, which the publishers of Leibniz's works have so far failed even to consider worthy of publication.»] G. VAILATI 1901a, p. 148–149.

Vacca, in his turn, identified the salient characteristics of Leibniz's logic starting from the autograph works published by Couturat. Peano's collaborator did not hesitate to affirm that Leibniz was in fact the first mathematician to have understood the importance of logic, attempting to build a complete, consistent symbolic system of it. Thus Vacca was able to conclude:

> «Da quanto precede appare quanto sia elevato il posto che Leibniz occupa nella storia della logica, e quanto da questo punto di vista sia stata utile la pubblicazione del Couturat. Egli dice, troppo modestamente, che la sua pubblicazione (p. III) «est un recueil de morceaux choisis, que parfois se réduit presque à un catalogue». Il volume edito dal Couturat di oltre 600 pagine supera di gran lunga le così dette edizioni complete di Gerhardt e di Erdmann. Se però questo volume è venuto in buon punto a soddisfare i più urgenti desideri degli studiosi di logica, non conviene però dimenticare che urge tuttavia la pubblicazione completa ed integrale delle opere di Leibniz. È un'opera colossale: si tratta di un centinaio di volumi in ottavo che occorrerà comporre, raccogliendo e decifrando con pazienza tutti i frammenti grandi e piccoli che Leibniz ha lasciato, e di cui il primo buon modello è quello ora datoci dal Couturat.»[57]

It was with interest that Couturat followed the work of promotion of his writings in Italy, several times thanking Peano and his collaborators.[58]

In addition to Peano's *Revue de Mathématiques* there was another scientific periodical, linked to the context of the University of Turin, which contributed to the spread of knowledge of Leibniz among mathematicians in Italy: this was the *Bollettino* under the direction of Loria. A friend of Peano since the time of his university studies, Loria contributed to the diffusion of the *Formulaire*, although he was perfectly aware of the criticisms that such a work would inevitably attract. He also shared the interest of Peano and his School in the history of logic, a subject to which he devoted the article *La logique mathématique avant Leibniz*, in which he wrote:

> «On voit, d'après ces lignes que nous avons empruntées à l'*Introduction au Formulaire de Mathématique* que vient de publier M. Peano dans sa qualité de directeur de la *Rivista di Matematica*, qu'un groupe de mathématiciens s'est proposé de rédiger une sorte de grande encyclopédie, une espèce de *repertorium* où l'on trouvera énoncés en symboles logiques les définitions et les théorèmes plus importants qui se rapportent aux différentes sciences exactes.

---

[57] [«From the above it emerges how elevated a place Leibniz occupies in the history of logic, and how useful from this point of view was the publication of Couturat. He says, too modestly that his publication (p. III) «est un recueil de morceaux choisis, que parfois se réduit presque à un catalogue». The volume published by Couturat, of more than 600 pages, is much greater than the so-called complete editions of Gerhardt and Erdmann. If, however, this volume has succeeded to a satisfactory degree in satisfying the most urgent wishes of logicians, we must nevertheless not forget that the complete and integral publication of the works of Leibniz is a pressing matter. It is a colossal work: it is a matter of about a hundred octavo volumes which will have to be put together, patiently gathering and deciphering all the fragments, small and large, that Leibniz has left, and the first good model of which is the one now given us by Couturat.»] G. Vacca 1903, p. 72–73.

[58] Cf. L. Couturat to G. Peano, 6.12.1903, E. Luciano, C.S. Roero (eds.) 2005, p. 54.

... Je ne veux pas discuter ici en détail ce projet, les critiques ... mon but ac-
tuel n'est pas d'essayer de déterminer la valeur de la méthode dont M. Peano
s'est fait un des avocats les plus actifs; je veux, au contraire, seulement faire
remarquer – ce qui semble avoir échappé même à Schröder – que l'entreprise
que j'ai signalée a été essayée, il y a deux cent cinquante années (c'est-à-dire
avant Leibniz), par le mathématicien français Pierre Hérigone.»[59]

As may be deduced, at this time Loria adopted the historiographic approach of Vacca
and of the *Formulaire*, namely the search for precursors. Hence his work may reasonably
be set beside that of Peano and of Vacca on the precursors of mathematical logic, or beside
Vailati's essay on G. Saccheri's *Logica demonstrativa*.[60]

The *Bollettino* soon became the journal of reference for the spread in Italy of the initiatives on Leibniz and for the updating of the secondary literature regarding the German
mathematician and philosopher.[61] In this journal – as we have said – Loria had accepted
for publication Vacca's article on Leibniz's manuscripts and Vailati's review of Couturat's *Logique de Leibniz*. These were important contributions, to the point that in 1906
Loria was to remark, with regard to the editing of volume XXI of the *Abhandlungen zur
Geschichte der Mathematischen Wissenschaften*:

> «[esso] porge un notevole contributo alla letteratura ed è nuovo sintomo del
> salutare risveglio di studi intorno al sommo filosofo-matematico Leibniz, ri-
> sveglio al quale il nostro *Bollettino* non è estraneo [...].»[62]

Furthermore, in the *News* section Loria published the announcements regarding the
plan for an international edition of Leibniz's works, information about the initiatives for

---

[59]G. Loria 1894, *La logique mathématique avant Leibniz*, Bulletin des sciences mathématiques (Darboux), s. 2,
18, p. 107–112, quotation at p. 108–109.

[60]G. Peano 1894e, *Un precursore della logica matematica*, RdM, 4, p. 120; G. Vacca 1899, *Sui precursori
della logica matematica*, RdM, 6, p. 121–125, 183–186; G. Vailati 1903, *Di un'opera dimenticata del P. Gero-
lamo Saccheri (Logica demonstrativa)*, Rivista filosofica, 5, 6, 4, p. 528–540. Cf. also G. Peano to G. Vacca,
15.11.1906, E. Luciano, C.S. Roero (eds.) to appear. This research on the 'precursors', whose value must obviously be put in perspective today, nevertheless justifies the choice of quotations selected by Vacca starting from
Leibniz's manuscripts. Vacca did not, in fact, use all the manuscripts he consulted, nor all those of Leibniz's
manuscripts that he considered interesting or important. He transcribed and published only those manuscripts
from which may be deduced a priority on Leibniz's part for anything regarding the propositions or proofs of
logic and arithmetic.

[61]On the subject of Leibniz's acquaintances in Italy, the name of Federigo Enriques cannot be omitted: from
1907 he directed the journal *Scientia, Revue internationale de synthèse scientifique* and, in 1922, published a
volume on the history of logic (*Per la storia della Logica. I principii e l'ordine della scienza nel concetto dei
pensatori matematici*, Bologna, Zanichelli, 1922). He had contacts with the School of Peano, and above all with
Vacca and Vailati, but his conception of logic and of historical research are very different. Enriques was interested above all in the philosophical aspects of Leibniz's writings, and gave space in *Scientia* to articles on and
reviews of Leibniz's philosophical works. Cf. for example G. Loria 1922 *J.M. Child, The Early Mathematical
Manuscripts of Leibniz ...*, Scientia, 31, p. 237–238. A review of Child's volume also appeared in *Archeion*,
another Italian journal on the history of sciences, contemporary with *Scientia* and published by Aldo Mieli. Cf.
E. Rufini 1924, *J.M. Child, The Early Mathematical Manuscripts of Leibniz ...*, Archeion, 5, 1, p. 68–69.

[62][«[it] offers a notable contribution to the literature and is a new symptom of the healthy revival of studies
concerning the supreme philosopher-mathematician Leibniz, a revival to which our *Bollettino* is not unrelated
[...].»] *Notizie. Abhandlungen zur Geschichte der Mathematischen Wissenschaften*, Bollettino di Bibliografia e
Storia delle Scienze Matematiche (Loria), 9, 1906, p. 94.

the centenary of his death and clarifications on the publication of his scientific correspondence.[63] In this case too the activity of promotion carried out by Peano and his School seems clear: thanks to the correspondence between Couturat and Peano it may be supposed, for example, that it was Couturat himself who sent Loria the information about the agenda for the edition of Leibniz's works, undertaken by an International Commission.[64] In addition, the publisher of the correspondence between Leibniz and Kochanski, repeatedly announced in the *Bollettino*, was Samuel Dickstein, who in his turn was in close contact with Peano.

5. International language and teaching

In 1903 Peano published his first article in *latino sine-flexione*, in which he illustrates a plan for an international language deducible from classical Latin, with the suppression of declensions and inflexional endings.[65] Thus a new chapter was opened in his intellectual biography and in his scientific output. Peano's studies on the international language, which he himself conceived as an application of mathematical logic, undoubtedly arose from his reading of Leibniz's manuscripts on the *Langue rationelle*, published by Couturat in the *Opuscules*. Peano remembers:

> «L. Couturat publica *Opuscules et fragments inédits de Leibniz*, qui contine studio de Leibniz, summe praetioso per constructione de Vocabulario philosophico. Libro, nunc edito, L. Couturat et L. Leau *Histoire de la langue universelle*, Paris a. 1903 p. XXI+571, expone 56 projecto de lingua artificiale. Si in futuro analyse et synthese invicem conveni, ut duo exercito de minatore, qui labora tunnel ex duo extremitate, tunc Lingua rationale et Characteristica universale de Leibniz fore idem.»[66]

From 1903 Peano developed his studies on Leibniz's philosophical language; Couturat, in contrast, progressively distanced himself from research on Leibniz, to the point that in 1905, when E. Borel asked him for an article for the newly founded *Revue du mois*, he answered:

> «Je te félicite sincèrement de l'idée de créer la *Revue du Mois* ; mais, pour le moment, je ne puis rien te promettre comme collaboration. Je t'ai dit combien je suis occupé. Je ne puis plus revenir sur Leibniz, que j'ai quitté depuis plusieurs années. Quand j'aurai un moment de loisir, je penserai à ta Revue.

---

[63]Cf. *Notizie. Carteggio di Leibniz*, Bollettino di Bibliografia e Storia delle Scienze Matematiche (Loria), 4, 1901, p. 127; *Notizie. Carteggio di Leibniz; Progetto d'un'edizione internazionale delle opere di Leibniz*, Bollettino di Bibliografia e Storia delle Scienze Matematiche (Loria), 6, 1903, p. 30, 126; AA.VV. 1906, *Notizie. Preparazione d'un'edizione completa delle opere di Leibniz; Abhandlungen zur Geschichte der Mathematischen Wissenschaften*, Bollettino di Bibliografia e Storia delle Scienze Matematiche (Loria), 9, 1906, p. 30, 94; *Notizie. Il II centenario della morte di Leibniz*, Bollettino di Bibliografia e Storia delle Scienze Matematiche (Loria), 19, 1917, p. 31–32.
[64]Cf. L. Couturat to G. Peano, 2.1.1903, E. Luciano, C.S. Roero (eds.) 2005, p. 49.
[65]G. Peano 1903d, *De latino sine- flexione*, RdM, 8, p. 74–83.
[66]G. Peano 1903d, *De latino sine- flexione*, RdM, 8, p. 80, 82. Cf. also L. Couturat to G. Peano, E. Luciano, C.S. Roero (eds.) 2005, p. 51–52.

Ce serait pour entretenir tes lecteurs des sujets qui m'occupent présentement,
à savoir la Logique formelle, d'une part, et la Langue internationale, d'autre
part.»[67]

Meanwhile, in 1908 Peano printed the definitive edition of the *Formulaire* which
for some time he had been using for his teaching of Analysis at the University of Turin.
In his lectures he liked to adopt the historical method, i.e., he proposed a reading of
passages from the classics of mathematics. So Peano devoted himself to going more
deeply into Leibniz's writings on differential and integral calculus, and transcribed parts
in his *marginalia* to the *Formulaire*, commenting them with his students.

The year 1910 was, perhaps, one of the most difficult for Peano, who was forced to
abandon the teaching of advanced Analysis, because of the criticisms of his colleagues
about his recourse to logic and to the *Formulaire* in his lectures. In February 1910 his
correspondence with Couturat was abruptly interrupted, but in any case Couturat was by
this time a fanatical interlinguist, devoting himself entirely to propagandizing the *Ido*. In
the same month, Peano and Vacca were involved in a controversy with Roderigo Biagini
in the journal of philology *Classici e Neolatini*.[68] The object of the dispute was whether
or not Peano's *latino sine-flexione* actually coincided with Leibniz's rational language,
though Biagini had recourse above all to the weapon of humor, trotting out the discovery
of Leibniz's unpublished works by Vacca in 1899 and Couturat's subsequent contributions. Both Peano and Vacca responded and Vacca, in particular, specified in irritated
tone:

> «Pubblicai, in più luoghi, il risultato di questi miei studi ed altri ne pubblicherò
> a tempo opportuno. In seguito a queste mie ricerche, che ebbero dagli studiosi
> competenti l'accoglienza che si meritavano, apparvero evidenti le insufficien-
> ze da me per primo segnalate. [...] Infine l'Associazione internazionale delle
> Accademie, soddisfacendo ad un voto espresso prima da me, poi dal Couturat
> e da altri, deliberò la pubblicazione integrale degli scritti di Leibniz [...]. Il
> Sig. R.B. ha mal garbo a dire che io ho pubblicato alcuni manoscritti i quali, o
> non erano stati veduti, o forse non voluti pubblicare. Erano stati veduti, perché
> pubblicati in parte dal Gerhardt; e non erano stati pubblicati perché gli editori
> precedenti non ne avevano capito l'importanza.»[69]

---

[67]L. Couturat to E. Borel, 23.11.1905, Fonds Borel Paris, c.p., in E. Luciano 2008, *Giuseppe Peano docente
e ricercatore di Analisi 1881–1919*, PhD Thesis, Torino, Dep. of Mathematics, vol. 2, p. 128. In France,
meanwhile, a harsh controversy broke out in the *Revue de Métaphysique et de Morale* on rigor and intuition,
involving mathematicians and philosophers such as Couturat, Russell, Poincaré, Boutroux, Borel, Winter and,
in Italy, Peano, Vacca, Pieri, Croce. Cf. E. Luciano, C.S. Roero (eds.) 2005, p. XXV-LX.

[68]Cf. R. Biagini 1910, *La lingua internazionale in servizio delle scienze e degli scienziati*, Classici e Neolatini,
a. VI, 2–3, p. 1–24; G. Peano 1911a, *A proposito della lingua internazionale*, Classici e Neolatini, 4, p. 281–
285; G. Vacca 1910, *A proposito delle edizioni delle opere di Leibniz*, Classici e Neolatini, 4, p. 286. Cf. also
G. Peano to G. Vacca, 17.12.1910, c. 1r, E. Luciano, C.S. Roero (eds.) to appear.

[69][«I published, in several places, the result of my studies and published others at an appropriate time. Following
my research, which were welcomed as they deserved by competent scholars, the inadequacies first pointed out by
me appeared evident. [...] Finally the international Association of the Academies, satisfying a vote expressed
first by me, then by Couturat and by others, resolved on the publication in their entirety of Leibniz's works [...].
R.B. has the impertinence to say that I published some manuscripts which either the editors had not seen, or did

After 1910 Peano was forced to give up lecturing on Higher Analysis and turned to the world of secondary teaching and of the interlingua.[70] His activities in the area of didactics were many, including the publication of manuals for high schools and professional institutes, but also initiatives intended to foster contacts between the university and the community of teachers.

In the School of Peano particular importance was given to the critical editions of original sources, and he had some collaborators who worked on the preparation of anthologies and translations, which in effect proved to be very useful to teachers and students, by making them familiar with the classics of mathematics such as Leibniz's works. It is not surprising, then, that the first two Italian translations of writings by Leibniz on mathematics should be owed to scholars linked to the School of Peano. In 1920 Alpinolo Natucci, a teacher in the upper schools in Pinerolo and a friend of Peano who shared his opinions on the pedagogy of mathematics, translated Leibniz's manuscript *De solidorum elementis*.[71] This is a text of Descartes on the polyhedral geometry and on figurative numbers, copied by Leibniz and then developed by him in the years he spent in Paris. Then, in 1927, Ettore Carruccio, a student of Vacca, translated the famous *Nova Methodus*.[72] This research was developed in relation to courses on the history of mathematics given by Vacca at the University of Rome. The translation appeared in a journal expressly addressed to teachers: the *Periodico di Matematiche*. Nor does the choice of manuscripts seem accidental, if we bear in mind that both solid geometry and differential calculus were subjects typical of the curricula for secondary education.

In 1924–25, Peano left the chair of Infinitesimal Analysis, transferring to that of Complementary Mathematics, which was precisely designed for the training of future teachers. In his courses he developed subjects concerning the logical and historical foundations of mathematics . Information about his lectures may be gathered from some articles by the students who attended them and published extracts from them, at his invitation. In 1928, for instance, Ugo Cassina presented an article devoted to π in the *Periodico di Matematiche*.[73] In this paper there is a careful examination of Leibniz's contributions on the series of the *arcotang*, reconstructed thanks to an examination of the writings and correspondence of Leibniz. Cassina's results were subsequently taken up by Fausta Audisio, another student of Peano's courses of *Matematiche complementari*, who wrote her degree dissertation on π, under his supervision, and also published a number of papers on this subject.[74] Audisio wrote:

---

not want to publish. They had been seen, because they were published in part by Gerhardt; and they had not been published because the earlier editors had not understood their importance.»] G. Vacca 1910, *A proposito delle edizioni delle opere di Leibniz*, Classici e Neolatini, 4, p. 286.

[70] Cf. E. Luciano, C.S. Roero 2008, *Giuseppe Peano Matematico e Maestro*, Turin, Dep. of Mathematics, p. 65–80.

[71] A. Natucci 1920, *Il De Solidorum Elementis di Leibniz*, Bollettino Mathesis, 12, 5–8, p. 117–127.

[72] E. Carruccio 1927, *Il «Nuovo Metodo» di Leibniz, con note storiche*, Periodico di Matematiche, s. 4, 7, p. 285–301.

[73] U. Cassina 1928, *Calcolo di π*, Periodico di Matematiche, s. 4, 8, p. 271–293.

[74] F. Audisio, *Il numero π*, Graduation Thesis in Mathematics, Turin, Historical Archives of the University, 1930, p. 1–48; *Calcolo di π in Archimede*, Atti della R. Accademia delle Scienze di Torino, 65, 1929–30, p. 101–108.

«La serie

$$\frac{\pi}{4} = 1 - \frac{1}{3} + \frac{1}{5} - \frac{1}{7} + \cdots$$

fu pubblicata da Leibniz, in «Acta Eruditorum», pp. 41–46, l'anno 1682: *De vera proportione circuli ad quadratum circumscriptum in numeris rationalibus* (Opera, III, p. 140). [...] Si sa che il resto di questa serie, dopo un termine qualunque, ha il segno ed è minore del primo termine trascurato. Quindi con questo criterio la serie è poco convergente e non atta al calcolo numerico. [...] Ma la regola precedente dà solo un limite superiore del resto. Il prof. Peano, in una sua lezione, indicò come si possa stimare il resto con maggiore approssimazione. Io espongo questo metodo e sviluppo il calcolo di $\pi/4$ con 5 decimali, servendomi di 14 termini della serie.»[75]

Cassina's and Audisio's works, like that of other students of Peano, which appeared in the last years of his life, are distinguished by the same structure: they arise from the reading of the classical sources, such as Leibniz; these sources are transcribed, translated and commented on, from a historical standpoint; finally, applying modern mathematical tools, among them the techniques of numerical calculus, these historical sources are the basis for the development of new results. The starting point is always the *Formulaire de Mathématiques*, so it is not surprising that the same passages from Leibniz's writings, both published and unpublished, should be repeated and taken up in exactly the same way by various students of Peano.

From the 1920s, Giovanni Vacca, for his part, interrupted his research activity on Leibniz. Friends and correspondents, Loria among them, contacted him several times, as the years went by, to ask for information about the existence of certain results or concepts in Leibniz and urged him to collate the notes he had made in Hanover, only partially published and divulged.[76] Nevertheless, Vacca had meanwhile become professor of History and Geography of Eastern Asia at the University of Rome, so he limited himself to a number of sporadic mentions, without taking up systematic studies on Leibniz. In an article on the geometry of folded paper, for example, he remarked incidentally:

«In un manoscritto di Leibniz abbiamo trovato la seguente lista: *Geometria est explicare figuras, quas natura et ars singulari quadam ratione producit, ita figurae cristallisationum, ecc.; Geometria sartorum; De artificio, puerorum,*

---

[75] «The series

$$\frac{\pi}{4} = 1 - \frac{1}{3} + \frac{1}{5} - \frac{1}{7} + \cdots$$

was published by Leibniz, in «Acta Eruditorum», pp. 41–46, in 1682: *De vera proportione circuli ad quadratum circumscriptum in numeris rationalibus* (Opera, III, p. 140). [...] It is known that the rest of this series, after any term, has the sign and is less than the first term ignored. Hence with this criterion the series is slowly convergent and not suitable for numerical calculus. [...] But the preceding rule gives only a limsup for the rest. Prof. Peano, in one of his lectures, indicated how the rest can be estimated with greater approximation. I expound this method and develop the calculus of $\pi/4$ with 5 decimals, using 14 terms of the series.» F. AUDISIO 1930, *Calcolo di π colla serie di Leibniz*, Atti della R. Accademia dei Lincei, Rendiconti, 6, 11, p. 1077–1080; *Il numero π*, Periodico di Matematiche, 4, 3, 1931, p. 11–42; *Ancora sul numero π*, Periodico di Matematiche, 4, 20, 1931, p. 149–150.

[76] Cf. for example G. Loria to G. Vacca, 26.7.1901, G. Loria to G. Vacca, 14.1.1903, P. NASTASI, A. SCIMONE (eds.) 1995, p. 97, 103–104.

quo fila digitis implicata educunt; De textoria arte; De geometria apum et aranearum.»[77]

One year earlier Peano had addressed himself to Vacca to obtain information on the developments of the publication of Leibniz's works. Peano had actually proposed a research on Wilson's theorem to his assistant Maria Cibrario. Vacca, as we have said, had undertaken his Leibnizian studies on precisely this theorem but, in 1929, he was no longer up to date and merely replied:

«Non so a che punto sia la pubblicazione dei manoscritti di Leibniz. Credo che sia stato pubblicato, intorno al 1914, uno o due volumi, di un catalogo completo dei manoscritti di Leibniz, ma io non l'ho visto. Per quanto riguarda il teorema di Wilson, di cui avevo dato notizia per la prima volta nel Bollettino di Loria, 1899, T. II, pag. 113, è comparso nel 1912, Bibliotheca Mathematica, 1912, vol. XIII, III Folge, Leipzig, Teubner, p. 29, un articolo di Dietrich Mahnke, *Leibniz auf der Suche nach einer allgemeine Primzahlgleichung*, il quale contiene uno studio sui manoscritti di Leibniz, più diffuso, e completo del mio. Il Mahnke riproduce i passi ricopiati da me, ed alcuni altri. Io ho ancora l'impressione che una diligente ricerca tra i libri della biblioteca di Leibniz, che si conservano ancora nella sua casa, ridotta a museo in Hannover, potrebbe dare risultati assai interessanti, ma io non ho modo di andare laggiù, e non so se altri, oltre il Mahnke, se ne occupino. Forse potrebbe occuparsene il Wieleitner, il quale scrive interessanti studi di storia della matematica.»[78]

However, Peano's intention was implemented: in 1929, once more in the *Periodico di Matematiche*, Cibrario published an article entitled *Teorema di Leibniz-Wilson sui numeri primi*.[79] In this work she retraces the history of the proof of this proposition, starting from the studies of Vacca – which by this time had become classic references and were even quoted in the *Encyclopédie des sciences mathématiques* – down to the more recent works by Dietrich Mahnke, which Vacca himself had suggested to Peano. Cibrario's article takes its cue from the *Formulaire*, from which the historical information is drawn. The subject is yet again chosen on the basis of its possible didactic implications, for Wilson's

---

[77]["In one of Leibniz's manuscripts we found the following list: Geometria est explicare figuras, quas natura et ars singulari quadam ratione producit, ita figurae cristallisationum, ecc.; Geometria sartorum; De artificio, puerorum, quo fila digitis implicata educunt; De textoria arte; De geometria apum et aranearum."] G. VACCA 1930, *Della piegatura della carta applicata alla geometria*, Periodico di Matematiche, 4, 10, p. 43–50.

[78]["I do not know what point the publication of Leibniz's manuscripts has reached. I believe that around 1914, one or two volumes were published, of a complete catalog of Leibniz's manuscripts, but I have not seen it. As to Wilson's theorem, which I first examined in Loria's *Bollettino*, 1899, T. II, pag. 113, there appeared in 1912, Bibliotheca Mathematica, 1912, vol. XIII, III Folge, Leipzig, Teubner, p. 29, an article by Dietrich Mahnke, *Leibniz auf der Suche nach einer allgemeine Primzahlgleichung*, which contains a study on Leibniz's manuscripts, more widespread and complete than mine. Mahnke reproduces the passages copied by me, and some others. I still have the impression that diligent research among the books in Leibniz's library, which are still kept in his house, now a museum in Hanover, might give very interesting results, but I have no way of going there, and I do not know whether others, besides Mahnke, are concerned with it. Perhaps Wieleitner, who writes interesting studies on the history of mathematics, could be interested in." G. Vacca to G. Peano, 25.1.1929, C.S. ROERO (ed.) 2001, *Giuseppe Peano Matematica, cultura e società*, Cuneo, L'Artistica Savigliano, p. 84.

[79]M. CIBRARIO 1929, *Teorema di Leibniz-Wilson sui numeri primi*, Periodico di Matematiche, s. 4, 9, p. 262–264.

theorem is included in the programs for competitive examinations for teachers of middle and secondary schools.

A tangible testimony to the success of the activity of making Leibniz's works known, carried on by the School of Peano, is found in the *Enciclopedia delle Matematiche Elementari*.[80] This encyclopedia, edited by L. Berzolari, D. Gigli and G. Vivanti, was a sort of *sussidiario* for students who were preparing for examinations for permanent teaching posts. There were numerous contributions from exponents of the School of Peano, which is why there are many cross-references in it to Leibniz's manuscripts and printed works on Logic, Arithmetic, *Analysis situs,* etc. These references often descend directly from the *Formulaire de Mathématiques* or from the publications of Peano and of Vacca; there are also, however, references to the works of Couturat. Leibniz's are almost the only manuscripts cited in the *Enciclopedia*.[81] Hence knowledge of them was considered an element for the cultural and historical training of future teachers of mathematics.

6. Conclusions

Today it seems almost inevitable to stand at some distance from a certain hagiographical literature on Peano which aims to give excessive emphasis to his links with Leibniz, in the end considering the *Formulaire* as the complete and immutable fulfillment of Leibniz's dream. In fact, the positions apropos this must be far more nuanced. Peano himself, as the years passed, became more prudent regarding the successes of algorithmic logic. His entourage – and above all those students who like Vacca had witnessed the fundamental developments of mathematical logic in the 1930s – did not hesitate either to make manifest a substantial skepticism, when faced with the enthusiastic initial declarations to the effect that, with the *Formulaire*, «Leibniz's dream had come true». Vacca, for instance, wrote in 1946:

> «É vero che in lavori precedenti G. Peano aveva considerato talvolta la logica matematica come la soluzione del desiderato di Leibniz [...]. Questo ideale di Leibniz, da lui non raggiunto, come non è raggiunto da nessun logistico moderno non era nemmeno raggiunto da chi lo aveva ingenuamente espresso nel Medioevo, da Raimondo Lullo, nella sua *Ars Magna*. Le citazioni di Leibniz negli scritti di Giuseppe Peano sono limitate effettivamente alle applicazioni della logica alla matematica.»[82]

[80]Cf. L. Berzolari, D. Gigli, G. Vivanti (eds.), *Enciclopedia delle Matematiche Elementari*, Milano, Hoepli, I, 1, 1930, p. 6, 7, 23, 26, 28, 30, 33, 34, 35, 36, 38, 39, 43, 47, 48, 49, 52, 57, 60, 64, 65, 68, 69, 70, 71, 90, 91, 102, 105, 193, 219, 280, 287, 422; I, 2, 1932, p. 5, 13, 75, 76, 109, 111, 116, 407, 421, 455, 456, 457, 458, 459, 460, 463, 471, 474, 477, 478, 481, 486, 488, 491, 498, 500, 502, 504, 506, 507, 515, 521, 525, 526, 527, 530, 533; II, 1, 1937, p. 53, 63, 291, 392, 432, 536, 572; II, 2, 1938, p. 50, 128, 148, 153, 164, 202, 382, 413, 429, 431; III, 1, 1947, p. 210, 211, 267; III, 2, 1950, p. 45, 55, 56, 108, 423, 578, 654, 690, 703, 707, 714, 715, 724, 727, 734, 736, 739, 740, 741, 749, 755, 762, 894, 900, 941.
[81]Cf. L. Berzolari, D. Gigli, G. Vivanti (eds.), *Enciclopedia delle Matematiche Elementari*, Milano, Hoepli, I, 1, 1930, p. 6, 7, 23, 26, 60, 287; I, 2, 1932, p. 75; II, 1, 1937, p. 291; III, 2, 1950, p. 894.
[82][“It is true that in earlier works G. Peano had sometimes considered mathematical logic the solution to Leibniz's dream [...]. This ideal of Leibniz, which he never achieved, just as no modern logician has achieved it, was not even reached by the man who had naively expressed it in the Middle Ages, Raimondo Lullo, in his *Ars*

However, there are some very peculiar features of the reception of Leibniz by Peano and his School. It certainly seems risky to include these scholars among the protagonists of the *Leibniz Renaissance* in Italy, since they did not carry out philosophical, logical or philological analyses on Leibniz. The importance of the School of Peano must therefore be assessed bearing in mind that these scholars were the only ones who, *in the Italian scientific community*, spread an exact knowledge of Leibniz's results in Logic, Arithmetic, Analysis etc.[83] Nor is this all: this knowledge was recovered from study carried out *on the sources* (printed and manuscript), in contrast with what happened, for example, in F. Enriques' work on the history of logic. Besides, Peano and his students were personally committed to the divulgation of the works of Couturat, of which, in Italy, they were the genuine promoters.[84]

On examining Peano's and his collaborators' works on Leibniz, one may identify the epistemological arguments that frame the historiographical research carried out in relation to the *Formulaire*. History contributes to the 'founding' of every mature discipline as mathematics, identifying the paternity of the concepts and theorems, the links among the theories and the stages that gradually led to its modern, rigorous structure. Peano dedicated scrupulous attention to questions of priority and to precursors' research. Equal attention is devoted to transcriptions, critical editions and so on, for it is not considered appropriate that historical research be separated from philological and etymological criticism.

These arguments became concrete in specific didactic demands. The teaching of mathematics, according to Peano and his collaborators, must be developed with the historical method. Information on the origins of concepts, rules, symbols and proofs in fact help the student to understand their usefulness. On a closer look, one of the most characteristic aspects of study activities on Leibniz by the School of Peano is their promotion in the domain of teaching, thanks to the quotations from Leibniz's manuscripts in the manuals for secondary schools, by means of the translations into Italian of his most important works, etc. Knowledge of Leibniz's writings was considered, as we have seen, an important element in the cultural formation of teachers of mathematics. Moreover, the emphasis on didactics is also one of the principal characteristics of the editorial policy of Peano's *Revue de Mathématiques* which, according to the statement at the beginning of the first volume:

Magna. The quotations from Leibniz in the writings of Giuseppe Peano are in effect limited to the applications of logic to mathematics."] G. Vacca 1946, *Origini della Scienza*, Roma, Partenia, p. 32.

[83]Cf. A. Padoa, *Il contributo di G. Peano all'ideografia logica*, Periodico di Matematiche, 4, 13, 1933, p. 16: «nel 1889, cioè quando il Peano negli *Arithmetices principia, nova methodo exposita* [...] riuscì primo ad esporre con linguaggio simbolico una teoria deduttiva (postulati, definizioni, teoremi e dimostrazioni), il contributo di Leibniz alla sua *Ars characteristica* era ancora in gran parte ignoto» [«in 1889, i.e., when Peano in the *Arithmetices principia, nova methodo exposita* [...] first succeeded in expounding with symbolic language a deductive theory (postulates, definitions, theorems and proofs), Leibniz's contribution to his *Ars characteristica* was still largely unknown ».]

[84]In 1930 Vacca was given the task of writing the entry *Leibniz* for the *Enciclopedia Italiana Treccani*.

«ha scopo essenzialmente didattico, occupandosi specialmente di perfezionare i metodi di insegnamento. Essa conterrà pure articoli e discussioni riferentisi ai principi fondamentali della scienza e alla storia delle matematiche e vi avrà parte importante la recensione dei trattati, e di tutte le pubblicazioni che riguardano l'insegnamento.»[85]

Certainly it may seem surprising to consider the works of Couturat on Leibniz, reviews of which were published in the *Revue de Mathématiques*, as "treatises and publications having to do with teaching". Nevertheless, the desire to make research on Leibniz available to the world of students and teachers seems to be a constant feature of the work of Peano.

On examining Peano's and his students' articles on Leibniz, one notes an indisputable convergence of intentions. Even more: the interweaving of cross-references and reciprocal quotations makes it difficult to establish the exact paternity of individual reflections. Often the same passages from Leibniz are analyzed and quoted in several works, by different authors. At other times we are face to face with a mechanism of inter-crossed collaborations, as happens in the case of the *Formulaire*. Notwithstanding, the convergence of quotations is not a case of 'plagiarism' as one might be led to believe. It seems to me symptomatic of the existence of a patrimony of knowledge considered 'common' to a School. Peano's students thought of themselves as members of a School, and often seemed to consider the results of their historical, mathematical or philological research as results 'of the School', without *distinguo* by the name under which they had been published. Thus is expressed one of the finest aspects of the personality of Peano, a Master who knew how to make collaboration one of the cardinal features of his scientific activity.

References

[Arrighi 1997] Arrighi G. (ed.), *Lettere a Mario Pieri* (1884–1913), Milano, Università Bocconi, Quaderni Pristem, N. 6. (1997).

[Barreau 1999] Barreau H., *Table ronde: portée et limites du formalisme leibnizien. Leibniz et le débat intuitionnisme-formalisme* in Berlioz D., Nef F. (eds.), *L'actualité de Leibniz. Les deux labyrinthes*, Studia Leibnitiana Supplementa, 34, Stuttgart, Steiner, 1999, p. 559–565.

[Carruccio 1970] Carruccio E., *Tre problemi di Leibniz nelle loro relazioni con la logica peaniana e post peaniana*, Accademia Nazionale di Scienze, Lettere ed Arti di Modena, 6, 12 (1970) p. 197–209.

[Couturat 1901] Couturat L., *La Logique de Leibniz d'après des documents inédits*, Paris, Alcan, 1901.

[De Zan-Luciano-Roero to appear] De Zan M., Luciano E., Roero C.S., *G. Vailati – G. Vacca, Carteggio 1899–1909*, to appear.

[Ferrari 2006] Ferrari M., *Vailati, Leibniz e la 'Rinascita leibniziana'*, in Ferrari M., *Non solo idealismo. Filosofi e Filosofie in Italia tra Ottocento e Novecento*, Le lettere, 2006, p. 165–204.

---

[85][«has essentially a didactic purpose and it occupies especially with perfecting teaching methods. It will also contain articles and discussions referring to the fundamental principles of science and to the history of mathematics and an important place will be given to reviews of treatises, and of all publications having to do with teaching.»] [G. PEANO], *Ai lettori*, RdM, 1, 1891, p. nn.

[Lanaro 1971] Lanaro G. (ed.), *Giovanni Vailati. Epistolario 1891–1909*, Torino, Einaudi, 1971.

[Lolli 1985] Lolli G., *"Quasi alphabetum": logica ed enciclopedia in G. Peano*, in G. LOLLI, *Le ragioni fisiche e le dimostrazioni matematiche*, Bologna, Il Mulino, 1985, p. 49–83.

[Luciano 2006] Luciano E., *The influence of Leibnizian ideas on Giuseppe Peano's work*, VIII Internationaler Leibniz-Kongress, *Einheit in der Vielheit, Vorträge*, Hannover 2006, p. 525–531.

[Luciano to appear] Luciano E., Roero C.S., *Giuseppe Peano – Giovanni Vacca, Carteggio 1894–1932*, to appear.

[Malatesta 1997] Malatesta M., *On a Particular Meaning of the Principle of Duality: Leibniz, Boole, Peano*, Metalogicon, 10, 1 (1997), p. 1–12.

[Mugnai 2000] Mugnai M., Pasini E. (eds.), *Scritti filosofici di G.W. Leibniz*, vol. I, *Scritti giovanili, Elaborazioni private, Il nuovo sistema*, Torino, UTET, 2000.

[Nastasi 1995] Nastasi P., Scimone A. (eds.) , *Lettere a Giovanni Vacca*, Palermo, Quaderni Pristem, N. 5 (1995).

[Pasini 1966] Pasini E., *Corpo e funzioni cognitive in Leibniz*, Milano, F. Angeli, 1966.

[Pasini 1995] Pasini E., *Segni e algoritmo nell'analisi leibniziana*, in M. Panza, C.S. Roero (eds.), *Geometria, flussioni e differenziali*, Napoli, La città del sole, 1995, p. 385–412.

[Patzig 1969] Patzig G., *Leibniz, Frege und die sogenannte 'lingua characteristica universalis'*, Studia Leibnitiana Supplementa, III, *Akten des internationalen Leibniz-Kongresses Hannover 14–19.11.1966, III, Erkenntnislehre Logik, Sparchphilosophie, Editionsberichte*, Wiesbaden, Steiner, 1969, p. 103–112.

[Peano 1894g*] Peano G., *Notations de Logique Mathématique (Introduction au Formulaire de Math.)*, Turin, tip. Guadagnini, 1894 [BDM Milano, coll. Op. I 46].

[Peano 1895aa**] Peano G., *Formulaire de Mathématiques, tome 1 publié par la Rivista di matematica*, Torino, Bocca, 1895 [BDM Milano, coll. Op. A 138].

[Peano 1895aa*] Peano G., *Formulaire de Mathématiques, tome 1 publié par la Rivista di matematica*, Torino, Bocca, 1895 [BDM Milano, coll. Op. I 46].

[Peano 1895aa] Peano G., *Formulaire de Mathématiques, tome 1 publié par la Rivista di matematica*, Torino, Bocca, 1895.

[Peano 1897b*] Peano G., *Formulaire de Mathématiques*, t. II § 1, *Logique mathématique*, s.l., s.e., 1897 [BDM Milano, coll. Op. A 140].

[Peano 1897b] Peano G., *Formulaire de Mathématiques*, t. II § 1, *Logique mathématique*, s.l., s.e., 1897.

[Peano 1898d] Peano G., *Additions et corrections à F_2*, RdM, 6, 1898, p. 65–74.

[Peano 1898f] Peano G., *Formulaire de Mathématiques*, t. II, § 2 *Arithmétique*, Turin, Bocca-Clausen, 1898.

[Peano 1899b*] Peano G., *Formulaire de Mathématiques*, t. II, n. 3, Turin, Bocca-Clausen, 1899 [BDM Milano, coll. Op. I 46].

[Peano 1899b] Peano G., *Formulaire de Mathématiques, publié par la Revue de Mathématiques*, II, 3, *Logique mathématique. Arithmétique. Limites. Nombres complexes. Vecteurs. Dérivées. Intégrales*, Turin, Bocca-Clausen, 1899.

[Peano 1900a*] Peano G., *Formules de logique mathématique*, RdM, 7, 1900, p. 17–22 [BDM Milano, coll. Op. I 46].

[Peano 1901b*] Peano G., *Formulaire de Mathématiques*, t. III, Turin, Bocca-Clausen, 1901 [BDM Milano, coll. Op. I 46].

[Peano 1901b] Peano G., *Formulaire de Mathématiques*, III, Turin, Bocca-Clausen 1901.

[Peano 1903f*] Peano G., *Formulaire mathématique, édition de l'an* 1902–03 (*tome IV*), Turin, Bocca-Clausen, 1903 [BDM Parma, coll. Per 0831709 999653].

[Peano 1903f] Peano G., *Formulaire mathématique, édition de l'an* 1902–03 (*tome IV de l'édition complète*), Turin, Bocca-Clausen 1903.

[Peano 1906g*] Peano G., *Formulario mathematico Editio V, Indice et Vocabulario*, (*Proba de* 100 *exemplare*), Torino, Bocca, 1906 [BDM Milano, coll. Op. A 139].

[Peano 1906g] Peano G., *Formulario mathematico editio V, Indice et Vocabulario*, (*Proba de* 100 *exemplare*), Torino, Bocca, 1906.

[Peano 1908a*] Peano G., *Formulario Mathematico, Editio V.* (*Tomo V de Formulario completo*), Torino, Bocca, 1908 [BDM Milano, coll. Op. A 130].

[Peano 1908a] Peano G., *Formulario Mathematico, Editio V.* (*Tomo V de Formulario completo*), Torino, Bocca 1908.

[Peano 1901d] Peano G., Cantoni E., Ciamberlini C., Eneström G., Padoa A., Ramorino A., Stolz O., Vacca G., *Additions et corrections au Formulaire*, t. III, RdM, 7, 1901, p. 85–110.

[Peano 1901i] Peano G., Arbicone A., Boggio T., Cantoni E., Castellano F., Vacca G., *Additions au Formulaire*, 1901, RdM, 7, 1901, p. 173–184.

[Rodriguez-Consuegra 1991] Rodriguez-Consuegra F.A., *The mathematical Philosophy of Bertrand Russell: Origins and development*, Basel, Birkhäuser, 1991.

[Roero 1999] Roero C.S., *I matematici e la lingua internazionale*, Bollettino dell'UMI, 8, 2-A, 1999, p. 159–182.

[Roero 2008] Roero C.S. (ed.), *L'Opera omnia e i Marginalia di Giuseppe Peano*, dvd-rom N. 3, Torino, Dipartimento di Matematica, 2008.

[Roero 2008a] Roero C.S., "The Formulario between mathematics and history", in *Giuseppe Peano between Mathematics and Logic. Proceedings of the Torino Conference*, 2–3 October 2008, Milan, Springer, p. 83–132.

[Serfati 2005] Serfati M., *La révolution symbolique. La constitution de l'écriture symbolique mathématique*, Paris, Pétra, 2005.

[Serfati 2006] Serfati M., *La constitution de l'écriture symbolique mathématique. Symbolique et invention*, SMF Gazette, 108, 2006, p. 101–118.

[Styazhkin 1969] Styazhkin N.I., *History of Mathematical Logic from Leibniz to Peano*, Cambridge, MIT Press, 1969.

[Vacca 1901] Vacca G., *Additions au Formulaire*, RdM, 7, 1901, p. 59–66.

[Vacca 1903] Vacca G., *La logica di Leibniz*, RdM, 8, 1903, p. 64–74.

[Vailati 1901a] Vailati G., *L. Couturat, La logique de Leibniz d'après des documents inédits*, RdM, 7, 1901, p. 148–159.

[Vailati 1901b] Vailati G., *L. Couturat, La logique de Leibniz d'après des documents inédits, Paris 1901*, Boll. Bibl. e Storia Sci. Mat. (Loria), 4, 1901, p. 103–110.

[Vailati 1903] Vailati G., *Aggiunte alle note storiche del Formulario*, RdM, 8, 1903, p. 57–63.

[Vailati 1905] Vailati G., *Sul carattere del contributo apportato da Leibniz allo sviluppo della logica formale*, Rivista di filosofia e scienza affini, 7, 1 (12), 5–6, 1905.

Erika Luciano
Università di Torinom, Dipartimento di Matematica 'G. Peano'
via Carlo Alberto 10, I-10123 Torino, Italia
erika.luciano@unito.it

Couturat's Reception of Leibniz

Anne-Françoise Schmid

Introduction

Louis Couturat (1868–1914) was a French philosopher and mathematician[1]. He mainly wrote works on the philosophy of mathematics and the international auxiliary language project. He was involved in the beginnings of the *Revue de Métaphysique et de Morale* and the founding of the French Society of Philosophy. He is known for his polemic with Henri Poincaré on the "logistic" question. His extensive correspondence (1897–1913) with Bertrand Russell, published in 2001 (Russell, 2001), contains exchanges on their respective interpretations of Leibniz. We will be making use of this still relatively unexploited document in the course of this paper.

Understanding Couturat's reading of Leibniz is a complex undertaking given that Couturat's work covered many fields: philosophy, logic, geometry and the international auxiliary language. He never presented himself as an inventor in any one of these disciplines, but he was very brilliant, and completed his studies in philosophy, mathematics, and later in linguistics, with the highest distinction. Couturat's particularity is that he dedicates all of these disciplines to the universality of thought and the cooperation of mankind for this goal[2], both by defending contemporary learning in mathematics, and more generally in the sciences, which he systematically sought to understand and disseminate, and by applying this knowledge in a rationalist philosophical framework. The elementary philo-

---

[1] A first bibliography of Couturat was established by André Lalande (Lalande, 1915), reprinted in (Loi, 1983).

[2] Here is what Couturat says to mark his commitment to the dissemination of the international language: "I think I do not need to argue the interest of such a work. You know that it was the dream of several great philosophers, especially that of Leibniz; it is coming more and more to the attention of an informed public as international relations develop and the scientific and civilized world expands through the accession of new peoples and the extension of commercial and intellectual exchanges. Conversely, the establishment of an international language would contribute greatly to multiplying these relations and exchanges, to increasing understanding and unity in the academic world, and finally to developing sympathy and the communion of ideas between different peoples and creating the "consciousness of mankind", which, according to Mr. Boutroux, is to be the work of the twentieth century." Couturat was a pacifist and an advocate of the socialist Jean Jaures; he was an influence in turning Russell away from his imperialist stance during the Boer War, and was a follower of the various treaties of arbitration that preceded the First World War.

sophical requisites of this position are the admission of a rational order and the idea that experience does not suffice to understand knowledge, that one also requires reason.

Today Louis Couturat is still known as the man who introduced the logic of his time into France. He is one of the landmark authors one cites to substantiate the idea that logic in France met with a tragic fate due to the premature deaths of its pioneers (Jacques Herbrand, Jean Nicod), and thus did not acquire the institutional status it deserved. Indeed, Couturat died abruptly, knocked down by a military vehicle during the mobilisation, on 3 August 1914. Yet, at the time, he had for all intents and purposes given up logic, and his last publications on the subject point to a partial return to his earlier readings of Schröder and the algebra of logic.

However Couturat is also known as the author of one of the three books on Leibniz (Couturat, 1901a) written almost simultaneously at the beginning of the 20th century, the two others being those of Bertrand Russell and Ernst Cassirer. Couturat saw his work both as a historical study and a way of connecting with contemporary debates on the foundations of mathematics and logic.

Only the first objective was really fulfilled if one is to believe the reviews and discussions generated by Couturat's book, much read and admired for its "lucidity". Couturat was no doubt disappointed by the reception, if one is to judge by the manner in which he defends his positions in subsequent texts in the *Revue de Métaphysique et de Morale* (Couturat, 1902b), as well as in the preface to *Opuscules et fragments inédits* (Couturat, 1903a).[3] Two years separate *La Logique de Leibniz* from the *Opuscules et fragments inédits*. Ubaldo Sanzo (Sanzo, 1991) interprets this as Couturat's reaction to the criticisms to which his book gave rise, the *Opuscules et fragments inédits* being presented as a set of fragments *intended* to clarify and justify Couturat's presentation. It is probable that what bothered Couturat was not only the criticism of his book, but the status attributed to him, which was based on an ambiguity. This ambiguity was compounded by Couturat's standpoint. In taking on Leibniz, he considered himself to be presenting the studies of a pioneer of many scientific and philosophical problems that were entirely in keeping with his time. Thus, by returning to Leibniz through the *Universal Algebra* (Whitehead, 1898), Couturat considered himself to be writing a work that would have its place in the most topical debates in the sciences of his day. But his book was understood first and foremost as a study on the history of philosophy, which significantly reduced its specifically philosophical and scientific impact. His thesis on the role of logic in Lebiniz's work was not followed. Even Russell saw it as a very beautiful book on the history of philosophy (Russell, 1945).[4] As a result, the book wasn't read in keeping with the author's full intentions. Yet one must remember the place Couturat sought to occupy in the scientific and philosophical world: that of a defender of a rationalism given fresh purpose by the most recent discoveries of the sciences.

---

[3] See Ubaldo Sanzo, 1991. He analyses the reviews of *La Logique de Leibniz*, and shows how the article from the *Revue de Métaphysique et de Morale* on "Leibniz's metaphysics" (Couturat, 1902), and then the publication of *Opuscules et fragments inédits de Leibniz* (C), were a defence of the theses presented in *La Logique de Leibniz* (Couturat, 1901).
[4] Russell, 1945, p. 568 et 571, and Russell, 1903, pp. 177–191.

We shall seek to understand this semi-failure through the way in which Couturat read Leibniz and intervened in logic of his time. To do so, we are going to present the principal features of Couturat's philosophical thinking, notably drawing on the considerable correspondence with Bertrand Russell between 1897 and 1913.

Couturat's rationalism

Couturat almost always worked on several subjects at the same time, except perhaps for the last years of his life, which were almost exclusively devoted to the defence of the international auxiliary language. He simultaneously tackled studies on the philosophy of science (principally on the infinite, set theory and geometry), logic (Boole, Schröder, Peano, Whitehead, Russell) and linguistics (in relation to the international language project; he was Meillet's student). This diversity attests to the fact that for Couturat there exists what one could call a "perpetual parallelism", to borrow a Leibnizian expression, not between God's goodness and omnipotence, but between philosophy and science. This parallelism is given concrete expression by a rationalist thesis that Couturat was never to abandon, and that one can already observe in the argumentation of his major doctoral dissertation *De l'infini mathématique* (Couturat, 1896a): a given scientific concept corresponds to a philosophical idea. One finds a certain number of statements similar to the following one:

> "...the mathematical idea of relation is simply the arithmetic translation of
> the philosophical idea of relation" (Couturat, 1896a, p. 426).

This thesis is rationalist, because it posits the relation between philosophy and science independently of the relation to the empirical. This rationalism assumes the idea of a universality of thought, whereby it has to be possible to relate any special concept to a philosophical idea. This correspondence between philosophy and science is, according to Couturat, the guarantee against scientific nominalism.

Here we find two demands inherent to Couturat's work. The first consists in justifying mathematical constructions through their relationship to philosophical thought, a position that is evident in his conception of truth:

> "...truths, whether logical or scientific, are relative to the thinking mind, and
> are true only insofar as they are consistent with rational principles" (Couturat,
> 1896a, p. IX).

This position is very present in the quoted work where the construction of sets of numbers is concerned, the latter being considered as schemas for orders of magnitude (Couturat, 1896a, p. 211), but also in all later philosophy of science debates that Couturat was to support with several French mathematicians.

In this way, Couturat was able to distinguish between what is "logically" constructed and that which is the "rationally" constructed (Couturat, 1896a, p. 134 or p. 425). The "logical" is that which can be constructed without philosophical consideration in a particular science, and the "rational" that which can correspond with philosophical ideas.

Couturat goes further still, by adhering to the principle whereby "all philosophy consists in choosing" between two equally logical alternatives the one which is most rational (Couturat, 1896a, p. X), a position which Bertrand Russell, in his review, qualified as "intellectual capitulation" (Russell, 1897b, p. 114).

Then there is the other demand, that which requires philosophy to also remain informed of new scientific knowledge and not to engage in rearguard battles. The ambition of *La Logique de Leibniz* is to show the importance of this attitude, to the point of criticising Leibniz when his attachment to Aristotle or Euclid prevented him from formulating ideas that he came very close to inventing. This is all explicit in Couturat's project:

> "My goal was to clarify and constantly comment on Leibniz's philosophical works through his mathematical works. You can easily understand the usefulness of such a study" (Russell, 2001, Couturat 30 June 1900, p. 189).

This suggests that, for Couturat, logic resides in the articulation of mathematics and philosophy. Like Russell, Couturat considers it as a discipline of its own; as a result of accepting Cantor's theory and transfinite sets, he attributed a place to classes and sets and developed an explicit treatment for them. Or again, speaking about Leibniz he states: he doesn't dabble in

> "... straggling philosophers who ignore the contemporary scientific movement and waste their energies in the study of past authors and the adoration of the past" (Russell, 2001, Couturat 31 October 1903, p. 314).

Couturat always takes sides, and does not try to balance opposite positions, except perhaps in politics where he greatly favoured any arbitration which delayed the declaration of the First World War, while actively campaigning for peace and socialism. Nonetheless, he conceded that "philosophical discussions are generally sterile" in his opening lecture at the Collège de France (Couturat,1906b, p. 340). Couturat therefore had the strengths and the weaknesses of someone who was always committed whatever the subject of dispute – mathematical, philosophical, or to do with the international language. There is always a more rational, more just position than another, there is no place for doubt in this rationalism, whatever the theme he is writing on. This is perhaps one of Couturat's greatest difficulties, synthesizing all the aspects of his knowledge when commitment is always militant, managing the balance between disciplines and methods and choosing between subjects worthy of interest. There is little room for fluctuation, hesitation, or paradox. He devoted all his strength to reason, and ended up turning against the criticist positions, first of all progressively, then very markedly; to him they appeared not only to mix knowledge and faith, but to give precedence to the latter. Couturat's whole intellectual attitude was driven by reason and he was convinced of the validity of all aspects of his activities, even in politics, where he was a pacifist and welcomed all the arbitration treaties in a very Leibnizian way.

From a philosophical point of view, this signified that Couturat was never sceptical. The drawback is that he has, in his own words, *"bêtes noires"* (Russell, 2001, Couturat 7 June 1898, p. 63), which one can list as follows:

1) Nominalism[5] (*passim*, against Edouard Leroy, (Couturat, 1900b), against Poincaré, against mathematicians who do not found their constructions rationally).

2) Symbolism that does not give rise to algorithms[6] (Russell, 2001, Couturat 3 January 1901, p. 216). Couturat wrote two articles on Peano (Couturat, 1899b and 1901d). His predilection for the algebra of logic led him to prefer the "algorithms" of algebra and logical equations to Peano's symbolism and the system of implications between propositions. For Couturat, the logic of classes came before the logic of propositions, and what mattered was their equivalence.

3) Aristotle and the scholastics[7] (Russell, 2001, Couturat 7 June 1898, p. 63, and Couturat 26 June 1904, p. 410).

4) The systematic search for paradox[8], because "a logical system must produce everything which *good sense* produces"[9]. Couturat characterises "common sense" as our spontaneous logic[10], rationality compatible with a rationalism whereby any principle must have intuitive clarity.

5) Commentators[11].

6) The adoration of the past[12] (Russell, 2001, Couturat 31 October 1903, p. 314).

Bolstered by his rationalism, his aim became to acquire knowledge of new ideas and to spread them. He acted as a sort of international "letterbox", corresponding with Italians (including Peano, Burali-Forti, Cremona, Vacca and Vailati), English philosophers (including Russell and Whitehead), Americans (including Veblen), and Germans (including Frege, Cassirer and Vaihinger). He played the role of a sort of new Mersenne. He

---

[5]To Russell: "...I am delighted to agree with you on most points, especially concerning philosophical principles, and to have found a precious ally against the nominalist tendency which reigns among mathematicians." (Russell, 2001, Couturat 3 December 1899, p. 146).

[6]On Peano's symbolism: "...I do not believe that the symbolism applied has great utility, nor that it is *essential* and indispensable. Furthermore, from the point of view of *formal* logic, I find it inconvenient; and in Mathematics, I see it as not much more than a stenography enabling to write formulas briefly, but in a horribly obscure and complicated form, for all intents and purposes *illegible*." (Russell, 2001, Couturat 3 January 1901, p. 216).

[7]"...I protested, saying that the Aristotelian tradition is dead, and that the living tradition is that of Pascal and modern mathematicians. That is what we have come to! We still have to fight Aristotle's authority! It's like having to kill dead men." (Russell, 2001, Couturat 26 June 1904, p. 410).

[8]To Russell: "...paradox is your *péché mignon*, and I acknowledge that *most* of your paradoxes are just and suggestive." (Russell, 2001, Couturat 12 May 1903, p. 290).

[9]Couturat states this in relation to the multiplicative axiom: "And then, seeing that you cannot prove something which seems so simple and so easy, I wonder if it is not the fault of the definitions or the principles of your system. You understand and excuse this attitude of a critical mind: but I believe that a logical system must be able to produce everything which *good sense* produces." (Russell, 2001, Couturat 10 July 1905, p. 512).

[10]"It would be unfortunate for Logic should there be, in its foundations, so complete a divorce between rigour and obviousness. (As you know, this criticism was once directed at Euclid's Geometry.) It seems that the basic elements of mathematical thought and all thought should present themselves to the mind with dazzling clarity, and not resemble subtle and thankless puns." (Russell, 2001, Couturat, 4 June, 1903, p. 295).

[11]To Russell: "...I have always had little taste for history, and I abhor commentators, who generally obfuscate their author. If I have found anything new by studying Plato and Leibniz, it was by reading the texts themselves, without the fuss of interpreters." (Russell, 2001, Couturat, 31 October 1903, p. 314).

[12]Couturat asserts his position against the "...straggling philosophers who ignore the contemporary scientific movement and waste their energies in the study of past authors and the adoration of the past." (Russell, 2001, Couturat, 31 October 1903, p. 314).

didn't consider himself to be on the side of the inventors, and the discussion of ideas on an international level seemed to him a pledge of progress and a partial guarantee of peace. This same posture led him to devote more and more of his time to perfecting and promoting the international language.

Justifications of the historical work

In a context such as this, what might the historical study of an author signify for Couturat? He presented his conception of the historical work to Russell, explaining that the aim was to reconstruct and interpret the system and to reconstruct its historical order:

> "I know very well that it is difficult to reconcile these two orders; however, I believe that it is indispensable with an author as "fluent" as Leibnitz" (Russell, 2001, Couturat 20 October 1900, p. 202).

Why this work, which apparently contradicted his declarations stating his disagreement with the study of the past? He sees Leibniz as a "precursor"[13] and that is what matters, as much as the links forged by the philosopher between the sciences and philosophy. These criteria were sufficient for Couturat to never vary his judgement on a few authors of the past, principally Plato, on whom he wrote his secondary thesis, as well as Descartes and Leibniz. For the same reasons, at first he admitted Kant's philosophy, at least partially, then vigorously rejected it. His reading of Russell's "popular" little article in *The International Monthly* entitled "Recent Work on the Principles of Mathematics" (Russell, 1901) and his exchanges with him certainly played a determining role in his criticism of Kant[14]. Another important cause is that the mathematicians at the time who were opposed to "logistics" (including Henri Poincaré and Pierre Boutroux) themselves belonged to a Kantian school of thought.

Couturat's preferences in logic

Before dealing with Couturat's reception of Leibniz, it is necessary to highlight the presuppositions through which Couturat read logic at the time he was undertaking his work on Leibniz. He wrote his main thesis *De l'infini mathématique* (1896a), where he defends Cantor, at a time when Russell had not yet acknowledged the latter's studies. He was therefore, as was said at the time, an "infinitist". He read logic as an algebraist, "...having learned logic through Schröder" (Russell, 2001, Couturat 16 November 1903, p. 330). His intellectual encounter with Russell incited him to more clearly distinguish between algebra and logic, but in his final presentation on the principles of logic (Couturat, 1913) he returns to very Schröderian formulations. As mentioned above, he is very wary of symbolism, saying of Frege that he had "an invincible repugnance for his symbolism" (Russell, 2001, Couturat 11 February 1904, p. 350), and of Peano that "I have no taste

---

[13] This is the term used by Couturat: "...Leibnitz, this genius precursor in all things..." (Russell, 2001, Couturat, 1 October 1900, p. 198).

[14] "You will find analogous or parallel opinions in the Conclusion of my *Leibniz...*" (Russell, 2001, Couturat, 21 July 1901, p. 257).

for his symbolism" (Russell, 2001, Couturat 1 October 1900, p. 199 and Russell, 2001, Couturat 3 January 1901, p. 217). He was to reconsider these judgements (Russell, 2001, Couturat 5 July 1901, p. 251), however, despite it all, the latter article shows that it is possible to present logic with a very sparing use of symbols[15]. Couturat's rationalism led him to assume an almost inevitable passage from the use of symbolism to nominalism. For him this at least represented a danger that one must always be attentive to, but which most mathematicians did not avoid. Obviously, such positions were to be moderated by everything he knew about the importance of symbolism in the work of Descartes, Leibniz, Whitehead and Russell. But he understood symbolism principally as serving algorithms.

It was above all his reading of mathematicians which led Couturat to turn towards Leibniz. Whitehead's *Universal Algebra* (Whitehead, 1898) was decisive as well as Grassmann's *Geometric Analysis* of 1847: "...I am not too surprised by our parallel work on Leibniz: mine was in part suggested by Whitehead's study and, subsequently by Grassmann, who, as you know, sided (after the fact) with Leibniz in his *Geometric Analysis* of 1847" (Russell, 2001, Couturat 13 May 1900, p. 179). At the very beginning of *Opuscules et Fragments inédits*, Couturat says that the *opportunity was presented to him* by Giovanni Vacca, in a conversation where the latter pointed out the importance of the unpublished Hanover papers (Couturat, 1903a).

Couturat and Russell as readers of Leibniz

We are once again going to refer to the correspondence between Couturat and Russell, because for a long time they argued over the interpretation of Leibniz. Indeed, Russell and Couturat did not take an interest in Leibniz for the same reasons. In Russell's work it is the (Russell, 2001, Russell 21 June 1900, p. 182) "...result of a purely philosophical class", where he replaced McTaggart, and where the subject had been chosen by the latter. Russell was interested in Leibniz's "philosophical logic" more than in his logic strictly speaking. In one letter Couturat stated:

> "I am convinced, like you, that Leibniz's logic is the centre and the heart of his system; it is what I show in my Preface, and what I set out to prove throughout the entire volume." (Russell, 2001, Couturat 13 May 1900, pp. 179–180)

Russell wrote in reply:

> "The logic I speak of is a logic in a less formal sense: what one could almost call epistemology: the analysis of judgements, etc. Nevertheless, what Leibniz says about symbolic logic is very remarkable" (Russell, 2001, Russell 21 June 1900, p. 182).

Unlike Couturat, Russell did not come to Leibniz through his studies. His interest in logic in this book is to do with the meaning, construction and consistence of the system.

---

[15] One must add a more material circumstance. Couturat, who liked technological innovations, very soon began writing his letters to Russell on a typewriter, which limited the use of mathematics and logical signs.

This difference is highly significant. The two authors do not deal with necessary truths and contingent truths in the same way. This difference is related to their respective conceptions of the place of logic in Leibniz's system. And finally it enables us to understand the progressive distance between Couturat's and Russell's attitudes in their understanding of the role of logic.

There is nonetheless a sort of coincidence, for a few years, between Russell's and Couturat's approaches. It has to do with their criticism of nominalism, with the project to develop a logic other than Aristotelian, with the criticism of Euclid and the search for a more "a priori" foundation of geometry through projective geometry, with the idea that one can deduce metaphysics from logic, and therefore with the fact their entire work has a systematicity which, in all its variations, remains quite stable[16].

Couturat's interpretation of Leibniz

Couturat speaks of his work on Leibniz in the following terms: "It is certain that I placed myself in a purely historical point of view, and that I did not introduce any preconceived theory into this study" (Russell, 2001, Couturat 12 May 1903, p. 291 and Couturat 27 March 1902, p. 276). Yet, he also states: "...the true Leibniz, not only has not been published, but has never been written" (Russell, 2001, Couturat, 20 October 1900, p. 202).

Couturat's interpretation is therefore also founded on that which was never written by Leibniz, the "guiding thread" of his system. We are going to follow the decisive aspects of this reconstruction.

In *De l'Infini mathématique* (Couturat, 1896a) Couturat already quotes Leibniz, for his idea of the actual infinite and the principle of continuity, and *De Arte Combinatoria* (1666), to which he attributes a special importance, as it is in accord with his rationalist and algebraic positions. His interpretation of Leibniz focuses principally on logic, and Couturat's interpretation of Leibniz develops in relation to the following theses:

1) Logic is the centre, the source and the heart of his system (Russell, 2001, Couturat 13 May 1900, pp. 178–179).
2) His metaphysics is a "panlogism" (Couturat, 1901a, p. XI and Russell, 2001, Couturat 4 October 1901, p. 261).
3) Logic is hidden in the system, acting as a "guiding thread".
4) No particular science (geometry, mechanics, theology) can be understood as the source of his system, even if each of them offers a point of view on the whole (Couturat, 1903c, p. 98).

Logic therefore cannot be treated as one or the other of these disciplines, it is as if "submerged"; for Couturat the aim is not to assume that it is a more important discipline than the others, however its central positioning results in the search for a "method" capable of accounting for the importance of each of the other scientific and philosophical

---

[16]See Daniel Garber's text, "Leibniz and the Foundations of Physics: The Middle Years", (Kathleen Okruhlik & James Robert Brown, 1985, pp. 72–73) which clearly shows the concomitant effects of the two works by Russell and Couturat on the interpretation of Leibniz.

disciplines. From there, it is partially unjust, at least in a first approach, to say that Couturat wasn't attentive to the role of mechanics or theology in the constitution of Leibniz's system. Thanks to the idea of a guiding thread, dear to Leibniz, then to Couturat, one can see logic as a method more than as a discipline which is already constituted.

Leibniz as a "precursor"

In his interpretation, Couturat constructs the opposition between a "precursor" Leibniz and a Leibniz who is too tied to tradition. Furthermore, the many fields in which Leibniz can be considered as a precursor – the term used by Couturat[17]– always involve something of his point of view on logic. According to Couturat's study on Leibniz and his later book *Les Principes des mathématiques* (Couturat, 1905d), this is principally observable where the following questions are concerned: the criticism of the syllogism, the universal language, the characteristic, logical calculus, geometric calculus, logic of relations, the logical definition of numbers (Couturat, 1905d, p. 27).

These objects are to a lesser or greater degree linked to logic, although in different ways. The criticism of the syllogism is one of the important passages in the development to a more mathematical logic. However, Couturat reproaches Leibniz for not going far enough in this direction, because his attachment to tradition prevented him from abandoning subalternation and partial conversion (Russell, 2001, Couturat 1 October 1900, p. 199). The universal language, the characteristic (articulation of symbolism and the encyclopaedia) both assume a relationship to logic in the sense that it acts as the negative condition for each. For the following areas, they are the development of logic, or assume a "translation" of logic in terms of number or magnitude. In all of these aspects, Leibniz always appears, in Couturat's view, as a precursor, advancing ideas that had not been formulated previously, and yet he had not been able to pursue these to their conclusions due to his respect for his predecessors. This was a consequence of (Couturat, 1901a, pp. 438–440):

1) his attachment to Aristotle (Russell, 2001, Couturat 01 October 1900, p. 199); and
2) his attachment to Euclid (on this point also Couturat appreciated Russell's "popular" little article in *The International Monthly* (Russell, 1901)).

The *Organon* and the *Elements* are, according to Couturat, too perfect as works to allow for invention, emphasising how far behind mankind is when such perfected monuments are published in a given domain (Couturat, 1901a, Conclusion).

The contrast between the inventor of new ideas and the philosopher who is too respectful of the works of the past plays an important role in Couturat's understanding of Leibniz. Rather than see this as a banality, one must consider that it is inherent to his conception of the links between philosophy and science. This contrast is to be related to that which one also quite regularly finds in his writings, where there are texts on the logic of the system (*De Arte combinatoria*, 1666, *Discours de Métaphysique*, 1686, *Opuscules et Fragments* (Couturat, 1903a) and others which act as its "novel" (*The Monadology* – "late and almost unintelligible summary of the doctrine" (Russell, 2001, Couturat 20

---

[17]Cf. note 13.

October 1900, p. 202), *The Principles of Nature and Grace*, ...). Michel Serres (Serres, 1968) clearly identified this division made by Couturat in Leibniz's work. He shows that it evened out the infinite variations of the system by opposing something as the centre of a partially formulated system and perspectives as important as they were unattainable, such as the Encyclopaedia for example, which for all intents and purposes only remained a project and a great "dream". Couturat wrote about Leibniz at a time when the critics knew well what a philosophical "system" was, but still weren't very sensitive to ideas equivalent to that of "architectonics".

Readings of Plato and Leibniz by Couturat

There are analogies between Couturat's readings of Leibniz and Plato. Indeed, the reading of Plato, *De Platonicis Mythis* (Couturat, 1896b), consisted in providing a way to distinguish the mythical from the heart of the system in the work. André Lalande (Lalande, 1915, p. 652) relates that Couturat had the intention of writing a work entitled *Plato's System in its Historical Development* and therefore of placing himself in a very similar perspective to the one adopted in his Leibniz seeking out the principle of the system by tracking its formulations chronologically.

How did Couturat think he could escape from the point of view of commentators on the history of philosophy and that of the worshipers of the past, to use his own expression, while seeking to describe the historical meanders of doctrines? It would appear that the decisive criterion is the relationship between the reading of the original texts and his conception of relations between philosophy and the sciences. A philosopher can be a precursor, this is true of all disciplines, not only philosophy, however philosophical ideas can be an obstacle to "new ideas".

"All truths are analytic"

Couturat attributed to Leibniz the thesis that "all truths are analytic" (Russell, 2001, Couturat, 20 October 1900, p. 202, and Couturat, 5 July 1901, p. 250). Nonetheless, he also stated that Leibniz "did not possess the notion of *analytic*" (Russell, 2001, Couturat 12 May 1903, p. 290). This summary of Leibniz's philosophy, which is sometimes erroneously attributed to Leibniz, given that Couturat's book was so widely read, is a quasi-Kantian formulation of a Leibnizian problem. It is not to be found in Leibniz, and was subsequently to be abandoned by Couturat (Lalande, 1914, p. 657, note 1).

Couturat presents this thesis by adding that it is a form of the principle of reason. Due to its universality, Couturat affirms that the principle of reason applies to every kind of truth. This formula accounts for the "praedicatum inest subjecto" of the *Discours de métaphysique* in relation to individual substances, but it articulates this subject-predicate logic with an idea of analycity linked to that of combinatorial analysis. Necessary truths and contingent truths are analytic (Couturat, 1901a, pp. 210–212). However, only the former require the principle of contradiction to be proved.

This form of the principle of sufficient reason is the converse of the principle of identity (Couturat, 1902b, p. 8) "Any identical proposition is true". This essential idea is presented as "hidden and as if submerged" (Russell, 2001, Couturat, 5 July 1901, p. 250) in most of the texts (Couturat,1902b, p. 8). The proposition "All truths are analytic" constructs a link between the theory of truth as predicate in the subject and the principle of sufficient reason (Couturat, 1902b), a link which was already highlighted by R.C. Sleight (Sleight, 1990, p. 90).

The importance of this explicit rendering of the principle of reason is that it does not differentiate between necessary propositions and contingent propositions. But, as Michel Fichant (Fichant, 2004, p. 18) points out, it also places individual substances and simple substances on the same level. Here a knot forms between logic, ontology and metaphysics, which becomes the subject of a long and quite obstinate discussion between Russell and Couturat in their exchange of letters, especially in 1900 and 1901, where, alongside their conception of the distinction between the analytic and the synthetic, they also discuss the distinction between the necessary and contingent, between that which has mathematical value and that which has philosophical value, and finally their ideas on the field of application of the principle of sufficient reason. Couturat states that there is no difference between existence and other predicates (Russell, 2001, Couturat 4 October 1901, p. 261), whereas Russell admits this difference (Russell, 2001, Russell 2 October 1901, p. 259), Russell finally accepts Couturat's position on the basis of a quote[18], but in his *History of Western Philosophy* (Russell, 1945), he is again very prudent on this question, and attributes Couturat's thesis to the esoteric doctrine[19].

The principle of reason is universal according to Couturat (Russell, 2001, Couturat 4 October 1901, p. 261). This is what distinguishes necessary truths from contingent truths according to Russell (Russell, 2001, Russell 2 October 1901, p. 259). The "universal characteristic" is the "root" and the "seed" of Leibniz's system for the former (Russell, 2001, Couturat 20 October 1900, p. 202). It is a mathematical idea for the latter (Russell, 2001, Russell 21 June 1900, p. 182). These differences express a divergence between the two philosophers in their way of seeing relations between logic and mathematics.

---

[18]Russell said himself to be convinced by the following text, which, as it happens, presents the problem highlighted by Couturat: "Atque ita arcanum aliquod a me evolutum puto, quod me diu perplexum habuit, non intelligentem, quomodo praedicatum subjecto inesse potest, nec tamen propositio fieret necessaria. Sed cognitio rerum geometricarum atque analysis infinitorum hanc mihi lucem accendere, ut intelligerem, etiam notiones in infinitum resolubiles esse". It can be translated as follows: "Thus I believe that I have explained something mysterious which for a long time left me perplexed and which I did not understand, namely how a predicate can be included in the subject and yet how the proposition thus formed is not necessary. But knowledge of geometry and the analysis of the infinite have brought me light and have allowed me to understand that notions can also be resoluble to the infinite" (Couturat, 1902b, p. 11, note 3).

[19]"What exactly Leibniz meant by the principle of sufficient reason is a controversial question. Couturat maintains that it means that every true proposition is "analytic", *i.e.*, such that its negation is self-contradictory. But this interpretation (which has support in writings that Leibniz did not publish) belongs, if true, to the esoteric doctrine. In his published works he maintains that there is a difference between necessary and contingent propositions, that only the former follow from the laws of logic, and that all propositions asserting existence are contingent, with the sole exception of the existence of God. Though God exists necessarily, He was not compelled by logic to create the world; on the contrary, this was a free choice, motivated, but not necessitated, by His goodness" (Russell, 1945, p. 568).

In Couturat's work, logic is a sort of rational foundation for mathematics, whereas for Russell logic is in a way more experimental than foundational[20], seeking what is most elementary in mathematics. On this point, they were later to be in disagreement, as Russell attested to indirectly in a article that appeared in the *Revue de Métaphysique et de Morale* (Russell, 1906b). For Couturat, the difference between synthetic and analytic has a universal value, whereas, for Russell, it has no importance in mathematics, but only in philosophy (Russell, 2001, Russell, 1 January 1905, p. 463). In Russell, the principle of reason enables to *differentiate* necessary truths and contingent truths, and even to give a definition of necessary truths, just as the principle of induction will later make it possible to give one of the definitions of the finite number[21]. Russell is a great deal more attentive to the architectonics of the different disciplines, whereas Couturat remains faithful to his rationalism, which assumes an almost term-to-term relationship between philosophy and mathematics. All of these discussions lead into another: the distinction between comprehension and extension in logic. Couturat maintained the standpoint of *Ars Combinatoria* where extension is concerned. Later we will see the importance of this point of view for his interpretation of Leibniz.

This discussion between Couturat and Russell is all the more difficult in that there is always a point where the difference between what is synthetic or analytic, of what is in comprehension or in intension gives rise to a problem. If one admits that all truths are analytic, how does one deal with this proposition? Is it itself analytic? If so, how can one admit that the principle is quite distinct from the principles of identity and contradiction? Similarly, one can finally admit that logic must operate in extension. But then, how does one consider the principal operator of a proposition? These are questions which one or the other of these authors develop, and which are integral to their way of theorising relations between logic and mathematics. Either logic is a sort of foundation, as in Couturat, and a synthetic aspect must indeed be present in its elements – for example the synthetic character of "all truths are analytic", or else, one admits with Russell that there may be a synthetic aspect, or an analytic aspect, in logic as much as in mathematics, a position which partially invalidates the relevance of this distinction. According to the latter, it has no importance in mathematics. In the end, for Russell, logic will not be a foundation, but the equivalent of an experimental science whose object is mathematics[22].

Placing the contingent and the necessary on the same level implies, for Couturat, at the same time a "mystery" of the system, and a fundamental problem.

---

[20] In an article of the journal *Philosophia Scientiae,* we highlighted Russell's criticisms of the idea of "basis" or "foundation" (Schmid, 2005, p. 175).

[21] This difference became very important during the polemic on the principle of induction with Poincaré, and on what was later called in logic the "final clause".

[22] "The Logistic method is essentially the same as in any other science. It contains the same fallibility, the same uncertainty, the same blend of induction and deduction, and the same necessity to contest, in order to confirm principles, so that they concur with results calculated through observation. Its object is not to ban "intuition", but to check and systematise its use, to eliminate the errors which its unchecked use gives rise to, and to discover the general laws from which one can, through deduction, obtain results never contradicted by intuition and, in crucial cases, confirmed by it. In all of that, Logistics is exactly on the same footing as astronomy for example, except that, in astronomy, verification takes place not through intuition but through sense." [Russell, 1906, 630].

The weak point or "mystery" is that the existence of non-realized possibles implies the contingency of the realized possibles, which are not opposed to them through contradiction, but are incompossible to them (Russell, 2001, Couturat 4 October 1901, p. 261). This is a way of defending contingency for "moral and *above all* theological" reasons (Russell, 2001, Couturat 4 October 1901, p. 262).

Given his interpretation, Couturat formulates the problem of Leibniz's system as follows: how is it possible that there are "analytic truths" that are not necessary and that, in other words, are contingent (Russell, 2001, Couturat 4 October 1901, p. 261)? The difference is in the idea of infinite analysis, which only distinguishes necessary truths from contingent truths:

> "*Contingentiae radix est infinitum. Veritas contingens est, quae est indemonstrabilis*" ("The root of contingency is the infinite. A truth is contingent when it is not subject to demonstration") (Hanover text quoted by Louis Couturat (Couturat, 1901a, p. 212)).

The criticisms of Couturat's interpretation

To begin with, let us point out that most critics considered the interpretations by Russell and Couturat in the same terms, and thought that both wrote about Leibniz because of his relations to logic. We saw that this is in part a confusion. Even Couturat was drawn to Leibniz more by mathematical readings (Whitehead, Grassmann) than by the logical ones (Schröder).

We have seen that while Couturat's study on Leibniz is generally very admired, few readers admitted his principal thesis on logic, and this is what Couturat sought to respond to both in the article he published in the *Revue de Métaphysique et de Morale* (Couturat, 1902b) and in the preface of his edition of *Opuscules et fragments inédits* (Couturat, 1903a). His major problem would be proving his standpoint: "When the fragment *Phil. VIII; 6–7* becomes known, everyone will be obliged to acknowledge my interpretation" (Russell, 2001, Couturat 5 July 1901, p. 250). This is the fragment published in the *Revue de Métaphysique et de Morale* (Couturat, 1902b, pp. 2–7).

The more recent reinterpretations of Couturat's thesis seek to show that Leibniz's texts were somewhat "forced" by Couturat. Kathleen Okruhlick points out that there are passages where Leibniz admits that there are essences which do not exist (Okruhlik, 1985, p. 190 *sqq.*), which obviously affects Couturat's thesis that existence is a predicate like any other. Similarly, individual substances (*Discours de Métaphysique*) and Monads (*Monadology*) are identified too quickly: "*The monad is the logical subject raised to the level of substance; its attributes become accidents 'inherent' to the essence of the substance*" (Couturat, 1902b, p. 99)[23]. These criticisms highlight an aspect which Couturat's interpretation leaves in obscurity.

---

[23]Couturat, (1902b) p. 9. Cf. M. Fichant, 2004, "Introduction" (p. 18) à son édition de *Discours de Métaphysique, Monadologie*, Folio Essais, Gallimard.

Other critics insist on the fact that some of the "special sciences"– mechanics, theology, metaphysics– were not understood in their function of elaborating the system, for example Martial Guéroult (Guéroult, 1967, 1)[24]. Perhaps one should remain more measured
in commenting the consequences of Couturat's logical interpretation. All differences of
perspectives are taken seriously by Couturat, he does not deny them, even where Dynamics is concerned, contrary to what Guéroult says, but it is as if he reduced them rationally
on an infinite surface whose organisation is entirely logical. One could reproach Couturat
less for having neglected such and such a perspective, than for not having developed the
full scope of the variations of Leibniz?s system. It is not a question of providing a classification of the sciences, rather a true architectonics, in a sense which we probably have not
finished discovering. But can one reproach Couturat with this failure? This architectonic
approach – in a somewhat different sense to Guéroult's architectonics, and closer to what
Michel Serres (Serres, 1968) says in his book on Leibniz – wasn't common in the history
of philosophy at the time. It was only later that it acquired some consistency.

But there is another structural criticism of Couturat's interpretation of Leibniz,
which concerns the "illatio duplex". It is the double assumption of an inference of implication between propositions and an inference of involution between the notions, which
connects the universal propositions "per se notae" and the infinite analysis of individual substances. It is this same double assumption that establishes the distinction between
mathematics and logic in Leibniz. This criticism was developed in particular by André
Robinet (1986). Leibniz has two discourses on the question of substantial forms, and it
depends on the point of view, in extension or in comprehension ("Illatio duplex", as clarified by André Robinet, 1986, is in the first text of *Opuscules et Fragments* 1–3: *Theol.*
VI, 2, f.11 (2p. in 4°) (Couturat, 1903a). Nevertheless, Couturat distinctly identifies the
tension in Leibniz's work between comprehension and extension. But he thinks that Leibniz's attachment to comprehension (Aristotle) caused him to neglect the algebra of logic
and the logic of relations (Couturat, 1901a, p. 376). For Couturat, logic is extensional
(algebraic conception of the system), that which is in comprehension is problematic, and
"piques his curiosity", he tells Russell, (Russell, 2001, Couturat, 12 May 1903, p. 291).

Couturat's admitted preferences are not without consequences for his reading of
Leibniz. They tend to reduce the difference between the implication among propositions
and involution in the development of a notion: "Yet the demonstration itself..., takes
place through the decomposition of the terms of the proposition to be demonstrated, so
that the analysis of the truths is reduced to the analysis of the concepts, that is to say the
definition" (Couturat, 1901a, p. 184).

Attempt at an interpretation

According to Couturat, there is a inversion in the order of the metaphysical theses (RMM,
1902, p. 9) between the *Monadology* and the texts in which the monad is the conclusion

---

[24]"Apart from the fact that these unilateral interpretations are in danger of strongly altering the true perspectives
of the doctrine, they go against what we know of Leibniz's activity since his earliest youth" (p. 2).

of a long deduction. This inversion can be "explained by the late date of this opuscule (1714)". (See also Couturat, 1901, p. 210 note 1.)

When Couturat states that there is only one source to Leibniz's system, logic, his argument as to the place of the other disciplines is often the following: the texts which seem to draw on sources of the system other than logic quite generally have a "polemical and negative value" (Couturat, 1902b, p. 22). There are only "negative determinations of the essence of substance" (Guéroult, 1967, p. 172).

Our hypothesis is that Couturat "replaces" the "illatio duplex" with a positive argumentation (that of substantial forms) and an argumentation that fuels the polemic, which remains entirely negative (that of phenomenalism; Couturat, 1902, p. 22). Thus, the other disciplines "found" the system by confirming the principles of logic (Couturat, 1902, p. 24). This resonates with Leibniz's maxim: "Systems are true by what they assert and false by what they deny" (quoted in Couturat, 1905e, p. 307). Couturat sought to make logic in Leibniz's work into a generator of the "true", and to consider the other scientific, theological and metaphysical disciplines, as true in themselves, without founding or generating the system. The use of the disciplines concerning the founding of the system served Couturat as a "polemical" argument or is considered by him as a "unilateral" approach. There is always a dissymmetry in how Couturat deals with Leibniz, which explains the difficulty he experienced in having his work seen as both historical and scientific. Couturat brings out the variational richness of Leibniz's thinking in the contrasts between completed work and dream, between the "precursor" and admirer of past works, between works that are consistent with logic and romanticised presentations of the system, between a close interpretation that articulates *De Arte Combinatoria* and the *Discours de Métaphysique* and a more Aristotelian interpretation, taking comprehension and the modalities into account without really reducing them.

How does Couturat treat his predecessors?

In France, among Couturat's contemporaries it is Emile Boutroux who published the most on Leibniz and German philosophy, editing the *Monadology*. Couturat barely quotes Boutroux, but acknowledges that he saw Leibniz's hesitation on the scope of the principle of contradiction and the principle of sufficient reason (Couturat, 1901a, p. 216 note 1).

Couturat's relationship with Boutroux is not a simple one. Emile Boutroux was Poincaré's brother-in-law and father of the mathematician Pierre Boutroux, with whom Couturat had engaged in polemics on the reception of "logistics". It should be pointed out that, in the name of the struggle against backward-looking philosophies, Couturat turned against almost all of his teachers, in particular those in mathematics. His Leibniz could not be placed on the same footing as the studies produced by the Kantian school in France at the time. Its place lay in the "perpetual parallelism" between the sciences and philosophy. Through his publication on Leibniz, Couturat was acting as an advocate and defender of modern ideas more than accomplishing the work of a historian. In a way, Leibniz enabled Couturat to turn against the Kantian thinking dominating French philosophy at the time.

But not only was this objective not recognised by his readers, in a way it ended up causing problems for Couturat.

Thanks to the correspondence with Russell we know that Couturat had completed two significant works. These were not projects, but fully elaborated works. In 1904, *La Logique Mathématique*, to be published by Naud in its "Scientia" collection, which became *Manuel de Logistique* (Alcan), or *Traité de Logistique*. In more or less the same period, he had completed a *Histoire de la logistique* (500–600 pages)[25]. What happened? Difficulties with the publishers? Discouragement before the technical difficulties of Russell's work? Excessive commitment to the dissemination of the international auxiliary language? All of these reasons must have had an influence. Couturat was probably tired out, as he mentioned several times in his correspondence with Russell, less by the struggle for "modern ideas", but by the constant adjustments being made by those who, like Russell, worked as inventors. These manuscripts remain lost for the moment, but they must exist somewhere[26]. In his following studies, the principle of reason returned in the form of the principle of abstraction ("it permits us to reduce every symmetrical and transitive relation to the equation of an abstract element (often itself imperceptible)" (Couturat, 1913, p. 178).

In his last presentation on logic ("The Principles of Logic", *Encyclopedia of the Philosophical Sciences*, London, Macmillan, 1913), Couturat principally returns to Schröder. Russell is very rarely quoted in this article.

There is therefore a history of Couturat's reading of Leibniz, both in Couturat's work and in that of his interpreters. Couturat's story is that of progressive weariness and disgust as regards the technical complications of the new logic. For his commentators, this is linked to his philosophical relations with Russell, where the positions of the two philosophers are almost always too quickly identified. Couturat's guiding thread, the part of the system which wasn't written, was often too hastily interpreted as a doctrinal corpus in his writings. Nonetheless, by abandoning his works, in a way Couturat did nothing to modify the interpretations of his commentators.

References

[Couturat 1896a] Couturat, Louis, *De l'infini mathématique*, Alcan, Paris 1896.

[Couturat 1896b] Couturat, Louis, *De Platonicis Mythis*, Alcan, Paris 1896.

[Couturat 1896 and 1897] Couturat, Louis, "Compte-rendu critique de Hannequin "Essai sur l'hypothèse des atomes dans la science contemporaine"." *Revue de Métaphysique et de Morale* 4 (1896), 778–797 and 5 (1897), 87–113, 220–247.

[Couturat 1898a] Couturat, Louis, "Sur les rapports du nombre et de la grandeur." *Revue de Métaphysique et de Morale* (1898), 422–447.

[Couturat 1898b] Couturat, Louis, "Compte-rendu critique de B. Russell, "Essai sur les fondements de la géométrie"." *Revue de Métaphysique et de Morale* 6 (1898), 354–380.

---

[25] See Russell, 2001, p. 661 and p. 662.

[26] The Poincaré Archives hold two of Couturat's manuscripts, an older one on logic, inspired above all by algebraic logic, and notes on Kant, probably for his 1904 article (Couturat, 2010).

[Couturat 1899a] Couturat, Louis, "Lettres à Brunetière, sur le pacifisme de Kant." *Le Temps* (27 March and 1 April 1899).

[Couturat 1899b] Couturat, Louis, "La Logique mathématique de M. Peano." *Revue de Métaphysique et de Morale* 7 (1899), 616–646.

[Couturat 1899c] Couturat, Louis, "Compte-rendu de B. Russell, "An Essay on the Foundations of Geometry"." *Bulletin des Sciences Mathématiques* (1899), 54–62.

[Couturat 1900a] Couturat, Louis, "Sur une définition logique du nombre." *Revue de Métaphysique et de Morale* 8 (1900), 23–26.

[Couturat 1900b] Couturat, Louis, "Contre le nominalisme de M. LeRoy." *Revue de Métaphysique et de Morale* 8 (1900), 87–93.

[Couturat 1900c] Couturat, Louis, "Sur la définition du continu." *Revue de Métaphysique et de Morale* 8 (1900), 157–168.

[Couturat 1900d] Couturat, Louis, "L'Algèbre universelle de M. Whitehead." *Revue de Métaphysique et de Morale* 8 (1900), 323–362.

[Couturat 1900e] Couturat, Louis, "Les Mathématiques au Congrès de Philosophie." *L'Enseignement Mathématique* 2 (1900), 397–410.

[Couturat 1900f] Couturat, Louis, "Compte-rendu du 1er Congrès International de Philosophie ; Section III : "Logique et Histoire des sciences"." *Revue de Métaphysique et de Morale* 8 (1900), 503–598.

[Couturat 1901a] Couturat, Louis, 1969, *La Logique de Leibniz*, Alcan, Paris, Hildesheim, Olms 1901.

[Couturat 1901b] Couturat, Louis, "Note sur les bases naturelles de la géométrie d'Euclide – contre M. de Cyon." *Revue Philosophique de la France et de l'Etranger* 52 (1901), 540–542.

[Couturat 1901c] Couturat, Louis, "Troisième volume du Congrès International de philosophie de 1900 : "Logique et Histoire des sciences"." Armand Colin, Paris 1901.

[Couturat 1901d] Couturat, Louis, "Compte-rendu critique de Peano, "Formulaire de mathématiques"." *Bulletin des Sciences Mathématiques* (1901), 141–159.

[Couturat 1902a] Couturat, Louis and Ladd-Francklin, Christine, "Symbolic Logic." *Dictionary on Philosophy and Psychology* (1902), J.M. Baldwin, ed., 640–645 and 650–651.

[Couturat 1902b] Couturat, Louis, "Sur la métaphysique de Leibniz (avec un opuscule inédit)", *in*: *Revue de Métaphysique et de Morale* 10 (1902), 1–25.

[Couturat 1903a] Couturat, Louis, *Opuscules et Fragments inédits de Leibniz*, Alcan, Paris 1903.

[Couturat 1903b] Couturat, Louis, and Leau, Léopold, *Histoire de la langue universelle*, Hachette, Paris 1903.

[Couturat 1903c] Couturat, Louis, "Compte-rendu d'Ernst Cassirer, "Leibniz System in seinen wissenschaftlichen Grundlagen"." *Revue de Métaphysique et de Morale* 11 (1903), 83–99.

[Couturat 1904a] Couturat, Louis, "La section de logique et de philosophie des sciences au Congrès de Genève." *Revue de Métaphysique et de Morale* 12 (1904), 1037–1077.

[Couturat 1904b] Couturat, Louis, "Russell (Bertrand) – "The Principles of Mathematics", Cambridge, Cambridge University Press, 1903." *Bulletin des Sciences Mathématiques* 28 (1904).

[Couturat 1904c] Couturat, Louis, "Délégation pour l'adoption d'une langue auxiliaire internationale." *L'Enseignement Mathématique* 6 (1904), 140–142.

[Couturat 1905a] Couturat, Louis, *A Plea for an international language*, George J. Henderson, London 1905.

[Couturat 1905b] Couturat, Louis, "Sur l'utilité de la logique algorithmique." Compte-rendu du 2e Congrès International de Philosophie, Kündig, Genève 1905, 706–712.

[Couturat 1905c] Couturat, Louis, "Les Définitions mathématiques." *L'Enseignement Mathématique* 7 (1905), 27–40.

[Couturat 1905d] Couturat, Louis, "Définitions et Démonstrations mathématiques." *L'Enseignement Mathématique* 7 (1905), 104–121.

[Couturat 1905e] Couturat, Louis, *Les Principes des mathématiques, avec un appendice sur la philosophie des mathématiques de Kant*, Alcan, Paris 1905.

[Couturat 1905f] Couturat, Louis, 1965, 1980, *L'Algèbre de la Logique*, Gauthier-Villars, réed. Georg Olms, puis Réed. Albert Blanchard, Paris, Hildesheim, Paris 1905.

[Couturat 1906a] Couturat, Louis, "Pour la logistique (réponse à M. Poincaré)." *Revue de Métaphysique et de Morale* 14 (1906), 208–250.

[Couturat 1906b] Couturat, Louis, "La Logique et la Philosophie contemporaine." *Revue de Métaphysique et de Morale* 14 (1906), 318–341.

[Couturat 1912] Couturat, Louis, "Die Prinzipien der Logik." *Enzyclopädie der philosophischen Wissenschaften*, Mohr, Tübingen 1912, 137–201.

[Couturat 1913] Couturat, Louis, "The Principles of Logic." *Encyclopedia of the Philosophical Sciences*, Macmillan, London 1913, 136–198.

[Couturat 1916] Couturat, Louis, "De l'abus de l'intuition dans l'enseignement mathématique." *Revue de Métaphysique et de Morale* 24 (1916), 879–884.

[Couturat 1917a] Couturat, Louis, "La Logique algorithmique et le calcul des probabilités." *Revue de Métaphysique et de Morale* 24 (1917), 291–313.

[Couturat 1917b] Couturat, Louis, "Sur les rapports logiques des concepts et des propositions." *Revue de Métaphysique et de Morale* 24 (1917), 15–58.

[Couturat 1973] Couturat, Louis, *De l'infini mathématique*, A. Blanchard, Paris 1973.

[Couturat 1979] Couturat, Louis, and Leau, Léopold, *Histoire de la langue universelle, suivi de: Les nouvelles langues internationales*, G. Olms, Hildesheim, New York 1979.

[Couturat 1980] Couturat, Louis, *Les principes des mathématiques: avec un appendice sur la philosophie des mathématiques de Kant*, A. Blanchard, Paris 1980.

[Couturat 1980] Couturat, Louis, *L'Algèbre de la logique*, Albert Blanchard, Paris 1980.

[Couturat 2010] Couturat, Louis, eds. Schlaudt, Oliver and Sakhri, Mohsen, *Traité de logique algorithmique*, Basel, Birkhäuser, "Publications des Archives Poincaré", 2010.

[Fichant 2004] Fichant, Michel, "Introduction" to his edition of the *Discours de Métaphysique, Monadologie*, Paris, Gallimard, Folio Essais 2004.

[Garber 1985] Garber, Daniel, "Leibniz and the Foundations of Physics: The Middle Years", in: Okruhlik, Kathleen & Brown, James Robert eds., 1985.

[Guéroult 1967] Guéroult, Martial, *Leibniz. Dynamique et Métaphysique*, Paris, Aubier-Montaigne 1967.

[Lalande 1915] Lalande, André, "L'œuvre de Louis Couturat", *Revue de Métaphysique et de Morale* 22 (1915), 644–688.

[Loi 1983] Loi, Maurice ed., *Louis Couturat ... de Leibniz à Russell...*, Paris, Presses de l'Ecole Normale Supérieure, "Philosophie" 1983.

[Okruhlik 1985] Okruhlik, Kathleen & Brown, James Robert eds., *The Natural Philosophy of Leibniz*, Dordrecht, Boston, Lancaster, Tokyo, D. Reidel 1985.

[Robinet 1986] Robinet, André, *Architectonique disjonctive, automates systémiques et idéalité transcendantale dans l'œuvre de G.W. Leibniz*, Vrin, Paris 1986.

[Russell 1897] Russell, Bertrand, "Review of L. Couturat, "De l'infini mathématique"." *Mind* 6 (1897), 112–119.

[Russell 1900] Russell, Bertrand, *A critical Exposition of the Philosophy of Leibniz*, Cambridge University Press, Cambridge 1900.

[Russell 1901] Russell, Bertrand, "Recent Work on the Principles of Mathematics." *International Monthly* 4 (1901), 83–101.

[Russell 1903] Russell, Bertrand, "Recent Work on the Philosophy of Leibniz", *Mind* 12 (1903), 177–201.

[Russell 1945] Russell, Bertrand, *A History of Western philosophy, and its connection with political and social circumstances from the earliest times to the present day*, Simon and Schuster, New York 1945.

[Russell 1906] Russell, Bertrand, "Les Paradoxes de la logique", *Revue de Métaphysique et de Morale* 14 (1906), 627–650.

[Russell 2001] Russell, Bertrand, *Correspondance sur la philosophie, la logique et la politique avec Louis Couturat (1897–1913)*, ed. Anne-Françoise Schmid, transcription Tazio Carlevaro, 2 vol., 735 p., Paris, Kimé 2001.

[Sanzo 1991] Sanzo, Ubaldo, *L'Artificio della lingua. Louis Couturat (1868–1914)*, Milan, FrancoAngeli 1991.

[Schmid 2005] Schmid, Anne-Françoise, "Perspectives hétérodoxes de Russell sur les fondements", in *Philosophia Scientiae*, Cahier spécial 5: "Fonder autrement les mathématiques" (2005), 175–198.

[Schmid 2009] Schmid, Anne-Françoise, "La Controverse entre Bertrand Russell et Henri Poincaré", in: Jean-Yves Béziau, Alexandre Costa-Leite eds.: *Dimensions of Logical Concepts*. Coleção CLE, volume 54, UNICAMP, Campinas, Brazil (Centre de Logica, Epistemologia et Historia de la Ciencia) 2009, 99–126.

[Serres 1968] Serres, Michel, *Le Système de Leibniz et ses modèles mathématiques,* Paris, P.U.F. 1968.

[Sleigh 1990] Sleigh, R.C. JR, *Leibniz and Arnauld. A Commentary on their Correspondence*, New Haven and London, The Yale University Press 1990.

[Whitehead 1898] Whitehead, Alfred North, *A Treatise on Universal Algebra*, Cambridge University Press, Cambridge 1898.

Anne-Françoise Schmid
33, rue de Fontarabie
F-75020 Paris, France
afschmid@free.fr

Russell and Leibniz on the Classification of Propositions

Nicholas Griffin

1. Background

We owe Russell's book on Leibniz to a very improbable series of events, of which surely the most improbable of all was that McTaggart was getting married. It was McTaggart who was scheduled to give the lectures on Leibniz that Russell ended up giving at Trinity College, Cambridge during Lent Term of 1899 and it was McTaggart's marriage that prevented him from giving them. Now getting married (even for McTaggart) would not normally be so traumatic an event as to prevent one from lecturing, but McTaggart's bride-to-be was a New Zealander. So McTaggart's nuptials in 1899 took him away from Cambridge for a very long time while he travelled to the other side of the globe in pursuit of love – the relation which, in his metaphysics, appropriately enough, held the Absolute together. In his absence, Trinity College looked around for a replacement lecturer, and offered the job – another unlikely event – to Russell. Why they would have chosen Russell is not clear. He was not known to be interested in Leibniz, or even very much in the history of philosophy. And there were other people at Cambridge who might well have been available: G.F. Stout regularly taught early modern philosophy there. But given that they did offer the lectureship to Russell, it is another mystery why he accepted it. His own work was at that time in a state of rapid transition which might have been expected to keep any normal workaholic more than fully occupied.[1] It was not that he needed the money: he was not a wealthy man, but he already had a Fellowship stipend from Trinity which he was giving away to the London School of Economics. He did not enjoy lecturing, being still quite painfully shy on public occasions. Nor would the subject have

---

[1] During the period in which he prepared and delivered the Leibniz lectures (summer 1898 to summer 1899) Russell worked on four different book projects on the philosophy of mathematics, completing drafts of two of them, in addition to working on the axiomatization of projective geometry and the concept of order, as well as on many smaller projects. Even more importantly it was a time of dramatic change in his philosophy, the period in which he rejected neo-Hegelianism and embraced analysis, the one real revolution in his thought he subsequently maintained (*MPD*, p. 11).

seemed particularly attractive to him. Describing his attitude to Leibniz at the start of his research, he said that the *Monadology* appeared to him as 'a kind of fantastic fairy tale, coherent perhaps, but wholly arbitrary' (*POL*, p. xiii). This hardly seems like a recommendation of the subject.[2] Nonetheless, Russell did accept the invitation and the result of this once-only teaching assignment was not just a book on Leibniz, but a very important book, which, as Nicholas Jolley says, exerted 'a huge influence on Leibniz studies ... for much of the twentieth century'.[3]

Russell probably accepted the task in the early summer of 1898, for his reading list[4] records him reading Langley's recent translation of the *Nouveaux essais* [NE] in June 1898, a work which otherwise he would have had no reason to read at that time. Thereafter, Leibniz works begin to appear regularly in his reading list for the next nine months, though it can hardly be said he was studying Leibniz with any great urgency. The next works to appear are two recent English translations of selections of Leibniz's writings: Duncan's *The Philosophical Works* [D] and Latta's *Monadology and Other Writings* [RL], in August and October respectively, followed by the *Théodicée* (in an unknown edition) in November. It may seem odd that Russell should have read Latta's selection after Duncan's, since everything Latta includes is in Duncan, but the reason is presumably that he was looking for better translations than Duncan's, of which he complains in his Preface (*POL*, p. xv).[5] Russell's main source for Leibniz text was Gerhardt's seven-volume *Philosophischen Schriften* [GP], at that time the most extensive collection of Leibniz material available in print, and his equally magisterial, *Leibnizens Mathematische Schriften* [GM], also in seven volumes. Both are cited extensively in *POL*, the former on almost every page. Russell records reading the *Philosophischen Schriften* in February 1899, which was presumably when he finished a task begun some months before. The *Mathematische Schriften* is not recorded in 'What Shall I Read?', probably indicating that Russell read it only selectively. Russell's library in The Bertrand Russell Archives contains his copy of the *Philosophische Schriften*: it is quite extensively annotated by Russell, with passages marked for attention and occasional marginal comments. But Russell did not have his own copy of the *Mathematische Schriften*: he tells Moore (9 June 1900) that he is using the copy from the Trinity College Library.[6] Although there are no copies of Duncan and

---

[2] There was one topic on which a study of Leibniz might have appealed to Russell when it was first proposed to him in 1898, and that was Leibniz's views on infinity, a topic which was giving Russell immense trouble in the philosophy of mathematics as he struggled to come to grips with Cantor. There are, indeed, places in the manuscripts on infinite number that Russell wrote shortly after his work on Leibniz which show Leibniz's influence (e.g., POM/D, pp. 121–2; cp. GP, i, p. 388; GP, ii, pp. 304–5; GP, vi, p. 629), and these would be worth studying, though I shall not do so in this paper. Leibniz's views on infinity were, in fact, less helpful to Russell than might have been expected, for they tended to confirm him in resisting Cantor.

[3] Jolley [2005], p. 216; see also Adams [2010], p. 309.

[4] 'What Shall I Read?', *CPBR1*, pp. 347–65. Apart from what he tells us in *POL*, all our information about Russell's reading on Leibniz comes from this list.

[5] About half of Latta's book is taken up with Latta's exposition of Leibniz's philosophy, but it was not this that attracted Russell, who, so far as I can tell, makes no use of it anywhere.

[6] Gerhardt printed the works in the language in which Leibniz had written them – German, French, or Latin. Moore was helping Russell with Latin translations. Wherever possible Russell used an existing translation, preferring Latta's to Duncan's, but correcting Duncan's rather than making his own.

Latta in Russell's library, it is plausible to assume that he did have his own copies of these relatively cheap and easily obtainable books, but disposed of them subsequently (he was not an assiduous keeper of books). The only other Leibniz text he cites in the book is Foucher de Careil's *Réfutation inédite de Spinoza par Leibniz* (1854), which, however, is not listed in 'What Shall I Read?'.[7]

Turning to secondary materials, Russell says he learnt more from Erdmann's 'excellent account of Leibniz in his larger history' (Erdmann [1834–53]) 'than from any other commentary' (POL, p. xiii), though he complains that the book was written in ignorance of much important material that had been published only after it was written, in particular the correspondence with Arnauld.[8] It was reading this correspondence and the *Discourse on Metaphysics* that, Russell said, threw 'a flood of light … on all the inmost recesses of Leibniz's philosophical edifice' ([POL], pp. xiii–xiv). Unfortunately, we don't know when this illumination took place.[9] In the Preface he cites only two other studies of Leibniz: Dillmann [1891], which he treats rather dismissively, and Stein [1890], which he praises for its historical scholarship.[10] Works on Leibniz had been appearing with some regularity in German during the nineteenth century as previously unpublished manuscripts dribbled out of the archives,[11] but very little material had appeared in English before Russell's book. Latta ([RL], p. v) is able to cite only Mertz [1884] and Sorley's *Encyclopedia Britannica* article, of which only about half is devoted to Leibniz's philosophy (Sorley [1875]), as precursors. The catalyst for German work on Leibniz was undoubtedly the publication of Gerhardt's two massive editions of Leibniz's mathematical and philosophical writings, but little of this material had appeared in English beyond Langley's translation of the *Nouveaux essais* and the selections by Duncan and Latta. Russell's rather striking (and still very useful) device of including almost 100 pages of classified extracts from Leibniz in English translation as an appendix to his book was a response to this situation. Russell's book was thus well-placed to make an important contribution to English thinking about Leibniz, almost irrespective of the value of his interpretation.

In this paper, however, I do not propose to assess Russell's interpretation of Leibniz, nor his contribution to Leibniz scholarship[12] – both of which would require more

---

[7]Interestingly, he used the original French edition, published in a very limited edition by the *Institut de France*, rather than the much more readily available English translation (Foucher de Careil [1855]), presumably because the former was available in the Cambridge University Library. He seems to have made no use of two other collections of Leibniz material edited by Foucher de Careil, [1857] and [1859–75]. The latter was the most extensive edition of Leibniz's writings to be published before Gerhardt's edition. It, too, was available in the Cambridge University Library, but was of less interest to Russell since it contained mainly political, historical and theological writings. Loemker criticizes the texts as inferior to Gerhardt's and Erdmann's (L, i, p. 103).

[8]The book does not appear in 'What Shall I Read?', no doubt because Russell read only the sections on Leibniz. Russell's library does contain a finely bound copy of Leibniz's *Opera Philosophica*, edited by Erdmann (Erdmann [1840]), which belonged to Russell's former brother-in-law, Frank Costelloe. It gives no sign of having been read.

[9]Russell would have read them in [GP], vols. i and iv, respectively.

[10]'What Shall I Read?', which includes Stein but not Dillmann, adds also Mertz [1884], vol. ii of Kuno Fischer [1889], and Selver [1885].

[11]Kvêt [1857], Caspari [1870], Class [1874], Fischer [1889a], Wernicke [1890], Thilly [1891], Werckmeister [1899] may be mentioned as of potential relevance to Russell, in addition to those already cited.

[12]There is a mass of useful information about the book, its composition and its reception, in O'Briant [1979].

knowledge of Leibniz and Leibniz scholarship than I possess. Rather, I want to look at one aspect of the contribution that Russell's foray into Leibniz scholarship made to the development of his own philosophy, in particular his philosophy of logic. Russell's interpretation of Leibniz was famous for treating almost the whole of Leibniz's philosophy as the outcome of his logical doctrines. This was not a view that Russell had when he started his study of Leibniz, but an interpretation he arrived at in the course of his work. It was a view that had a clear attraction to Russell and one which was consonant with Russell's own recently acquired position on how philosophy should be conducted; a view which placed logic at the beginning of philosophy, and the analysis of propositions at the beginning of logic.[13] This was a view that Russell thought he shared with Leibniz, though some prominent Leibniz scholars dissented (e.g., Duncan [1901]). After a summary of Leibniz's premisses, Russell begins his account with the following bold declaration: 'That all sound philosophy should begin with an analysis of propositions, is a truth too evident, perhaps, to demand a proof. That Leibniz's philosophy began with such an analysis, is less evident, but seems to be no less true' (*POL*, p. 8). Russell and Leibniz differed, however, on how propositions were to be analyzed. It is Russell's treatment of Leibniz's account of propositions, and its relation to Russell's own account, on which I want to concentrate in this paper. It will require a little background on where Russell stood philosophically in 1899.

Russell's study of Leibniz came at a critical time in the development of his own philosophy. In 1898 he had been working on *An Analysis of Mathematical Reasoning* [*AMR*], a major book which, as he told Couturat[14], was an attempt to answer the Kantian question: How is pure mathematics possible? A question to which he expected to give a purely Kantian reply. Despite this, Kantian elements were less in evidence in this work than in Russell's previous efforts at dealing with the special sciences, which had been dominated by transcendental arguments. In *AMR*, by contrast, large parts of the extant manuscript are taken up with the classification of propositions and their analysis into their component terms, work which was entirely of a piece with the type of philosophical analysis that Moore and Russell started publishing the following year.[15] Even so, *AMR* is a genuinely transitional work. All the analytic labour can be seen as being undertaken in support of a transcendental deduction, this time of Boolean algebra from the possibility of judgment. Writing *AMR* brought to a head the problems facing Russell's earlier idealist approach to philosophy, and the next extended work of which we have substantial fragments, *The Fundamental Ideas and Axioms of Mathematics* [FIAM], written in 1899,[16] is radically different in tone. Neo-Hegelianism has been entirely abandoned, and so too has

---

[13]In a letter of 5 May 1900 Russell told Couturat, then at work on his own book on Leibniz's logic (Couturat [1901]), that, while *POL* was on the whole of Leibniz's philosophy, it was the logic that interested him and he decided to write the book only after he discovered that the whole of Leibniz's system was based on his logic ([BRLC], i, pp. 170–1).

[14]3 June 1898 (*SLBR1*, p. 188).

[15]For more information on Moore's and Russell's work at this time, see Griffin [1991], pp. 296–309; and Griffin [forthcoming].

[16]In between there were at least two other book length projects on philosophy of mathematics in this exceptionally busy period of Russell's life: *On the Principles of Arithmetic* (of which only two chapters survive; *CPBR2*, pp. 247–60) and *An Inquiry into the Mathematical Categories* (of which we have only an outline; *CPBR2*, pp. 26–7).

the use of transcendental arguments. There is not even a lingering vestige of deference to Kant, whose appeals to intuition are now declared unnecessary (FIAM, p. 270) – worse abuse was in store for them in subsequent writings.

Russell gave his Leibniz lectures between the writing of *An Analysis of Mathematical Reasoning* and that of *Fundamental Ideas and Axioms*, and it is not difficult to find the impact of Leibniz on *FIAM*. Clinton Tolley, in a paper entitled 'Humanizing Logic: Kant's Departure from Leibniz' (Tolley, forthcoming), argues that Kant developed his view that logic should be anchored in the human understanding as an explicit reaction against Leibniz's overtly theological conception of logic. In the period 1898–9, Russell was going in the opposite direction: starting from Kant, and seeking to dehumanize logic, or at least to depsychologize it. Although Russell had always been opposed to psychologism, by embracing transcendental arguments in his early analysis of the sciences he was sailing close to the wind. He acknowledged that Kant's own philosophy of space was psychologistic and he sought to avoid the same complaint being laid against his own by insisting that his use of 'a priori', unlike Kant's, was purely logical (*EFG*, p. 3). Whether this was so (and even what he meant by it) is not so easy to discern. A failure to avoid psychologism was one of the chief complaints that Moore made against Kant in his second dissertation, a study of Kant's moral philosophy,[17] and he came to apply it also to Russell in a strongly critical review of Russell's *Essay on the Foundations of Geometry* (Moore [1899]). By the time the review appeared Russell agreed with his criticisms. There is a certain appropriateness, then, that Russell, in reacting against Kant, should turn to the philosopher against whom, on Tolley's account, Kant had reacted. It was not that, by studying Leibniz, Russell came to appreciate the perils of psychologism, nor even that he learnt how to avoid them: Russell would have no truck with Leibniz's theological basis for logic, and there were points on which Russell thought that Leibniz himself was prone to make the same mistakes as Kant.[18] It was rather that Leibniz would confirm him in the direction he was already taking.

When *A Critical Exposition of the Philosophy of Leibniz* first appeared in 1900, it was almost the first work to be published that represented the new, anti-idealist philosophy that Russell and Moore had begun to develop at the end of the nineteenth century. Moore's important but obscure 'The Nature of Judgment' [1899a] had appeared in *Mind* at the end of the previous year and his somewhat less important but even more obscure 'Necessity' [1900] was published in *Mind* in 1900. Russell, himself, though he had been active giving papers, had previously published nothing of the new ideas. One can imagine, then, the degree of incomprehension that greeted the book: reviewers had very little idea of the basis of Russell's criticisms. Nonetheless, the book was quite widely reviewed,[19] though most of the reviews were quite short. They noted the book's difficulty (*The Saturday Review*, 12 Jan., 1901), the arcane nature of its subject matter (*The Westminster Review*, Dec., 1900), and that it was for specialists only (*The Guardian*, 3 April, 1901). The anonymous

---

[17] His most effective statement of the charge is in Moore [1903].

[18] He criticizes Leibniz's treatment of innate ideas in *New Essays*, e.g., on exactly this point: it is 'more like Kant's doctrine', he says, 'than it has any right to be' (*POL*, p. 163).

[19] The Russell Archives contains fifteen reviews of the first edition. See O'Briant [1979], pp. 186–91 for more detail about the reviews.

Guardian reviewer, however, rather presciently had this to say: 'it is impossible not to see in Mr. Russell's work elements of real originality and great power of argument, which together may lead to striking results in the future'. *The Literary Review* (1 July, 1902), by contrast, declared in tones verging on bewildered horror: 'one gathers that Mr. Russell is quite unorthodox'. Still, that was better than James Tufts, who in the *American Journal of Theology* said the book had 'comparatively little value' (Tufts [1902]).

Two prominent English-language Leibniz scholars, however, reviewed the book at length in prominent journals: George Martin Duncan in *The Philosophical Review* (Duncan [1901]) and Robert Latta in *Mind* (Latta [1901]).[20] Duncan, in a hostile review in which he took offense at Russell's criticisms of his own translations of Leibniz, simply ignored Russell's treatment of the analysis of propositions, which forms the heart of his book, maintaining merely that Leibniz did not arrive at his philosophy in that way (Duncan [1901], pp. 296–7). Latta, by contrast, genuinely attempted to come to grips with Russell's main contention, but was constantly frustrated by the fact that the point of view from which Russell was criticizing Leibniz was both completely unfamiliar and not fully explained. Indeed, Latta noted that Russell's criticism would apply, not just to Leibniz, but to 'the fundamental principles of all the great modern systems'. 'It is a pity', he went on, 'that in making so comprehensive a charge Mr. Russell has not given us a more complete account of his own position, for if his contention be just, his relational theory of the proposition must be of incalculable importance to philosophy' (Latta [1901], p. 527). One suspects more irony than generosity in this remark, but there was justice in his complaint. 'Mr. Russell has a logic of his own', Latta said, 'which he indicates from time to time, but which he no where fully expounds and justifies' (p. 526). Even with the advantages of hindsight, and of vast masses of Russell material not available to Latta, it is still no easy matter to determine precisely what Russell's logic was at the time he wrote his book on Leibniz.

Russell summed up Leibniz's position with admirable concision as follows:

> Every proposition is ultimately reducible to one which attributes a predicate to a subject. In any such proposition, unless existence be the predicate in question, the predicate is somehow contained in the subject. The subject is defined by its predicates, and would be a different subject if these were different. Thus every true judgment of subject and predicate is analytic – i.e., the predicate forms part of the notion of the subject – unless actual existence is asserted. Existence, alone among predicates, is not contained in the notions of the subjects which exist. Thus existential propositions, except in the case of God's existence, are synthetic, i.e., there would be no contradiction if the subjects which actually do exist did not exist. Necessary propositions are such as are analytic, and synthetic propositions are always contingent. (*POL*, pp. 9–10)

I shall examine the various parts of this account in what follows. The view that the predicate is contained in the subject is dealt with in § 2. The thesis, which I shall call (SP)

---

[20]Russell had corresponded with Latta while writing the book. See O'Briant [1979], pp. 169–71, for Latta's letters to Russell. Russell's letters are lost.

that all propositions are of subject-predicate form (SP-propositions, for short) is considered in § 3. The vexed question of whether this means that all propositions are analytic occupies § 4, and the question of whether any propositions are synthetic is considered in § 5. The remaining two sections deal with Russell: § 6 with the relation between the containment thesis and Russell's conception of analysis and § 7 with what Russell might have learnt about relations from his study of Leibniz.

2. Decompositional analysis and the containment principle

Russell's first extended work employing the new 'logic', at least the first one that was free from Kantian and Hegelian influences, was *The Fundamental Ideas and Axioms of Mathematics*. Russell's approach in the new manuscript is mereological: the business of philosophy is primarily the intellectual analysis of complex unities (propositions) into their constituent parts. There is even an attempt to understand inference in terms of the containment of the conclusion in the premises: 'Symbolic Logic as the Calculus of Whole and Part pure', as Russell put it (FIAM, p. 267). The fundamental ideas of mathematics are the simple elements from which complex items can be constructed; the fundamental axioms are those propositions from which the others can be derived. This is a style of analysis that Michael Beaney [2003] has usefully called 'decompositional analysis': the decomposition of complex entities into their constituents.

Russell's study of Leibniz would have encouraged his use of decompositional analysis, for, as Beaney [2003] says, Leibniz was 'pivotal' in ensuring the dominance of the decompositional conception of analysis in early modern philosophy. In Leibniz, the doctrine was encapsulated in what Beaney calls 'the containment principle'. It is stated in a letter to Arnauld of 14 July 1686:

> In every affirmative true proposition, necessary or contingent, universal or singular, the concept of the predicate is included in that of the subject, *praedicatum inest subjecto*. (GP, ii, p. 56)[21]

and in notes on one of Arnauld's letters that Leibniz made around the same time:

> In consulting the notion which I have of every true proposition, I find that every predicate, necessary or contingent, past, present, or future, is comprised in the notion of the subject. (GP, ii, p. 46)[22]

Russell gives the latter statement as the first of the 'leading passages' which he translates in the appendix to his book (*POL*, p. 205); it is the lynchpin of his interpretation of Leibniz.

---

[21]Cf. also: GP, ii, 52; GP, iv, pp. 432–3; GP, vii, pp. 199–200, 300; *NE*, IV, xvii, 8; C, pp. 85, 388, 518–19. Russell cites the last passage (which was not available to him in 1899) in his 1937 preface to the second edition of *POL* (p. v). From it, Russell maintains, Leibniz deduced the 'main doctrines' of the *Monadology* 'with terse logical rigour', thereby validating Russell's claim that Leibniz's philosophy 'was almost entirely derived from his logic' (*POL*, p. v).

[22]That Leibniz restricts the principle to affirmative propositions in the letter but not in the notes is perhaps puzzling but probably not important, since negative propositions are readily converted to affirmative ones, as Parkinson ([1995], p. 200) notes, citing C, p. 86.

With the containment principle came a method of proof. In the case of some true propositions, e.g., 'a rose is a rose' or 'a red rose is a rose', the inclusion of the predicate in the concept of the subject is explicit; these Leibniz calls 'identities' (GP, v, p. 343). In the case of other propositions, e.g., 'a crimson rose is red', where the inclusion is not explicit, Leibniz maintains that it can be made so by the successive substitution of definitional equivalents.[23] So, in principle at any rate, the truth of a true proposition can be demonstrated by reducing it to an identity. Leibniz's best-known example is his demonstration that $2 + 2 = 4$ (NE, IV, vii, 10). He starts by defining 4 as $3 + 1$, 3 as $2 + 1$ and 2 as $1 + 1$, then successive substitutions give:

$$2 + 2 = 4,$$
$$2 + 2 = 3 + 1,$$
$$2 + 2 = (2 + 1) + 1,$$
$$2 + 2 = 2 + (1 + 1),\text{[24]}$$
$$2 + 2 = 2 + 2,$$

which is our identity. Now, as Beaney [2003] points out, the importance of identities for Leibniz is that they are 'true per se' (GP, vii, p. 300), their truth is 'self-evident' (GP. vii, p. 309), or 'known through itself' (C, p. 372 = P, p. 62), or recognized 'immediately' (NE, IV, ii, 1). Moreover, seeing that an identity is true does not require an immediate grasp (what Leibniz calls an intuition) of the concepts involved (though, of course, it may require knowing an abstract definition of them). Identities, Leibniz says, are 'true in virtue of [their] form' (L, i, p. 294 = GP, iv, p. 426); their truth is established syntactically rather than semantically. Moreover, the substitutional method of proof is itself mechanical: for every proposition, there either is a series of substitutions which result in an identity (in which case the proposition is true) or there is not (in which case it is false). The outcome does not depend, as Beaney puts it, upon 'the vagaries of our own mental processes' (cf. NE, I, i, 1; IV, vii, 9). The approach was nicely free of psychologism, but this was not enough to commend it to Russell, who rejected Leibniz's proof procedure as well as his logical doctrines (POL, p. 19).

Now it would seem patently clear from Leibniz's statement of the containment principle that it applies to all true (affirmative?) propositions. An immediate consequence of this is that all propositions are subject-predicate in form; or, strictly, that if there are any propositions not of subject-predicate form then none of them is true. Russell famously attributes the doctrine (which I shall label (SP)) that all propositions are SP-propositions to Leibniz – it is the first of the five theses on which he claims Leibniz's philosophy is based

---

[23] 'Now it is certain that every true predication has some basis in the nature of things and, when a proposition is not an identity, that is to say, when the predicate is not expressly contained in the subject, it must be included in it virtually. This is what the philosophers call in-esse, when they say that the predicate is in the subject. So the subject term must always include the predicate term in such a way that anyone who understands perfectly the concept of the subject will also know that the predicate pertains to it' (Discourse on Metaphysics, § 8: L, i, pp. 471–2 = GP, iv, 433). Cf. also GP, vii, p. 309; C, p. 519.

[24] This step, of course, goes beyond the resources Leibniz explicitly allows, since it requires the associativity of addition: $(2 + 1) + 1 = 2 + (1 + 1)$. Leibniz gives a different example at A, VI, 4B, pp. 1515–16 which does not require associativity. I'm grateful to Massimo Mugnai and Richard Arthur for information on this.

(*POL*, p. 4). On present showing he would seem to have good grounds, for it was surely not Leibniz's intention to countenance whole categories of propositions not of SP-form, all of which were false. The alternative is to deny the containment principle as stated. Nevertheless, Russell's attribution of (SP) to Leibniz, on which his most important criticisms of Leibniz depend, has been controversial. I shall consider it in more detail in the next section.[25] But first some more about the containment principle, some of which will provide further evidence that Leibniz held (SP).

To begin with, it is not immediately clear exactly what the containment principle amounts to. What is the *concept* (or, in Russell's translation, the 'notion') of the subject and the *concept* of the predicate? I suspect that the concept of the predicate is simply the concept which occurs in the predicate. So if the predicate is 'is a rose' then the concept of the predicate is 'rose'. But what of the concept of the subject? Leibniz makes it clear in his letter to Arnauld that the concept of an individual substance (he takes Adam as his example) contains 'everything that will ever happen to him', 'all his predicates', it is the knowledge God had of him before he decided to create him (L, i, pp. 508, 510 = GP, ii, pp. 48, 50).[26] An individual concept is thus complete. Thanks to Leibniz's principle of the identity of indiscernibles, one can be sure that there will be only one individual which has all of Adam's predicates. Thus an individual concept is a concept under which only one individual falls.[27] Other features of Leibniz's philosophy ensure that each individual concept will be infinitely complex (cf. Parkinson [1965], pp. 27–8).

It is notable that nowhere in *POL* does Russell discuss Leibniz's notion of the concept of the subject; he usually speaks of the subject where Leibniz would have spoken of the concept of the subject.[28] Parkinson, by contrast, takes great care with the distinction. The concept of the subject and the concept of the predicate, for Parkinson, are actual constituents of the proposition. He takes it as 'obvious' that the subject of a proposition, the thing which the proposition is about, cannot itself it be an constituent of the proposition ([1965], p. 8). Similarly, he says that it is 'clear that the quality of being a king [the predicate] is not in Alexander [the Great] in the sense that the concept of being a king is in the concept of Alexander' (p. 29). When Leibniz says that the predicate is in the subject, he means, Parkinson says, no more than that the predicate is a predicate *of* the subject (p. 30).

Russell's relative neglect of the distinction was not part of the appalling tendency to use/mention confusions of which he is so often (and so unjustly) accused. It is rather that

---

[25] Further texts which support the attribution include L, i, p. 362 = C, p. 51.

[26] There is evidence that Leibniz also wanted to include relations, e.g., GP, ii, p. 37; L, ii, p. 517 = GP, ii, p. 56. But the evidence is not conclusive. At GP, ii, p. 37, for example, Leibniz uses the French verb '*envelopper*' which is weaker than the English 'include'. Massimo Mugnai, to whom I'm grateful for pointing this out, suggests that Leibniz may be operating with two notions of an individual concept: narrow and broad. What is clear is that Leibniz intended to include in the individual concept the intrinsic 'relational accidents' (to be discussed in the next section) on which the relation depends.

[27] At least, that is, so far as the actual world is concerned. When other worlds are taken into account, the situation becomes very much more complex (cf. Cover and O'Leary-Hawthorne [1999], Ch. 3).

[28] Leibniz's usage is inconsistent (cf., e.g., GP, vii, p. 300) as Parkinson [1965], pp. 6–7 explains.

his own account of propositions at this time left no room for such a distinction.[29] Contrary to what Parkinson says is obviously impossible, Russell, between 1899 and 1907, held a direct realist view of propositions: Russellian propositions, except such as contained denoting concepts, actually contained the objects that they were about. Mont Blanc itself is part of the proposition that Mont Blanc is more than 3000 metres high, as Russell famously told an astonished Frege (Frege [1980], p. 169). If it were the concept of the subject, rather than the subject itself, that was contained in the proposition, then, Russell maintains, it would be inexplicable how the proposition could be about the subject, rather than about the concept of the subject. The case of denoting concepts shows that the problem was real enough: Russell was never able to offer an account of how the occurrence of a denoting concept in a proposition resulted in the proposition's being about the term (if any) denoted by the denoting concept, rather than about the denoting concept itself.

Now if Leibniz holds, as Russell asserts, that all propositions are of subject-predicate form, then Leibniz's account of the relation between the subject (-concept) and its predicates gives good support to the containment principle and its associated method of analysis by transformation into identities. Indeed, for SP-propositions, the containment principle itself is a tautology: if the concept of the subject S of an SP-proposition consists of all the concepts that S falls under (equivalently, all the predicates that S has) then the proposition will be true just in case the concept of the predicate is among the concepts which make up the concept of the subject. If (SP) is correct, then the containment principle will be true for *all* propositions, as Leibniz seems to have supposed. And if we think of specifying – either partially or (*per impossibile*) completely – the concepts which make up the individual concept of the subject as providing a partial or complete definition of the subject-concept, then Leibniz's method, of establishing the truth of propositions by reducing them, through the substitution of definitional equivalents, to identities, looks plausible. But does Leibniz hold (SP) as Russell maintains? The evidence so far strongly suggests that he does, but the claim is not uncontroversial.

3. Did Leibniz hold that all propositions were SP-propositions?

That Russell rejected the containment principle comes as no surprise. He rejected it because he rejected the doctrine (SP) that all propositions are SP-propositions. This was his most fundamental criticism of Leibniz and most of his other criticisms are consequences of it. Russell's rejection of (SP) was not part of his rejection of neo-Hegelianism, though when he came to reject neo-Hegelianism he came to think that neo-Hegelians were logically committed to it. It is doubtful if Russell ever held (SP), but, if he did, he had abandoned it by the time he started work on Leibniz, for it is rejected without comment in *AMR*. Against (SP) Russell noted two kinds of counter-example in *POL*:[30] numerical propositions of the kind 'there are three men' and (much more importantly) relational

---

[29] Shortly afterwards, when he introduced the notion of denoting concepts, he did admit a similar distinction in his own account of propositions, but only in a limited range of cases. Denoting concepts had a brief currency, terminated by the theory of descriptions in 1905.

[30] He had noted several more in *AMR* (pp. 167–73). See Griffin [1991], pp. 276–7, 281–5 for discussion.

propositions (*POL*, pp. 12–15). Both Ishiguro ([1972], pp. 72–3) and Parkinson ([1965], pp. 37–8) point out that, as regards the former, Leibniz came closer to an adequate account (in fact, closer to the account that Russell later promulgated) than Russell gives him credit for. In the passage they cite, Leibniz treats '*a*, *b*, and *c* are three *m*'s' as *a* is an *m*, *b* is an *m*, *c* is an *m*, and *a*, *b*, *c* are disparate'.[31] While impressive, this is still not adequate, since it is not equivalent to the original proposition: it expresses 'there are at least three *m*'s' not 'there are (exactly) three *m*'s'. More importantly, it does not reduce the original entirely to SP-propositions: '*a*, *b*, and *c* are disparate' is not an SP-proposition.[32]

The case of relations is very much more complicated. Relations play an important part in Leibniz's philosophy, notably in his theories of space and time, and he has a good deal to say about them in brief remarks scattered throughout his corpus. It is clear from the detailed study by Massimo Mugnai[33] that Leibniz's treatment of relations depends heavily on the scholastic tradition according to which there exist only substances and their absolute (i.e., non-relational) accidents (Mugnai [1992], p. 102). It was very widely held, especially in the sixteenth and seventeenth centuries (cf. Mugnai [forthcoming], § 3), that relations had to be grounded in their terms and that the foundations on which they were grounded could only be absolute accidents of the substances they related, for (as Aquinas put it) an accident 'never extends beyond the subject in which it inheres'.[34] In this, as Mugnai says, Leibniz 'conformed ... to the traditional doctrine' ([forthcoming], § 3). It is very doubtful that Russell knew much, if anything, of the scholastic tradition when he wrote his book on Leibniz, but that these were Leibniz's views he was able to discover, even though the textual evidence available to him was much less extensive than what is available now.

Russell's key piece of evidence – he says it is of 'capital importance' (*POL*, p. 13) – was a widely cited passage in § 47 of the 5th letter to Clarke:

> The ratio or proposition between two lines *L* and *M* may be conceived in three several ways; as a ratio of the greater *L* to the lesser *M*; as a ratio of the lesser *M* to the greater *L*; and lastly, as something abstracted from both, that is, as the ratio between *L* and *M*, without considering which is the antecedent, or which the consequent; which the subject and which the object.... In the first way of considering them, *L* the greater is the subject, in the second *M* the lesser is the

---

[31] Russell can hardly be blamed for not recognizing this, since the passage to which they both appeal to was first published by Couturat in 1903 (C, p. 240): Gerhardt's transcription of the manuscript breaks off just before the relevant passage with a note that the rest of the document is more in the nature of an exploratory study (GP, vii, p. 221). Russell acknowledged the importance of the material omitted by Gerhardt in his review of Couturat's [1901], but cites an adjacent passage to the one noticed by Ishiguro and Parkinson (*RWPL*, p. 548).

[32] '*a*, *b*, and *c* are disparate' cannot be construed as an SP-proposition with multiple subjects, a form which Russell admits – albeit briefly and as distinct from SP-propositions proper – in *AMR*, p. 169. '*a*, *b*, and *c* are disparate' is not at all of the same form as '2, 3 and 5 are prime', which will serve as an example of the type of proposition Russell had in mind in *AMR*. There are passages in Leibniz, however, in which he seems to concede that attributions of number to collections are not of subject-predicate form: cf. the passage from an unpublished text of 1703 quoted by Mugnai [1992], p. 30.

[33] In what follows I am heavily indebted to Mugnai [1992] and [forthcoming]. Mugnai describes Leibniz's position on relations as a 'moderate nominalism' ([forthcoming], § 4).

[34] *In quator Libros Sententiarum*, II, d. 27, q. 1, ar. 6, quoted in Mugnai [forthcoming], § 2.

subject of that accident which philosophers call *relation* or *ratio*. But which
of them will be the subject, in the third way of considering them? It cannot be
said that both of them, *L* and *M* together, are the subject of such an accident;
for if so, we should have an accident in two subjects, with one leg in one, and
the other in the other; which is contrary to the notion of accidents. Therefore
we must say that this relation in this third way of considering it, is indeed *out
of* the subjects; but being neither a substance, nor an accident, it must be a
mere ideal thing, the consideration of which is nevertheless useful.[35]

The passage makes it as clear as possible that relations cannot be considered in the third
way, as polyadic accidents – for that would be 'contrary to the notion of accidents' – and
thus relations, in that sense, cannot be real at all, but 'mere ideal thing[s]', or 'entities
of reason' (*NE*, II, xxv, 1).[36] 'We have a *relation*', Leibniz says, 'as soon as two things
are thought together' (A, VI, 4A, p. 28); he even coins the term '*concogitabilitas*' for the
possibility of thinking things together which gives rise to relation (A, VI, 4A, p. 866). In
a note still unpublished in the Leibniz Archives in Hanover (LH IV, 7c, Bl. 35v; quoted
by Mugnai [forthcoming], § 4),[37] Leibniz writes: 'A relation is the *concogitabilitas* of two
things'. In *New Essays* he talks about relations as supplied by the understanding (*NE*, II,
xii, 3; II, xxx, 4).

   That no accident can inhere in more than one substance links Russell's two counter-
examples to the claim that all propositions are SP-propositions, for both relational propositions and propositions that ascribe numbers to collections would violate this requirement. Leibniz explains the matter quite clearly in another unpublished passage Russell
could not have seen: '"ten" is not an attribute of anything. In fact one cannot predicate
"ten" of a single aggregate or of singular things. The same applies to a relation which
is common to many subjects, such as the similarity between two things' (LH IV, 7c, Bl.
75–8, quoted Mugnai [forthcoming], § 4).

   The possibility that two things can be thought together, and thus the possibility
that they can be related, depends upon the existence of the foundation for the relation
in each of them. This further scholastic doctrine about relations is also to be found in
Leibniz in passages noted by Russell. Thus 'There is no denomination so extrinsic as not
to have an intrinsic one for its foundation' (GP, ii, p. 240; *POL*, p. 205). And, rather more
explicitly, in his letter of 21 April 1714 to Des Bosses, Leibniz first repeats the point
quoted earlier, that an accident cannot inhere in two substances, as if the opposite opinion
were an obvious absurdity: 'I do not believe that you will admit an accident that is in two
subjects at the same time'. He then continues:

   My judgment about relations is that paternity in David is one thing, sonship
   in Solomon another, but that the relation common to both is a merely mental

---

[35]GP, vii, p. 401 = L, ii, p. 1147 as quoted by Russell *POL*, pp. 12–13. Russell quotes the passage again in
POM, p. 222.

[36]In discussing Leibniz's theory of space, Russell refers to these expressions as 'abusive epithets' (*POL*, p. 129).
They occur in several passages, e.g., GP, ii, pp. 486, 517.

[37]Another unpublished passage, also cited by Mugnai (§ 4), reads: 'If relations were real beings in things, endowed with a reality different from that arising from the fact that they are thought, then they would be accidents
in two subjects simultaneously, because a relation has the same right of being in both' (LH IV, 3, 5a, Bl. 23v).

thing whose basis is the modifications of the individuals. (GP, ii, p. 486 = L, ii, p. 992; *POL*, p. 206)

As Mugnai explains this passage, paternity in David and sonship in Solomon are two relational accidents, each of which inheres in one substance only.

> If the accident of being a father applies to David, then this means that some-where David must have a child; and if Solomon is a son, then he must have (or have had) a father. For Leibniz, however, what the expressions *father* and *son* denote are *absolute things* (David, Soloman) not some "extra-mental" relative object. In the "external world" there are only individual substances with their internal properties. (Mugnai [forthcoming] § 5)

Leibniz, indeed, is in something of a dilemma: on the one hand, relations have an important role to play in his philosophy; on the other, his views about the nature of sub-stance and accidents prevent him from taking them quite at face value. In various places, he makes different proposals.[38] For example, in *Generales Inquisitiones* he proposes re-ducing 'Caesar is similar to Alexander' to 'Caesar is similar to a thing [*res*] which is Alexander' (C, p. 357), which Mugnai ([1992], pp. 70–1) presents symbolically as a re-duction of '$S(a, b)$, either to '$S(a, x)\&A(x)$' or to '$S(a, x)\&x = b$'. It is not immediately obvious what advantages are to be derived from such an analysis, but it is possible that Leibniz thinks that '$S(a, x)$' is preferable to '$S(a, b)$' because only in the latter does an attribute have a leg in two distinct *substances*. But then the attribute in '$S(a, x)$' still has two legs and the second would seem to be in nothing at all. Since the variable (with im-plicit existential quantification) in '$S(a, x)$' ranges over substances it is hard to think that the supposed problems of two-legged attributes have been satisfactorily avoided.

The second proposal is limited, but more interesting. It reduces 'a is similar to b' to a pair of monadic predications 'a is c & b is c', where 'c' is a monadic property, possessed by both a and b and in respect of which they are similar. The reduction requires some tweaking, because the *analysans* contains more content than the *analysandum* (which does not identify c as the respect in which a and b are similar), but this can be dealt with by replacing 'c' by an existentially quantified variable, thus: '$(\exists c)(a$ is c & b is $c)$'. The procedure applies to all equivalence relations, and can be extended to other symmetrical relations with a bit more tweaking (so that 'a and b are different', e.g., becomes '$(\exists c)[a$ is c & $\sim(b$ is $c)]$'), but not beyond. Asymmetrical relations resolutely resist such an analysis, as Russell famously argued (*POM*, pp. 221–4).

Leibniz's third proposal is quite general. It would reduce relational propositions to pairs of SP-propositions related by various binary propositional operators, his favourites are '*quatenus*' ('insofar as'), '*qua talis*' ('as such') and '*eo ipso*' ('by that very fact').[39] Thus 'Paris loves Helen' becomes 'Paris loves and by that very fact (*eo ipso*) Helen is

---

[38] They are surveyed in detail by Mugnai ([1992], Ch. 4).

[39] Ishiguro ([1972a], p. 208) worries, quite reasonably, about what sort of propositional operators these will be. Certainly, they will be neither extensional nor modal. They will be some form of sociative operator, such that '*A quatenus B*', for example, holds only if there is some appropriate connection between *A* and *B*. A full Leibnizian theory of relations would evidently require a logic (yet to be supplied) of these operators. Cf. Sylvan [2000], Ch. 4 for a partial taxonomy of such connectives (of which relevant implication is the best known).

loved'; 'Caius is killed by Titus' becomes 'Insofar as (*quatenus*) Titus is murdering, Caius is murdered' (cf. Mugnai [1992], p. 72); and 'Titus is wiser than Caius' becomes 'Titus is wise and as such (*qua talis*) is superior, insofar as (*quatenus*) Caius *qua* wise is inferior' (cf. C, p. 280).[40] From Russell's point of view, which by 1899 regards propositions as complex terms, it could not be thought of as any improvement at all, as regards the reduction of relations, to replace a binary relation between two substances by a binary relation between two propositions. But even without taking a Russellian view of propositions, it is hard to see how relations are eliminated for (as Ishiguro [1972a], pp. 207–9 notes) in each case the *analysans* contains terms (e.g., 'inferior'/'superior' in the third example) which are implicitly relational, even though they occur grammatically as monadic predicates. Elsewhere, Leibniz devoted considerable ingenuity to eliminating *implicitly* relational expressions (cf. Mugnai [1992], pp. 58–69) which makes it difficult to suppose that he would have been content with leaving them in tact in the present analysis.

Ishiguro's claim is that Leibniz did not wish to reduce relations. Most importantly, she cites ([1972a], p. 203) a passage, first published by Grua (vol. ii, p. 542) and so not available to Russell, in which Leibniz maintains that 'prior' and 'posterior' are basic (i.e., simple) concepts which are incapable of further analysis.[41] If so, it is understandable why he should have been unconcerned about implicitly relational expressions remaining in his analysis of relational sentences, but it is not understandable why he should have proposed the analyses in the first place. With the possible exception of Ishiguro, Leibniz scholars seem to be generally agreed (1) that Leibniz, following the scholastic tradition, did not admit relations into his ontology and (2) that, in consequence, he attempted to replace relational propositions by propositions which involved only monadic predicates. They differ as to whether Leibniz held that (2) would be adequately carried out (as, e.g., Kulstad [1980] contends) if relational propositions were replaced by propositions which involve only monadic relational predicates, that is predicates of the form '$R(\_, a)$', usually known in the Leibniz literature as 'relational accidents'. Now Leibniz certainly distinguishes between relations proper and 'relational accidents', the former being 'outside the subjects' whereas the latter inhere in one subject only,[42] but if a reduction to relational accidents was all that was required, it is hard to see why it should have cost much effort, or, for that matter, been worth the effort it cost (cf. Rescher [1967], p. 72).

Against Rescher's specific charge that such a 'reduction' would be circular, McCullough ([1977], p. 37) argues that the relations to be reduced are universals while the relational accidents to which they are reduced are 'singular "properties"', or tropes, from which universal relations are abstracted. On this view, fatherhood in David and sonship in Solomon and fatherhood in Vasily and sonship in Yevgeny are four distinct singular properties. As an attempted elimination of relations, McCullough's suggestion hardly helps very much: the reduction may not be exactly circular as Rescher alleged, but the 'singular properties' are still relational, so the reduction has not been complete. But this

---

[40]Leibniz's suggested analysis sometimes vary: cf. the different treatment of 'Titus is more learned than Caius' given by Mugnai [1992], p. 72. I'm not sure if anything hangs on the differences, but I suspect not.

[41]Ishiguro's discussion of the issue in [1972a] is much more thorough than in her book [1972], which I previously criticized in [1991], pp. 344–5.

[42]Cf. GP, ii, p. 486 discussed above, and see Mugnai [1992], pp. 116–17 for comment.

is to concede the broader point McCullough wants to make: ultimately, he claims, Leibniz wished to reduce only universal relations, particularized relations (tropes) he had no wish to reduce at all. Indeed, the preference for particularized relations both conforms to some of the constraints that Leibniz places on his account of relations and also explains some features of it which are otherwise puzzling. In the first place, the tropes obviously inhere in only a single substance, though for each relational trope inhering in a given substance there is another trope – the converse trope, as it were – which inheres in a different one. Thus relational tropes satisfy Leibniz's scholastic requirements on the nature of attributes. Moreover, the elimination of universal relations, as things abstracted from the tropes, would explain exactly why Leibniz maintains that relations are mere ideal things, even though he is prepared to countenance inferences involving relations and a relational theory of space and time – these latter cases depend presumably upon tropes and not upon universal relations. McCullough's doctrine allows Leibniz to have his cake and eat it. Indeed, McCullough pushes his view further and claims that Leibniz held that *all* the properties of monads are relational properties, understood as tropes.

It would take us too far a field here to evaluate properly this radical proposal, but the main difficulty it seems to me with McCullough's approach is that it is difficult to reconcile it to the containment principle. If Leibniz intended to leave relational tropes in tact, it is difficult to see what can become of the containment principle. There is certainly no problem (as we've seen) in maintaining that relational tropes are accidents inhering in only one substance, but it is very difficult to believe that such relational accidents are *contained* in the substance in which they inhere. For part of a relational trope is the second term of the relation, whose first term is the substance in which the trope inheres. So if the trope is contained in the first substance, then so must be the second substance, and this would seem to be impossible.[43] If McCullough is right that Leibniz did not wish to reduce relational tropes, and thus was prepared to concede that some propositions were fundamentally relational, then it would seem that the containment principle applies only to a proper sub-set of propositions, viz. SP-propositions. But if McCullough is also right in thinking that all a monad's properties were relational properties, i.e., relational tropes, then it would seem that the containment principle will apply to no propositions at all, for all propositions, on that view, will be relational.

In general, it is quite difficult to see how Leibniz could have maintained the containment principle (at least in the form in which he states it) if he did not also believe that relations (including relational tropes) could be reduced. It is also difficult to see how any such reduction could be successful, for the reasons Russell gave (*POM*, pp. 221–6). Parkinson ([1965], pp. 39–55), after a very careful examination, comes to the conclusion that Leibniz held (erroneously) the reducibility thesis that every putatively relational

---

[43]One possible way out would be to take *containment* as it occurs in the containment thesis to be a non-transitive relation. In this it would resemble the *composition* relation as it occurs in Russell's principle of acquaintance: Every proposition which we can understand must be composed wholly of constituents with which we are acquainted (KAKD, p. 154). On Russell's principle it is not necessary to be acquainted with the constituents of the constituent of a proposition which one understands – they are not part of the composition of the proposition.

proposition is equivalent to one or more SP-propositions.[44] Whether this means that Leibniz thought there really are relational propositions, though each of them could be replaced by equivalent SP-propositions, or whether he thought that all propositions were really SP-propositions, even though some of them are expressed by sentences of relational form, is something that Parkinson says cannot be determined from Leibniz's writings, though he finds the former 'more probable'. In the end, I don't think it makes much difference, partly because the reduction claim itself is false and partly because, even if the reduction claim were true on Parkinson's first interpretation, so that there really were relational propositions, they would be left as essentially idle wheels in Leibniz's logical system, for they play no role in his proof procedure or his definition of truth, at least so long as the containment principle is taken seriously. They might, it is true, have some other role to play, possibly a psychological one: they might be the propositions which the mind grasps as proxies for the equivalent SP-propositions which capture the thought in its proper logical form.

In the end, I think that, given the textual evidence available to him, Russell's ascription of (SP) to Leibniz is warranted. Even with the much more extensive textual evidence available today, the ascription still seems plausible. The passage in Grua that Ishiguro quotes gives some grounds for doubt, but, even if it admits of no other interpretation than the one she puts on it, it would seem insufficient evidence to warrant retraction of the containment principle as Leibniz states it. The Grua fragment seems altogether less considered than the containment principle. And, more generally, if Leibniz did not believe that relational propositions could be re-expressed as SP-propositions, it is very difficult to understand how he could maintain the containment principle in the form in which he states it.

4. Analytic and necessary propositions in Leibniz

One apparent consequence of the containment principle is that it makes all true SP-propositions analytically true. This follows immediately given one of the two accounts traditionally given of analyticity, namely:

(A1) A proposition is analytic if its predicate (concept) is contained in its subject (concept).[45]

Moreover, if, as I argued in the last section, Russell was correct in attributing (SP) to Leibniz, it follows that *all* true propositions are analytically true – and this is generally, and I think quite rightly, taken to be a very troubling doctrine. Indeed, since Leibniz states the containment principle as applying to *all* propositions, not just SP-propositions, we do not have to add (SP) as a premiss to get this conclusion – it follows from the containment principle as stated. So the question arises: did Leibniz really believe, as Russell claims he did, that all true propositions are analytic?

---

[44]Rescher [1967], Ch. 6; [1979], pp. 59–61, agrees.
[45]I state it as a sufficient condition, rather than a definitional equivalence, to leave open the option of combining it in a disjunction with the other traditional account, namely that a proposition is analytic if it follows from the law of no contradiction alone. A sufficient condition is all that is needed here.

There is a problem, as Russell notes (*POL*, p. 16), in discussing Leibniz's views on the distinction between analytic and synthetic propositions since Leibniz himself never used these terms. Leibniz frames his discussion entirely in terms of necessary and contingent propositions. Russell, of course, having previously worked in the Kantian tradition, was long accustomed to having two distinctions here and to regarding the propositions of mathematics, for example, as both synthetic and necessary. Watling ([1970], pp. 15–19) thinks both that Russell assumed that Leibniz thought the two distinctions were coextensive and that this led him into errors of interpretation of Leibniz. I'm not persuaded that either charge is correct, but, certainly, it would be rash to assume that Leibniz identified the necessary with the analytic and the contingent with the synthetic (though, no doubt, even rasher to assume that he distinguished them).

This, however, makes very little difference to the problem at hand, at least if we stick to the conventional wisdom about analyticity, which holds that no analytic proposition is contingent. It then follows from (A1) and the containment principle as stated that no proposition is contingent. Assuming that the necessary/contingent distinction is both exclusive and exhaustive, it follows that all true propositions are necessary. Now this is certainly a conclusion Leibniz doesn't want. According to Parkinson ([1995], p. 202) there is general agreement that Leibniz wanted an objective distinction between necessary and contingent propositions, that is, a distinction which does not depend on human ignorance.[46] All we have achieved so far, therefore, is to convert the problem of allowing for synthetic propositions into the problem of allowing for contingent propositions. We need to consider how it may be avoided.

If this is to be done at all, some of the conventional wisdom has to give. Leibniz commits himself to an idiosyncratic account of propositions. But he does not commit himself to any doctrines about analyticity, and we may well avail ourselves of the flexibility this allows. The first thing to note is that (A1) itself says nothing about necessity – the concept of containment is not a modal notion, there is nothing in it to suggest that one thing may not be contingently contained within another. This, of course, applies to the containment principle as well. Accordingly, (A1) together with the containment principle entails that all propositions are analytic, but it does not entail that all propositions are necessary. To get that conclusion we must add the premiss that all analytic propositions are necessary, and this does not follow from the definition of 'analytic proposition' alone, at least not as so far given. If we wish to maintain the containment principle as stated and that there are some true, contingent propositions, it is necessary to suppose that there are some true propositions in which the predicate is only contingently contained in the concept of the subject. There is nothing untoward in this. But if we also wish to maintain the standard definition of analyticity (A1), it follows that there are true propositions which are both analytic and contingent. If this is an oxymoron too strong for our stomachs, we must make adjustments. But it is important to realize that our options are limited. We can give up the containment principle, but since that was the centre-piece of Leibniz's logic, abandoning it should be the last resort of a sympathetic exposition. Or, we can give up

---

[46]Thus, e.g., Rescher ([1967], p. 35) 'There is no question that Leibniz *wished* to find a place for contingence'. Cf. also Jolley [2005], p. 216.

(A1) as a characterization of analyticity. If we refuse to give up either of these, we end with the conclusion that all propositions are analytic and thus that either all propositions are necessary or that some analytic propositions are not necessary. There are no other ways out.

It may help at this point to ignore the concept of analyticity, and consider just the containment principle. Suppose that S is a subject the concept of which comprises the concepts $F_1, \ldots, F_j, \ldots, F_n$. Then the proposition that S has the predicate F_j might seem to be a necessary proposition, at least if 'a red rose is red' is necessary. But the proposition that S is F_j is not identical to the proposition that an F_j is F_j, even if S is an F_j. Whether 'S is F_j' is a necessary truth will depend on the relation of the subject to the predicates which comprise its concept. If the concept of a subject comprises, as Leibniz says it does, *all* those concepts under which the subject falls and if, moreover, as Leibniz also claims, there are *some* contingent SP-propositions, i.e., there are some concepts under which the subject contingently falls, then, the concept of the subject will include those concepts under which the subject only contingently falls, and thus it will be only a contingent truth that $F_1, \ldots, F_j, \ldots, F_n$ comprise the concept of the subject S. In this case, unless F_j happens to be a predicate which is necessarily true of S, the proposition that S is F_j will be a contingent truth. The upshot is this: either the notion of containment used in (A1) and the containment principle is a modal notion or not. If containment is a modal notion then any proposition asserting that a predicate is contained in a subject is, if true, necessarily true. *Obviously*, if we treat containment in this way then the containment principle leaves no room for contingent truths. But, *equally obviously*, if containment is not a modal notion in the containment principle then it cannot be a modal notion in (A1) either (assuming that the two principles are to work together), and thus the concept of analyticity itself carries no modal force and contingent analytic propositions are admitted. It seems to me that the second of these two options is clearly to be preferred.[47] I shall refer to it as the contingent containment solution to the problem of contingent propositions.

There are places where Leibniz seems to embrace it:

> [I]n any proposition: A is B, B is always in A itself, or its notion is in some way contained in the notion of A itself; and this either with absolute necessity, in propositions of eternal truth, or with a kind of certainty, depending upon a supposed decree of a free substance, in contingent things; and this decree is never wholly arbitrary and destitute of foundation, but always some reason for it (which however inclines, and does not necessitate), can be given .... Truths which are inclined but not necessitated are contingent, and not necessary, truths (GP, vii, 300; quoted *POL*, p. 33).[48]

---

[47] If one insists that some truths are contingent but cannot countenance contingent analytic ones, then one must abandon either (A1) or the containment principle. Perhaps the most natural move would be to strengthen (A1) to

(□A1) A proposition is analytic if its predicate (concept) is necessarily contained in its subject (concept).

Since Leibniz does not address the concept of analyticity we are not in a position to attribute (□A1) to him, but the thesis is not without it's attractions.

[48] Cf. also the *Discourse on Metaphysics*, § 13: '[T]here are two kinds of connection or sequence. One is absolutely necessary, for its contrary implies a contradiction.... The other is necessary only *ex hypothesi*, and by

Indeed, immediately after stating the containment principle, Leibniz suggests that Arnauld's opposition to it was based on his false belief that Leibniz held the relation between the predicate-concepts and the subject-concept 'to be intrinsic and necessary at the same time'. Leibniz, however, asserts that he holds it 'to be intrinsic, but in no way necessary' (GP, ii, p. 46; cf. also GP, ii, p. 56).

It may seem that the real difficulty for the distinction between necessary and contingent propositions in Leibniz's system comes from his proof procedure: a proposition is true just in case it either is or can be reduced to an identity. And identities, since they are true in virtue of their form, are necessarily true. Moreover, as we have seen, Leibniz maintains that true propositions which are not identities can *always* be reduced to identities by the substitution of *definitional* equivalents. If these definitional equivalences are also necessary truths, then *all* true propositions will be necessary, for one cannot start with contingent propositions, substitute necessary equivalents within them, and end with necessary propositions. If the definitional equivalences are necessary, then either one starts with a necessary proposition and ends up with a necessary identity or one starts with a contingent proposition and ends up with a contingent identity. But this result depends upon our two assumptions. The first of these – that identities are necessary – has, no doubt, the stronger claim upon us (and Leibniz). But the second – that definitional equivalences (at least as Leibniz understands them) are necessary – would seem to have little warrant given what (as we've seen) Leibniz tells Arnauld in defence of the containment principle. If an individual concept is defined by the predicates it contains, and if (as Leibniz asserted to Arnauld) some of these predicates are contained only contingently, then some of the definitional equivalences involved in the subject-concept will be only contingently true.

Unfortunately, Leibniz does not avail himself of this opportunity. In the first place, as Russell notes (*POL*, p. 13), Leibniz distinguishes two types of SP-propositions. There are those, like 'This is red' and 'Socrates is wise', which relate individual and species (ISP-propositions), and those, like 'Red is a colour' and 'Wisdom is a virtue', which relate species and genus (GSP-propositions) (cf. L, i, p. 509 = GP, ii, p. 49). Now this distinction offers help with the necessity problem in a number of different ways for there is some plausibility in supposing that GSP-propositions are necessary and that ISP-propositions are (in general) contingent. In the first place, that Socrates has the property of being wise depends, according to Leibniz, upon the 'free decree of God' – God may have willed him otherwise – but that wisdom is a virtue, depends upon essences which 'are in the divine understanding prior to any consideration of the will' (L, i, p. 509 = GP, ii, p. 49). Now one could interpret this as claiming that GSP-propositions were necessary in the strong sense that God could not have made them otherwise, whereas ISP-propositions were necessary in the weaker sense that their truth depended upon God's will.[49] Russell, in notes (on Duncan [D], p. 170) written while he was preparing his Leibniz lectures, has this to say:

---

accident, so to speak, and this connection is contingent in itself when its contrary implies no contradiction' (L, i, p. 476 = GP, iv, p. 437).

[49] Alternatively, the necessary truths depend upon God's understanding, the contingent ones on his will, as Leibniz suggests in the *Monadology* (L, ii, p. 1051 = GP, vi, p. 614) – a position Russell scornfully demolishes (*POL*, pp. 178–81).

> When God has chosen a possible world, everything (in it) is comprised in his choice, and has that sort of necessity which can now be ascribed to things future, which is *hypothetical* or *consequent* necessity. This does not destroy the contingency of things, or produce that absolute necessity which contingency does not allow. [Leibniz means, by *consequent* necessity, the necessity of what follows from a contingent truth – i.e., the kind of necessity alone recognized by Bradley.][50]

If this is correct, then in the weaker sense of 'necessity', a truth's being necessary does not preclude its being contingent. ISP-propositions are necessary in this weak sense; GSP-propositions have 'absolute necessity' and are not contingent.

Now this seems hardly a satisfactory solution to the problem of contingent propositions. For one thing, it is way more seriously theological than it ought to be. For another, it is not clear what it amounts to. Is God's understanding bound by the strongly necessary truths, or are they strongly necessary because of God's understanding of them? But one of the supposed strongly necessary truths is that God exists, and, as Russell remarks (*POL*, p. 179), God's existence can hardly depend upon his understanding of it. Moreover, it is far from clear why God's will is any less strongly necessitating than his understanding.

Leibniz, at one point, poses the problem of contingent propositions this way: '[I]f, at any particular time the concept of the predicate inheres in the concept of the subject, how can the predicate ever be denied of the subject without contradiction and impossibility?' (L, i, p. 405). But this formulation of the problem can be resolved along the lines of the contingent containment solution. Obviously, it is a contradiction to maintain that the concept of a subject contains a predicate at a given time and that it does not contain it at that time, but from that it does not follow that, if the concept of the subject contains a predicate at a given time, then it is impossible for it not to contain that predicate at that time: $p\&\sim p$ is impossible; $p\&\diamond\sim p$ is not.[51] Once again there does not seem to be a genuine difficulty in admitting contingent truths along with the containment principle, so long as proper care is taken with modalities.

Leibniz, however, goes on to offer a solution to the problem of which he seems inexplicably proud (L, i, p. 407; C, pp. 18, 388–9; P, pp. 77–8), for it seems utterly fantastic, indeed, to be no solution at all. He argues that an individual concept (unlike a specific concept) will comprise infinitely many predicates. Accordingly, he claims, in order to reduce an ISP-proposition to an identity one will require infinitely many definitional equivalences. He then claims that necessarily true SP-propositions are those which either

---

[50] BR, notes on Duncan's translations of the *Philosophical Works of Leibniz*, in Russell's 'Mathematical Notebook' (RA 230. 030000-F1), p. 123.

[51] One should not be anachronistic here. Leibniz was not working within a modern modal logic and it is possible that admitting the contingency of $p\&\diamond\sim p$ would require significant adjustments in his logic. In an interesting article on Leibniz's logic as a precursor of modern combinatory logic, Lorenzo Peña argues that Leibniz's logic of 1686 is flawed chiefly because it yields the 'disastrous' inference rule $\diamond p, \diamond q \vdash \diamond(p\&q)$ (Peña [1991]). An instance of the disaster yields the inference: $\sim\diamond(p\&\sim p) \vdash \sim\diamond p \lor \sim\diamond\sim p$. Clearly such a logic is going to have difficulty distinguishing contingent from necessary truths. But, as Massimo Mugnai has pointed out to me, since Peña's disastrous inference rule is directly at odds with Leibniz's concept of compossibility there must be some doubt as to whether a logic which produces it can be attributed to Leibniz.

are identities or can be reduced to identities in a finite number of steps; all the others are contingent. But this seems to reduce a modal distinction to an epistemological one, for it makes the modal distinction depend upon the fact that we do not have infinite knowledge: for God, who knows all the definitional equivalences, all subject-predicate truths will be necessary. For the rest of us, the necessary propositions will be those which we can prove; the contingent ones, those which we can't.

It might be objected that Leibniz's distinction is not based on whether anyone (apart from God) could actually perform the reduction to identities, but whether there *was* such a reduction or not. The idea, it might be claimed, is that if a proposition required an analysis that could not be completed in a finite number of steps then the analysis could not be completed at all: the analysis would be, as it were, a potential analysis not an actual one. But if we take this view of infinite analysis we invalidate Leibniz's proof procedure, for we acknowledge that there are true propositions, viz. contingent ISP-propositions, which cannot actually be reduced to identities. The reduction, in these cases, is merely a potential one. Either infinite analyses are genuine analyses or they are not. If they are not, then the propositions for which they are required cannot be shown to be true by Leibniz's proof procedure. If they are then, while these propositions cannot be *shown* (in a finite number of steps) to be true by the proof procedure, nonetheless, as far as the proof procedure is concerned, these propositions will be true in exactly the same way as those propositions which require only a finite analysis. In this last case it is difficult to see how the two kinds of propositions could have different status with respect to the analysis alone; any purported difference in the status of the propositions would have to depend on something else. The only reasonable suggestion seems to be to make the distinction depend upon whether the analysis is actually carried out. But this is to confound the distinction between necessary and contingent propositions with an epistemological distinction between propositions which have a proof which can be completed and propositions which have an equally valid proof but one which can't be completed.

Moreover, there is no guarantee that Leibniz's approach will ensure that ISP-propositions are contingent and GSP-propositions are necessary. Getting that result would require, in particular, that no GSP-proposition had an infinitely complex subject-concept. One might want to argue that if it did, humans would not be able to grasp it. Whatever the virtues of this argument (and it looks to me like just more psychologism), they are not such as Leibniz can avail himself of, since he has admitted that individual-concepts are of infinite complexity and yet can be grasped. It will not do, for example, for Leibniz to argue that we can refer to an individual without having to grasp all the concepts under which it falls, for why cannot the same claim be made about an infinitely complex species-concept in a GSP-proposition?

Russell (*POL*, pp. 60–3) considers this approach – citing passages from GP, iii, p. 582; vii, p. 309 for it – but denies that Leibniz uses the distinction between finite and infinite analyses to make the distinction between necessary and contingent propositions. He acknowledges that Leibniz holds that all contingent propositions have infinitely complex analyses, but denies that it is their infinite analysis which constitutes their contingency. To suppose otherwise, he says, is to confuse 'the general character of all contingents … and the meaning of contingency itself' (*POL*, p. 61). The confusion, he thinks, has 'led many commentators to think that the difference between the necessary and the contingent has an essential reference to our human limitations, and does not subsist for God' (*ibid.*). Russell would certainly be reluctant to ascribe to Leibniz any such psychologistic distinction between necessary and contingent propositions. It is obvious that he takes an infinite analysis to be just as genuine an analysis as a finite one, and thus that, even if the subject (concept) of a true proposition is infinitely complex, the predicate will be contained within it. If, then, what makes a proposition contingent is simply that it has an infinitely complex subject (concept), it follows that contingent propositions, like necessary ones, are analytic. It is because Russell was unable to accept that Leibniz held that contingent propositions could be analytic that he rejected the idea that Leibniz held that the difference between finite and infinite complexity in the subject (concept) constituted the difference between necessary and contingent propositions.[52]

But on this he was to change his mind, as a result of evidence uncovered by Couturat. Russell and Couturat had discussed the matter intermittently since 1900 when Couturat first read Russell's book: it was an important point on which their interpretations of Leibniz disagreed. The first mention of it occurs in a letter by Couturat on 20 October 1900 where Couturat says that in his view Leibniz held that 'all truth (and not merely all necessary truth) is analytic' (*BRLC*, i, p. 202). Couturat returns to it on 5 July 1901, as he is finishing his own book on Leibniz, when he mentions for the first time previously unpublished manuscripts he has found which he thinks establish his interpretation incontrovertibly. Russell, however, was not immediately convinced. Writing after he had read Couturat's book (2 October, 1901), he comments that he had given due emphasis to the containment principle (which is certainly true) and that he had quoted all the passages Couturat cites (except for those which Couturat was publishing for the first time) but he thinks that all of them (including the new texts) can be interpreted along the lines he suggested in *POL*. He urges Couturat to re-read *POL*, §§ 14, 26 (*BRLC*, i, p. 259). Couturat, however, stuck to his guns and in the end it was Russell who gave way. What convinced him was not any of the texts Couturat had cited in his book (Couturat [1901]) but a fragment he published the following year in the *Revue de métaphysique et de morale* (Couturat [1902]). The passage, which Russell quotes (in Latin) in his review of Couturat [1901] reads:

> Thus I believe I have uncovered a mystery which perplexed me for a long time; for I did not understand how the predicate could be contained in the subject, and yet the proposition not be necessary. But knowledge of geometry and infinitesimal analysis let me see the light, so that I understood that concepts also may be resolved to infinity.[53]

This passage makes it explicit that Leibniz held that the predicate was contained in the subject, even when the subject had to be 'resolved to infinity', and thus that the proposition was analytic even though it was contingent. As Russell told Couturat ahead of the publication of the review:

---

[52] He is explicit about this in a letter to Couturat on 23 March 1902 (*BRLC*, i, p. 272).
[53] Couturat [1902], p. 11n, [RWPL], p. 543. Translation from *CPBR4*, p. 657.

I quoted several texts in my book that are hardly capable of any other interpretation, but I could not suppose that anyone could take analytic judgments as contingent. For this reason, it is the quotation that you give ([Couturat [1902]], p. 11, note) beginning 'Ita arcanum aliquod' that finally persuaded me of the correctness of your theory.[54]

So we have now to consider the question how Russell, in his book, thought that Leibniz distinguished necessary from contingent propositions in such a way as to prevent contingent ones from being analytic – i.e., in such a way as to protect contingent propositions from the containment principle. Ironically, the solution he came up with was also one for which the textual evidence improved after he wrote. His suggestion (*POL*, p. 16) is based on the distinction between ISP- and GSP-propositions. Russell proposes to exempt existence from the containment principle. He claims, then, that all existential propositions, except that which asserts the existence of God, are contingent. Now this is only a limited concession to contingent propositions. True, GSP-propositions remain necessary since they are not existential, and this is the result we want. But most ISP-propositions, all that do not assert existence, remain necessary as well – and this is a result Leibniz wishes to avoid. It can be avoided, however, if we suppose that all ISP-propositions are implicitly existential in that their truth depends upon the existence of their subject. Then ISP-propositions will be contingent, for a proposition the truth of which depends on a contingent proposition must itself be contingent.

This approach is nicely elaborated by Mates [1972] and Mondadori [1975] in terms of Leibniz's possible world semantics. If wisdom is one of the attributes of Caius then, assuming that for Leibniz individuals are identified (across worlds as well as within them) by means of their individual concepts, Caius is wise in every world in which he exists. But this does not make 'Caius is wise' a necessary truth, for there are worlds in which Caius does not exist. So if it were Leibniz's view that propositions with unexemplified subject-concepts are all false (a widely held view which I shall call the existential assumption), 'Caius is wise' would be false in those worlds in which Caius does not exist and thus, in the actual world, would be only a contingent truth.

This approach has a lot to commend it, but there are problems. In the first place, it seems to involve denying that existence is a predicate (for how else do we warrant excluding existence from the containment principle?), a view that is associated with Kant, not Leibniz. Moreover, Leibniz surely is committed to the view that existence is a predicate,[55] since he thinks the ontological argument is valid, a criticism made by Couturat ([1902], p. 12). Secondly, Russell himself suggests (*POL*, p. 50) that this approach runs into conflict with Leibniz's principle of the indiscernibility of identicals. Russell takes the latter to involve a confusion of a substance with the sum of its predicates, but in this case, Russell claims, 'predications concerning actual substances would be just as analytic as those concerning essences or species' (*POL*, p. 50). Russell evidently assumes some form

---

[54]Russell to L. Couturat, 23 March 1902 (*BRLC*, i, p. 272). Translation by Jolen Galaugher.

[55]For textual evidence, cf. *NE*, IV, i, 8. Mugnai, however, maintains that the issue (like so much else in Leibniz) is not so simple. Mugnai suggests that after 1690 Leibniz regarded existence as a kind of second level predicate, a concept which supervenes on other concepts. If that is the case, it would seem natural to exclude it from the scope of the containment principle.

of mereological essentialism for his argument; without that his conclusion doesn't follow, even if Leibniz does identify a substance with the sum of its predicates. Of course, as we've just seen from Russell's post-publication exchange with Couturat, this conclusion may have been exactly the one Leibniz was happy to draw. Nonetheless, the conclusion can be avoided by means of the contingent containment thesis, even if the identity of indiscernibles is adopted: all that is required is that contingent identities also be admitted. Even if the identity of indiscernibles involves the identification of substance with the sum of its predicates, the conclusion can still be avoided by the contingent containment thesis, provided mereological essentialism is rejected.

Thirdly, there is only indirect evidence that this was Leibniz's position in the texts available to Russell.[56] The closest Leibniz comes to it in the texts Russell used is the following, which Russell translates in his Appendix of leading passages:

> The notion of a species involves only eternal or necessary truths, but the notion of an individual involves, *sub ratione possibilitatis*, what is of fact, or related to the existence of things and to time, and consequently depends upon certain free decrees of God considered as possible; for truths of fact or of existence depend upon the decrees of God.[57]

This is less clear than one would wish, and returns us to the old contingency-as-dependence-on-God's-will defence. Though these two methods of defending contingent propositions need not be incompatible, as Leibniz's combining them shows, nonetheless it is unsatisfactory to have a relatively clear distinction between existence and other properties made to depend upon a murky distinction between God's will and his understanding.

Nonetheless, as Curley says ([1972], p. 83), Russell's 'instincts about what Leibniz would say were surely sound' for existence *is* clearly made an exception to the containment principle in works published after Russell's, notably in the following passage:

> [A]ll truths about contingents, i.e., about the existence of things, depend upon the principle of perfection. All existences, the existence of God alone excepted, are contingent. But the reason why one contingent thing exists, rather than others, is not sought from its definition alone, but from comparison with other things. For since there are infinitely many possibles which nevertheless do not exist, the reason why these exist rather than those must not be sought from the definition (otherwise, not to exist would imply a contradiction, and the others would not be possible, contrary to the hypothesis), but from an extrinsic principle, which is that these are more perfect than the others. (Grua, p. 288; quoted Curley [1972], pp. 83–4.)

Or the following, which is appealed to by Mates and Mondadori:

---

[56]I rely here on Curley's claim ([1972], p. 83), though as Curley notes 'negative existential propositions about what is in Leibniz are always dangerous'. Russell himself certainly provides no textual evidence.

[57]GP, ii, p. 39, as given at *POL*, p. 209. Russell quotes part of the passage at p. 26. Cf. also the other indirect evidence he assembles there: *N.E.* IV, x, 1; IV, xi, 13 = GP, v, pp. 414, 428–9. Better evidence is given by Mates and Mondadori: C, p. 393 = LP, p. 82.

This however presupposes denying every proposition in which there is a term which does not exist. In order, namely, to keep [the principle] that every proposition is true or false, [I consider] as false every proposition that lacks an existent subject or real term. (C, p. 393, as translated by Mates.)

Unfortunately, this neat solution falls foul of a problem noted by Fried [1978]. In addition to contingent truths, Leibniz needs to admit contingent falsehoods: the negation of every contingent truth must be a contingent falsehood. As Fried points out, the Mates-Mondadori approach will not permit this. On the containment principle (assuming bi-valence), a proposition will be false when the predicate-concept is not contained within the subject-concept. If 'Caius is wise' is a contingent truth, 'Caius is foolish' should be a contingent falsehood. But, by the principle that all propositions with unexemplified subject-concepts are false, 'Caius is foolish' will be false in all worlds in which Caius does not exist; and since Caius has the same individual-concept in every world in which he exists and 'foolish' is not contained in that concept, 'Caius is foolish' will also be false in every world in which he exists. So 'Caius is foolish' is false in every possible world and thus is not a contingent falsehood. The upshot is that the existential assumption together with possible worlds semantics will not on their own solve the problem of contingent propositions. To solve it we still need to fall back on contingent containment. If a predicate's inclusion in a subject-concept is a contingent matter, then there will be some worlds *in which Caius exists* where 'foolish' will be included in Caius's subject-concept. 'Caius is foolish' will thus be true in these worlds and thus will be a contingent falsehood in the actual world. The contingent containment solution works well with possible world semantics, but it requires abandoning the idea that an individual has the same individual concept in every possible world. It is not clear, at least to me, whether this was Leibniz's view or, if so, how important it was to his philosophy. I shall not pursue these issues here. It is time to look at Russell.

Interestingly enough, synthetic propositions were not a problem for Russell. Indeed he thought that there were many more of them than Leibniz did; for Russell, all the propositions of mathematics and logic were synthetic. The problem for Russell was contingent propositions – he was not persuaded that there were any (*POL*, p. 25). His grounds for thinking so were only tentatively put forward: 'It may be questioned... whether... there is any sense in saying, of a true proposition, that it might have been false' (*POL*, p. 24). He was inclined to think that Kant, by admitting that the propositions of mathematics were both necessary and synthetic, had 'prepared the way for the view that this is true of *all* judgments', though 'it must be confessed that, if *all* propositions are necessary, the notion of necessity is shorn of most of its importance' (*ibid.*).[58] I shall not pursue this view of Russell's here, but the fact that it was Russell's view means that the inquiry into whether Russell was in a position to admit synthetic SP-propositions has to be pursued differently from the way it was pursued in the case of Leibniz, for we cannot assume, in Russell's system, that some propositions are synthetic because they are contingent.

---

[58] Russell subsequently came to think that necessity and possibility applied only to propositional functions (cf. N&P). In 1899–1900 he did not have the concept of a propositional function (nor, indeed, the modern notion of a variable on which it depends).

5. Analytic and synthetic propositions in Russell

At the time of his study of Leibniz, Russell had already embraced the method of decompositional analysis, but this on its own was hardly sufficient to get the results he needed, it was not until he discovered Peano's symbolic logic in the summer of 1900 that he found what he needed. In this methodological vacuum, one might have expected him to seize upon Leibniz's substitutional procedures. There were many things in Leibniz's methods that Russell embraced, but the substitutional proof-procedure was not among them. Most importantly, he accepted Leibniz's account of definition as 'the analysis of complex ideas into their simple constituents' (though this was a position he already had from reading Moore [1899a] ahead of publication).[59] Similarly, he found that Leibniz's 'search for the simple ideas which form the presuppositions of all definition' in the Universal Characteristic coincided with his own search for the fundamental concepts of mathematics (*POL*, pp. 18–19). In the *Monadology* (§ 35, quoted *POL*, p. 19) Leibniz wrote: '[T]here are *simple ideas*, of which no definition can be given; there are also axioms and postulates, in a word *primary principles*, which cannot be proved' (GP, vi, p. 612). It was *exactly* to identify these, in the special case of mathematics, that Russell had started writing *The Fundamental Ideas and Axioms of Mathematics*, the first of his works in which exclusively analytic methods were employed. But here again Leibniz was mainly influential in confirming Russell in the direction he was already taking. Where Russell differed from Leibniz was over the latter's claim that the primary principles were identities.[60]

At the time he was writing on Leibniz, Russell denied that the axioms of mathematics were identities because he regarded identities as tautologies 'and so not properly propositions at all' (*POL*, p. 17). This idiosyncratic view was a hold-over from his earlier allegiance to Bradley, and didn't last much longer. I think it must have been rejected as soon as Peano was taken on board. Russell hedges on whether he thinks this applies only to what he calls 'pure tautologies' (*POL*, p. 18), e.g., '*A* is *A*', or whether he intends to include more complex identities, e.g., '*AB* is *B*'. He lists examples of both types together and says '[m]ost of these... assert nothing; the remainder can hardly be considered the foundations of any important truth' (*POL*, p. 17). It is not necessary to identify Russell's position more precisely, since for either type of proposition his fundamental objection is essentially the same: identities, of either type, do not have enough content to form the basic axioms of mathematics.

Apart from the problematic status he accorded tautologies at this time, Russell's reasons for thinking that the axioms of mathematics are not analytic are more interesting and more complex. Russell is not explicit in his Leibniz book whether he thinks that any identities are among the axioms of mathematics, but he does argue that not all the axioms

---

[59]*POL*, p. 18. In other work written around the same time Russell distinguished this notion of *philosophical* definition (or analysis), from mathematical definition, in which a term was defined, not by analysis, but by its relation to other terms (AOG, pp. 410–12). Mathematical definition was, of course, not restricted to complex terms.

[60]Had he not done so, he might well have spent some time investigating Jevons' equational logic (Jevons [1874]) which was an attempt, with which Russell was certainly familiar, at formalization along the lines of Leibniz's substitutional method. Influenced, no doubt, by Whitehead's [1898], which he read before he started working on Leibniz, Russell in 1898–99 favoured Boolean algebra as the algebra of symbolic logic.

of mathematics can be of this type and that some at least of the axioms must be synthetic. In the case of analytic SP-propositions, which are not pure tautologies, the subject-concept is always complex, and Russell argues that the predicates which make up the subject-concept must be 'compatible or jointly predicable predicates' (*POL*, p. 18). Now this was a position that Leibniz held (cf. *NE*, IV, ii, 1) and Russell may well have got it from him. For in *An Analysis of Mathematical Reasoning*, which Russell completed shortly after reading Leibniz's *New Essays*, Russell held that two predicates could be combined (or synthesized) to form a third only if they were not 'mutually contradictory' (*AMR*, pp. 169, 186).[61] The point of the compatibility requirement is not that the subject-concept has to be configured in such a way as to ensure that there is a subject to which to attach the predicate – for a consistent subject-concept gives no guarantee of that in any case. Russell's point, as his discussion of SP-propositions in *AMR* makes clear, is rather that unless the constituent predicates are compatible they cannot be synthesized at all. For my part, I see no reason whatsoever to think that this is the case. But given that it's Russell's position, it will certainly apply equally to GSP-propositions. So if Russell's argument were taken to show that no ISP-propositions (beyond pure tautologies) are analytic, it would equally show that no GSP-propositions are (at least, none with a complex subject-concept). Yet, Russell seems to have been quite happy to allow GSP-propositions to be analytic. The consistency restriction that he placed on predicate formation in *AMR* gave him enormous difficulty when he tried to mesh his philosophical account of the elements of propositions with the formal requirements of Boolean algebra (cf. Griffin [1991], pp. 281–2, 287–8, 291–3) and in the end he apparently just abandoned it (*AMR*, pp. 193–4). But this had been done by the time Russell wrote his Leibniz book and it is not clear in the Leibniz book whether Russell would reject inconsistent subject-concepts altogether.[62]

Whether inconsistently synthesized predicates be philosophically suspect or not, there is a strong case to be made for Russell's insisting that the predicates which make up the subject-concept of a proposition must be compatible if the proposition is to be a fundamental proposition in any branch of knowledge; for, if they are not, nothing will fall under the subject-concept and, according to Russell, the proposition, in general, will be false. Given, then, that the analytic propositions that have any claim to play a foundational role in mathematics must have consistent subject-concepts, Russell argues that not all the axioms of mathematics can be analytic; for, since the compatibility of the predicates is presupposed by an analytic proposition, the proposition that the predicates are compatible cannot itself be an analytic proposition (*POL*, p. 18). Thus, Russell claims, 'there is always involved, in definition, the synthetic proposition that the simple constituents are compatible' (*POL*, p. 20).

---

[61] His ground for this seems to have been that what is contradictory is meaningless, and thus that propositions that contained an inconsistently synthesized predicate would not be genuine propositions at all.

[62] His refusal to distinguish between subjects and subject-concepts raises a further interpretational difficulty here, for obviously he would reject inconsistent subjects – there just are no such things – and it may be this claim that he had in mind when he said that the predicates must be consistent. If they make up the subject, they obviously must; but if they make up the subject-concept, the case is not so clear. Indeed, I think powerful grounds can be given for thinking that inconsistent subject-concepts have to be admitted.

This argument for the syntheticity of compatibility seems hardly compelling. Russell argues that such propositions are presupposed by analytic propositions and so cannot be analytic themselves. But all he has shown is that they are presupposed only by a certain type of analytic proposition, namely analytic SP-propositions. There is no reason why these analytic propositions should not presuppose analytic propositions of a different type.[63] Only if the elementary judgments of compatibility were judgments of SP-propositions, would Russell's argument be valid, and Russell (obviously rightly) holds that they are not, for they are relational propositions. Now obviously relational propositions are not analytic in the sense in which we have hitherto been using the word. But might they be analytic in the other traditional sense of the word, namely, that their truth follows from the law of no contradiction alone, i.e., that to deny them involves an immediate violation of the law of no contradiction? Russell rejects this possibility. Russell, in fact, thinks that *nothing* follows directly from the law of no contradiction, except the proposition that some proposition is true (*POL*, p. 22). Moreover, although he doesn't mention it in the Leibniz book, he has, as we shall see, further grounds for thinking that *no relational* propositions can be analytic.

Russell has an additional argument for his conclusion that the elementary propositions of compatibility must be synthetic. When Leibniz says that the concepts which make up the subject-concept of a definition must be compatible, he means that the subject-concept must not be self-contradictory (GP, iv, p. 424). But as Russell explains, two *simple* terms[64] cannot on their own yield a contradiction 'since mere analysis will not reveal any further predicate possessed by the one and denied by the other' (*POL*, p. 21), so 'any collection of simple ideas would be compatible, and therefore every complex idea would be possible' (*POL*, p. 19). If simple terms are incompatible, it is because they have a synthetic relation of incompatibility; just as, if they are compatible, they have a synthetic relation of compatibility. Either way, some synthetic propositions are logically prior to analytic propositions, so the fundamental principles of any science cannot consist entirely of analytic propositions. Hence Russell's conclusion: 'if there are any necessary propositions, there must be necessary synthetic propositions' (*POL*, p. 23).

If Russell is right about the prior judgments of compatibility and incompatibility, this conclusion is safe enough. But he draws from his premises another conclusion which warrants further consideration. He claims that, since even SP-propositions such as 'the equilateral rectangle is a rectangle' depend on synthetic propositions of compatibility and incompatibility, this shows that they are 'not wholly analytic', since 'they are logically subsequent to synthetic propositions asserting that the constituents of the subject are compatible' (*POL*, p. 22). Taken at face value, Russell's claim is a strange one: analyticity, surely, does not admit of degrees. It may be that Russell is here operating with an

---

[63]This problem, of course, would not arise if all propositions were SP-propositions, but this is something Russell denies. The argument may, of course, be a good *ad hominem* argument against Leibniz (assuming that Russell is correct to attribute (SP) to him), but it cannot (yet) form part of Russell's case for the synthetic a priori status of mathematical propositions.

[64]Russell gives *good* and *bad* as examples (*POL*, p. 20), obviously following Moore's yet-to-be-published views about the indefinability of *good*.

unarticulated notion, drawn from his decompositional style of analysis, that the propositions presupposed by a given proposition are in some way contained within it, so that the original proposition to some degree (but not wholly) shares their character. This would be an exceedingly murky doctrine, and we should probably not saddle Russell with it by taking his words too seriously.

But his remark does raise the question of whether the fact that a putatively analytic proposition presupposes a proposition which is not analytic impugns the analyticity of the former. It is clear that no analytic proposition can presuppose a contingent proposition, but Russell's prior judgments of compatibility are necessary, though not analytic. It seems as if he might be assuming an additional feature of analytic propositions, which he nowhere states, but which was commonly ascribed to them, namely that their truth could be established, not merely *a priori*, but upon immediate inspection of the proposition itself. If this is assumed, then it follows that the truth of an analytic proposition cannot depend upon the truth of *any* other proposition. It is not clear whether we should add this assumption on Russell's behalf, but, if we do, it will ensure that even paradigm analytic propositions (all except the pure tautologies) are not analytic at all – not that they are not wholly analytic.

We are not in a position to draw a very exact conclusion about which (if any) propositions Russell at this time thought were analytic: his texts are just not explicit enough to do so. But one thing is absolutely certain: analytic propositions are neither numerous nor important. As we have seen, Russell thinks that perhaps only one proposition is analytic in the sense of following from the laws of no-contradiction alone: the proposition that some proposition is true. As regards relational propositions, Russell, as already noted and as we shall see in more detail in the next section, denied that any of them were analytic. For SP-propositions, on the other hand, Russell (following Bradley) denied during the period 1898–1900, that those which were pure identities were propositions at all. Those which were not pure identities were genuine propositions, but it seems impossible to determine whether he thought that any of them were analytic. All of them, as we have just seen, depend upon prior synthetic propositions of compatibility of their subject-concept, and Russell, so far as I can tell, is just not sufficiently explicit to allow us to decide whether he thinks that this is sufficient to preclude their being analytic. Even if it does not, it still leaves Russell, as we shall see in the next section, with very few analytic propositions – far fewer then one would expect. The only clear example we have of an analytic truth from Russell during this period is the proposition that some proposition is true. To see this properly, we must return to the containment principle.

6. The containment principle in Russell

Given Russell's early realist view of propositions it is surprising to find him embracing as much of the containment principle as he does. While he rejected its universal applicability, he did embrace it for SP-propositions. Combining this with his realist view of propositions, it follows that he embraced the second of the views that Parkinson thinks obviously wrong (that the predicate is in the subject itself, rather than in the subject-concept) as well as the first (that the subject itself, and not the subject-concept, is in the proposition). That he did so cannot, I think, be established from the book on Leibniz, and indeed the position may have been one that he arrived at after his study of Leibniz. Russell in fact says almost nothing about it, perhaps not surprisingly because, during the period that he held it (which was until 1905), his comments on his analytic methodology were generally brief and almost always incidental to a discussion of some other issue. Nor, after 1898, did he have much to say about the nature of SP-propositions – we shall see why shortly. The doctrine, I believe, was one that he quite clearly held, and that he was also quite clear about holding it: it was not an unsuspected consequence of his other beliefs. But it was a position that he never explored in writing (at least not in any of the writings that have survived), largely I suspect because it quickly became part of the accepted background of his rejection of neo-Hegelianism. And then, in 1905, it was swept away in the aftermath of the theory of descriptions. Ironically, the one clear indication that we have of it is from just before it was abandoned.

In 1905 Russell was debating the nature of relations with the neo-Hegelian philosopher, Harold Joachim,[65] who believed that relations were grounded in what he called the 'natures' of their terms – very much as Leibniz, following the scholastic tradition, believed they were grounded in the 'intrinsic accidental denominations' of the subject. What exactly Joachim meant by the 'nature' of a term was very far from clear. In his published reply, Russell called it 'the ghost of the scholastic essence' ([NT], p. 530); though, as we shall see, he might as accurately have called it the ghost of Leibniz's subject-concept. In his correspondence with Joachim he suggested two possible meanings: it could mean 'all the prop[osition]s which are true of the thing' or it could mean 'the adequate analysis of the thing'.[66] A definition, he explained, should give an adequate analysis, but would not give all the propositions that are true of the thing defined. 'Hegelians', he went on, 'consider... that these two notions are indistinguishable. They give no reason for this view; but it follows from the principle that every proposition consists in the attribution of a predicate to a subject. I deny this principle, and maintain my distinction.'[67] Obviously, the position which Russell here ascribes to the Hegelians would also be one he would ascribe to Leibniz.[68]

Russell's remark is, I believe, of prime importance for understanding the general features of his philosophy through roughly the period 1898 to 1905, though we don't know the exact period during which the doctrines it implies were held. It is obvious from the context that Russell is speaking only about atomic propositions. With this restriction assumed, Russell claims that it follows from (SP) that there is no difference between the

---

[65]The debate went on over two years, before and after the publication of Joachim's book, *The Nature of Truth* (Joachim [1906]). For their correspondence during this period, see Connelly and Rabin [1996]; for discussion of the debate, see Griffin [2007b].

[66]The use of the definite article here is somewhat puzzling since it implies that there will, for each thing, be only one analysis which is adequate. I'm not sure that Russell actually intended to say this, but he probably held that only one analysis was (metaphysically) correct and may well have held that any incorrect analysis was *ipso facto* inadequate.

[67]Connelly and Rabin [1996], p. 137. I have discussed the passage at greater length in Griffin [2007a].

[68]Interestingly, the most famous argument of the neo-Hegelian arguments against the reality of relations, Bradley's infinite regress argument (Bradley [1893], pp. 17–18), was anticipated by Leibniz (A, VI, 3, p. 399).

set of propositions true of a thing and the set of propositions required for an adequate analysis of it. Now this inference would hold *only* if it is assumed that *all* the predicates of a thing (but not its relations) are to be included in an adequate analysis of the thing. The proof is as follows: Let A be the set of propositions which give an adequate analysis of a thing; let P be the set of true propositions attributing predicates to it; and let T be the set of true propositions about it. If, as Leibniz and the Hegelians suppose, every proposition attributes a predicate to a thing, then $T = P$. From this, Russell can show that $T = A$, only if he assumes $P = A$. So Russell is assuming that *all* a thing's predicates are to be included in its analysis. Since Russellian analysis at this time is plainly decompositional analysis, it follows that *all* a thing's predicates are contained within it. Since, on Russell's direct realist account of propositions, the thing itself is the subject of an SP-proposition in which it occurs, it follows that all the predicates are contained within the subject: *praedicatum inest subjecto*.

It is, perhaps, necessary to point out that this principle plays a very different role in Russell's philosophy from the role played by the containment principle in Leibniz's, for Russell insisted upon a sharp and unbridgeable distinction between SP-propositions and relational propositions. Thus, while Leibniz could regard the containment principle as providing truth-conditions, either directly or indirectly (if the truth conditions of relational propositions had to be given via the truth conditions for the corresponding SP-proposition – i.e., the proposition which ascribed an intrinsic relational accident to one of the terms), for all propositions, Russell took his principle to supply truth conditions for only a proper subset of propositions. Moreover, as we shall see, this subset turns out to be much smaller than might have been supposed.

A number of theses follow immediately from this position. Firstly, it follows that, in general, a term's relations will not be part of its analysis; some of them (e.g., the relation of the term to its parts) may be, but most of them will not. This comes as no surprise on a decompositional account of analysis, and is, in any case, to be expected given Russell's distinction between philosophical definition (or analysis) and mathematical definition by means of relations (cf. AOG, pp. 410–12). We can see now that mathematical definitions are going to be synthetic according to Russell – and so, too, will be the mathematical propositions which depend on them, assuming that it really is Russell's position that an analytic proposition cannot depend upon a synthetic one. But there were other reasons for him to hold this position. For a while at any rate, he regarded all relational propositions (and the overwhelming majority of mathematical propositions were relational) as synthetic. Obviously they are not analytic according to the traditional definition in terms of subject and predicates (A1), but neither did Russell think that any of them could be established as analytic by the other traditional definition, that they followed from the law of no contradiction alone, for he thought only the proposition that some proposition is true followed from the law of no contradiction alone (*POL*, p. 22). Moreover, in the case of relational propositions, Russell explicitly acknowledged this: 'it can never be self-contradictory to deny a relation' (FIAM, p. 279). Secondly, it follows from Russell's account of analysis that *all* a term's predicates are part of its analysis. Moreover, it would seem to follow from this that *all* true SP-propositions are analytic. This is a very much more surprising result, for it is the very position from which we have been trying to save

Leibniz. So a major issue is now whether we can save Russell from it or somehow make it palatable.

Although Russell made use of decompositional analysis in *AMR*, before he studied Leibniz,[69] his commitment to it was not wholehearted, and his initial solution to the problem of the analyticity of SP-propositions involved what amounted to a limitation on decompositional analysis, though he was perhaps less clear both about the solution and its effects on decompositional analysis than this remark suggests. In *AMR*, Russell denied that predicates are terms:

> The peculiarity of predicates is, that they are meanings. Now although it is impossible to speak of meanings without making them subjects..., yet meanings *as such* are the antithesis of subjects, are destitute of being, and incapable of plurality. When I say 'Socrates is human', *human* as used in this judgment does not have being and is not a logical subject. I am, in a word, not asserting a relation between two subjects. (*AMR*, p. 174)[70]

This leads to a satisfactory solution to the problem of the analyticity of SP-propositions; for the subject is a term which, if complex, is composed of other terms, so if the predicate is not a term it cannot be among the constituents of the subject. Thus the predicate will never be included in the subject, and no SP-proposition will be analytic. The damage done to decompositional analysis is not that we cannot, on this account, decompose the subject into constituent terms, including the predicate – there is no real loss there, since if the account is correct, the predicate is not among the constituents of the subject. The real loss is in the decompositional analysis of propositions: SP-propositions can no longer be analyzed into subject and predicate, for the predicate is not a term which could be identified in the analysis. This leaves it entirely unclear how, if at all, SP-propositions can be analyzed.

No doubt ways around this could have been devised, and relatively easily, if Russell had been firm in denying that predicates were terms. But he was not, for the doctrine led to a quite inextricable tangle. If predicates as such cannot be made logical subjects, there is no way to talk about them and thus no way to formulate a theory of them. In response to this, Russell claimed that '[e]very predicate may be made a subject' and that the distinction of subject and predicate is 'only a distinction of aspect' (*AMR*, p. 174). But if predicates are not terms, the difference between subjects and predicates seems far more than a mere distinction of aspect. Once a predicate like *human* is made a subject, it ceases to be mere meaning and becomes a term. 'As soon as I make *human* a term... I have added something, namely being, one-ness, and diversity of being from other terms,

---

[69] Or, more accurately, the relevant parts of Leibniz. As we have seen he read the *New Essays* at the same time as he was writing *AMR*, but this was well before he came upon the correspondence with Arnauld and the *Discourse on Metaphysics*.

[70] The problem here is more familiar in the form in which Frege faced it in his function-object analysis of propositions. Frege [1892] denied that the function can be treated as an object, i.e., made into the subject of a proposition, much more resolutely than Russell, who here tries the same approach.

which *human* as predicated did not possess' (*AMR* 174).[71] This gives predicates a peculiar chameleon-like quality, switching from meanings to terms according to their role in the proposition. This cannot be satisfactory. If terms, predicates, and propositions really are mind-independent items, as Russell's anti-psychologism required, then I cannot literally add being, one-ness and diversity to them by moving them around in a proposition; indeed, I cannot literally move them around in a proposition either, for the Russellian proposition is not something that I construct. Terms and propositions are what they are, and not another thing (to paraphrase Moore's epigraph from Bishop Butler).

At the time Russell wrote *AMR* it is not clear that he thought there was a problem of analytic SP-propositions. Analyticity does not figure in his discussion and he seems to have given no thought to whether SP-propositions were analytic or not. His concern was to try to understand the fact that predicates could play a dual role in propositions. It was probably his study of Leibniz that brought the question of analyticity vs syntheticity to his attention. Certainly Chapter II of his book on Leibniz contain his most extended early treatment of the topic. We have already considered Russell's case against Leibniz's claim that all 'primary principles' or axioms are identities and thus analytic. He rejected it on the grounds that such SP-propositions presuppose synthetic propositions asserting the compatibility of the predicates which make up the subject (-concept). But this only shows that not *all* first principles can be SP-propositions; it does not show that the SP-propositions are not analytic. It may be thought to run against the spirit of analytic propositions to have them presuppose synthetic ones. We not untypically think of analytic propositions as those whose truth can be grasped without further consideration beyond the proposition itself. But that is an epistemological consideration; from the point of view of Russell's (and Leibniz's) radical anti-psychologism, it is neither here nor there how we grasp the truth of a proposition. If we define analytic SP-propositions as those in which the predicate is contained in the subject, there seems no ground for supposing that they cannot presuppose synthetic propositions, provided the synthetic propositions are necessary, as Russell's judgments of compatibility among predicates clearly are.

Russell, as we've seen, is clearly troubled by the prospect that all SP-propositions are analytic. He refuses to countenance the possibility that an analytic proposition may be contingent – and, until Couturat persuaded him of the contrary, he refuses to countenance the idea that Leibniz thought they could be. Oddly enough, the last was not a problem for Russell himself, for he was prepared to maintain (albeit hesitantly) that all propositions were necessary – but he was not prepared to admit that all of them (or even a large number of them) were analytic. Accordingly, he seeks ways to save himself (and Leibniz) from this untoward result. He is (rightly) not impressed by Leibniz's distinction between finite

---

[71] Those familiar with Russell's early work will recognize here the first version of a familiar problem, better known when formulated in terms of relations: the relating relation and the relation as term are required both to be different, according to their role in the proposition in which they occur, but must also be identical, because otherwise we couldn't be talking about relating relations. This problem, one of a type which I've called 'double aspect problems' (Griffin [1993]), leads to major difficulties in Russell's early philosophy: the famous problem of the unity of the proposition, the notion that analysis always involves falsification, the difficulty of giving a coherent account of functions, and even the problem about denoting concepts that led to the theory of descriptions.

and infinite analyses; nor, of course, with any distinction between what is true by virtue of God's will versus what is true by virtue of his understanding. Moreover, the distinction between ISP- and GSP-propositions, though useful and important, does not seem to cut at the right joints to capture the distinction between analytic and synthetic propositions. At the time of writing the book on Leibniz, Russell's favoured way of avoiding the conclusion that all SP-propositions are analytic seems to be his argument that, except in the case of pure identities (propositions of the form *A* is *A*), all SP-propositions logically presuppose prior synthetic propositions. But this approach, as I have argued, is not convincing without a fuller account of analyticity which Russell nowhere adumbrates. To this point, therefore, the problem has no satisfactory solution.

In an earlier discussion of this issue (Griffin [2007a]), I argued that the view that all SP-propositions are analytic is in fact Russell's considered opinion, but that it is saved from absurdity by the fact that Russell held that there were far fewer SP-propositions than we would normally suppose. This is the approach that Russell took in 'The Classification of Relations', a paper written while he was giving his Leibniz lectures. There he argues that predication itself is a relation and predicates are terms – thus firmly rejecting his approach the previous year in *AMR*.[72] (The same position appears in *POL*, p. 15.) In 'The Classification of Relations' he leaves it as an open question whether there are any SP-propositions at all (COR, p. 141).

How does this policy work out in the case of the types of SP-proposition Russell has been considering in connection with Leibniz? Before his work on Leibniz, he maintained that GSP-propositions were relational and asserted 'necessary connections of contents' (*AMR*, p. 173). He thought this group included 'most of the axioms of mathematics' (*ibid.*). Shortly afterwards, once he had discovered Peano's work, he treated them, in the way that is now familiar, as universally quantified implications. It was the other group of putative SP-propositions, ISP-propositions, that required work. The key suggestion here came from Leibniz's attempts to ensure that ISP-propositions are in general contingent while GSP-propositions are in general necessary. In his notes for a reply to Arnauld, Leibniz writes: 'The notion of a species involves only eternal or necessary truths, but the notion of an individual involves, *sub ratione possibilitatis*, what is of fact, or related to the existence of things and to time' (GP, ii, p. 39; quoted *POL*, p. 26). Adapting this general idea but implementing it by means of relations, Russell, in writings immediately after *AMR*, treats all ISP-propositions as relational propositions involving the individual thing, the concept predicated in the proposition, existence, and particular regions of space and time.

So, while Russell is prepared to accept that all SP-propositions are analytic, his view is that there are very few of them, and maybe none at all. On the other hand, all relational propositions- and this includes the vast majority of propositions (including perhaps all that we would usually think of as SP-propositions) – are synthetic. In coming to these conclusions, Russell is not at all driven, as Leibniz was, by the need to allow for contin-

---

[72]This, of course, solved the double-aspect problem in its original form, but it came back immediately in connection with relations, which occur as meaning when they relate and as terms when they are occur as logical subjects.

gent propositions, for Russell is prepared to countenance (though with some hesitation) the view that all propositions are necessary. Similarly, he is prepared to countenance the view that all propositions are synthetic (though again, not without hesitation). This, of course, would require him to deny that there are SP-propositions, but it is not entailed by that denial and the doctrine that all relational propositions are synthetic, for there are other sorts of propositions. Whether any of them have a claim to be regarded as analytic is left unclear. The only proposition for which such a claim is made (by no means unequivocally) is the proposition that some proposition is true. This may well be the only analytic proposition Russell is prepared to allow. All of which serves to show how very unusual Russell's early theory of propositions is.

7. The question of relations

In *My Philosophical Development* Russell says, rather mysteriously, that he 'first realized the importance of the question of relations' when he was working on Leibniz (*MPD*, p. 61). But what exactly was 'the question of relations'? It was not the idea that relations were of considerable importance in the philosophy of the special sciences – for Russell was well aware of that long before he started work on Leibniz, it informs even his first philosophical book, *An Essay on the Foundations of Geometry*. His recognition of the importance of relations in mathematics had been growing since then. Moreover, Russell's concern with the formal properties of relations, which led to the distinguishing of transitive, symmetrical and reciprocal (i.e., reflexive) relations (in something like their modern senses), goes back to the second half of 1898 (cf. *CPBR2*, pp. 26–7) – also before his serious work on Leibniz had started. Indeed, by the time he came to write *AMR*, Russell had come to think that transitive asymmetrical relations were crucial to the whole of pure mathematics – and thereby hangs a problem. As a neo-Hegelian Russell had planned to work his way through the special sciences, making dialectical transitions from narrower sciences to broader ones in order to relieve the antinomies generated by the excessive abstraction of the narrower ones. One type of antinomy turned out to be ubiquitous: it appeared in geometry as the antinomy of the point (*EFG*, p. 189) and, later in the same year, in arithmetic as the antinomy of quantity (RNQ, p. 81). By the following, in *AMR*, it had become 'the contradiction of relativity' (*AMR*, p. 166) and had cast its shadow over the whole of pure mathematics. In *AMR* (pp. 225–6) Russell has an argument to show that the contradiction arises wherever you have transitive, asymmetrical relations – since such relations 'pervade almost the whole of mathematics', Russell concludes, the 'fundamental importance of this contradiction to Mathematics is thus at once proved and accounted for' (*AMR*, p. 226). To modern eyes the contradiction hardly looks very intimidating; it is just this: that we have two distinct terms without a difference in the conceptions applicable to them (*AMR*, p. 166). In other words, we have terms numerically distinct, but qualitatively identical. Thus, in the case of points, though each point is numerically distinct from all the others, all of them have exactly the same intrinsic nature. We have, as Russell puts it with a Hegelian flourish, a concept of their difference without there being a difference in the conceptions applicable to each (RNQ, p. 81). What makes this apparently innocuous result a genuine problem for Russell is the assumption, hitherto unstated, which underlies his entire theory of relations through to 1898; namely, the neo-Hegelian 'axiom of internal relations', that relations are always grounded in the (intrinsic) natures of their terms. At this point we can see clearly enough that if R is an asymmetrical relation holding between a and b, then a and b must have different natures, different conceptions must be applicable to each. For whatever it is about the nature of a that grounds its relation R to b, b must have a different nature for b neither has the relation R to a nor to itself. If relations supervene on the natures of their terms, as neo-Hegelians (and Leibniz) suppose, then asymmetrical relations are impossible. And yet for the mathematical sciences they are essential.[73]

In other words, with the doctrine of internal relations, Russell has a genuine contradiction permeating the whole of mathematics. Moreover, it is not a contradiction that is easy to remove by transition from one science to another – as Russell found through many fruitless efforts in the period 1896–98. The contradiction simply re-emerges in a new form in each new science. But if one rejects the doctrine of internal relations, the problem simply vanishes. At some point between 1898, when *AMR* was completed, and 1899, when he starts to write the 1899–1900 draft of *The Principles of Mathematics*, Russell does just this. In fact, he keeps the very sheet from *AMR* in which he presented the argument for the contradiction as pervasive in mathematics and simply changes the conclusion: instead a *modus ponens* argument proving the contradiction of relativity from the doctrine of internal relations, it becomes a *modus tollens* argument refuting the doctrine of internal relations *because* it entails the contradiction of relativity.[74] It is hard to over-estimate the importance of this move, for it constituted Russell's break from neo-Hegelianism, the one true revolution in his philosophical development (*MPD*, p. 11).

Up to this point there had been in Russell's philosophy absolutely no discussion of the doctrine of internal relations at all – in *AMR* it is not stated as a premiss of the argument for the contradiction of relativity. Indeed, it is not stated in pre-1899 works at all: it is stated only after it is rejected. Until then, it had functioned as an unrecognized assumption, and this, of course, is what enabled it to survive so long and do so much damage.[75] So the question now naturally arises, what brought it to light? Unfortunately, Russell says too little for us to be certain about this, but it is noteworthy that Russell's course on Leibniz was given between the completion of *AMR* in 1898 and the start of POM/D in 1899. It is certainly plausible to conjecture that it was Leibniz's tortuous, and ultimately unavailing, efforts to show how relations may be grounded on the intrinsic relational accidents of their terms that brought it to his attention. It was certainly important enough to warrant Russell's labelling it '*the* question of relations'. Moreover, he was

---

[73] I have examined this problem (and its ultimate resolution) in some detail in Griffin [1991], Ch. 8 and more briefly in Griffin [forthcoming].

[74] Compare *CPBR2*, pp. 225–6 with *CPBR3*, p. 93. The argument reappears, somewhat rephrased, at *POM*, p. 224. It may well be that the sheet of typescript that travelled from *AMR* to POM/D may have spent some time in the intervening work FIAM (cf. the editorial comment on the manuscripts, *CPBR2*, p. 262).

[75] Of course, in some sense it had to be recognized earlier, otherwise Russell would not have been able to recognize the contradiction of relativity as a problem. The point is that, at this earlier time, Russell took it to be simply part of what was involved in acknowledging relations at all. What came to be recognized around 1898–99 was that the doctrine of internal relations was a separate, additional assumption, not automatically embraced along with relations; that relations were not necessarily internal.

hardly in a position to give Leibniz more direct acknowledgement for it was, of course, the *rejection* of Leibniz's position which had been important to him.

There is no reason to suppose that, before his study of Leibniz, Russell knew much about the scholastic doctrine of relations on which Leibniz's views were based. A philosopher not imbued with the scholastic tradition will find it very striking that Leibniz simply takes the idea of an accident's inhering in more than one substance as obviously absurd. For such a philosopher, to see the assumption stated is to bring it into question. While Russell, of course, was entirely outside the scholastic tradition, his own tradition, that of the British neo-Hegelians, shared the same assumption. It is certainly not implausible to suppose that it was his study of Leibniz which persuaded Russell of the importance of the assumption. Finally that it derived from the scholastics no doubt made it easier to dismiss as an ancient error. But recognizing that it survived unchallenged among the neo-Hegelian philosophers of his day could easily have produced the view, to which Russell in subsequent years gave frequent expression, that it was a hitherto almost universal prejudice among philosophers of all persuasions. Most important, seeing that it did not have to be accepted showed Russell a way out of the impasse in which by 1898 his dialectic of the sciences found itself.

The best evidence we have for this is, alas, only circumstantial, but some of it is certainly suggestive. One of the clearer statements of Leibniz's insistence that relations be grounded in the individual concepts of the terms comes from the correspondence with Arnauld, which (as we've seen) shed a flood of light on Russell's understanding of Leibniz. In his long letter of 14 July 1686, Leibniz had this to say:

> [T]he concept of an individual substance includes all its events and all its denominations, even those which are commonly called extrinsic.... For *there must always be some foundation for the connection between the terms of a proposition, and this must be found in their concepts* (GP, ii, p. 56 = L, ii, p. 517).[76]

In his book on Leibniz Russell collects a number of key passages on the topic together in his Appendix of leading passage under the heading 'Are all propositions reducible to the subject-predicate form?'. The first of these: 'There is no denomination so extrinsic as not to have an intrinsic one for its foundation' (GP, ii, p. 240 = *POL*, p. 205) is in fact the first time that Russell puts a version of the doctrine of internal relations into print. But there seems no way we can determine whether, in citing this passage, Russell was coming across the axiom of internal relations stated explicitly, in full generality, for the first time; or whether he gave the statement its prominent position because he already recognized its supreme importance for his own philosophy. It is notable, also, that the examples Leibniz gives in *New Essays* of relations of comparison which concern disagreement (namely: cause and effect, whole and parts, situation and order) (*NE*, II, xi, 4 = *POL*, p. 206) are exactly the ones that Russell is concerned with in *AMR* and POM/D. Indeed, in both places he uses cause and effect as his main example, even though it might seem out of place in a work on pure mathematics.

---

[76]Russell has marked the passage in his copy of Gerhardt with a double vertical line in the margin and the words 'v[ery] imp[ortan]t'.

If this conjecture is correct, then Russell's study of Leibniz was of the very greatest importance to his own philosophical work. Of course, what he learned from Leibniz was not the lesson Leibniz wanted to teach. But Leibniz's efforts to state and defend an assumption that Russell had hitherto almost unconsciously shared with him made it clear to Russell how difficult the assumption was to defend and how easy and desirable it would be to abandon it. As with many philosophical doctrines, the attempt to state the doctrine of internal relations clearly – as Leibniz did – was to invite it to be called into question – which was what Russell did.

8. Conclusion

Of course there is vastly more to say about Russell's book on Leibniz. Even as regards their views on propositions, I have done barely more than scratch the surface. Many issues have been raised without being tracked to satisfactory conclusions. In this paper I have been able to do little more than scratch the surface of some of the deep connections between Russell's philosophy and Leibniz's. For example, I have not examined McCullough's suggestion that Leibniz sought to eliminate universal relations in favour of relational tropes, nor how Russell might have perceived these issues, nor the effect that this perception may have had on his own decision, a couple of years after his book on Leibniz was published, to eliminate relational tropes in favour of universal relations.[77] Nor have I examined the way in which Russell's early commitment to a direct, strongly realist view of propositions may have affected (and possibly distorted) his understanding of Leibniz's treatment of propositions. In particular, Russell entirely ignores Leibniz's distinction between the subject of a proposition and its subject-concept, perhaps because, being in the grip of a theory of propositions which left no room for the distinction, he did not notice it or perhaps because he noticed it, but thought it of no importance. It is not clear to me how important this oversight is, nor whether it led Russell into other errors of interpretation of Leibniz's theory of propositions. These issues, and many others, are worth further consideration.

Here my main concern has been to consider the ways in which Russell's study of Leibniz affected two positions Russell took up either immediately before or at the time he started his work on Leibniz, namely, his view that decompositional analysis was the proper method for philosophy and his rejection of the neo-Hegelian doctrine of internal relations. These views were central to Russell's rejection of the neo-Hegelian philosophy he had previously held and they paved the way for the major philosophical advances he made in the next few years. Russell got neither of these views from Leibniz – indeed, as we've seen, Leibniz himself, in company with the scholastics, accepted the doctrine of internal relations and Russell got decompositional analysis from Moore rather than from Leibniz. On the first of these issues, though it is impossible to be sure (Russell simply says too little about it to enable us to decide), it is quite plausible to suppose that it was coming across the doctrine of internal relations stated by Leibniz (perhaps the first time Russell

---

[77] At the time he wrote the Leibniz book, Russell accepted relational tropes (cf. *CPBR3*, Appdx. I.1), but in *POM* (pp. 50–2) he rejects them.

had seen it stated in full generality), but without any arguments in its favour (apart from deference to the scholastic tradition) being supplied, led Russell to see how easily it could be rejected. If so, then Russell's study of Leibniz was of the greatest importance to the development of his own philosophy. The chief argument for this is timing. For whatever reason, Russell came to identify and reject the doctrine of internal relations at about the same time he gave his Leibniz lectures. Unfortunately, in the works in which he makes the change he makes no reference to Leibniz, and in the work on Leibniz he makes no direct reference to the doctrine of internal relations.

By contrast, Leibniz embraced the decompositional method of analysis and this led him into serious difficulties in allowing for propositions that were either synthetic or contingent, in particular in allowing for synthetic or contingent SP-propositions. It may have been that Russell's study of Leibniz brought this problem to his attention, but in any case Russell was convinced that Leibniz, because he held that all propositions were SP-propositions, had no satisfactory solution to it. By contrast, Russell, adhering to a firmly realist view of relations, was able to deal with the problem, first by denying that the contingent/necessary distinction was applicable to propositions; secondly, and more importantly, by allowing that all true SP-propositions were analytic while insisting that all true relational propositions were synthetic. At first sight this seems hardly to give a satisfactory solution to the problem of synthetic propositions: far too many propositions, it would seem, remain analytic on this account. In fact, the problem was, if anything, the other way around. Russell was actually countenancing, albeit tentatively, the view that *all* propositions were synthetic, a view for which he thought that Kant had paved the way. If this is the case, then it follows that there are no SP-propositions – all plausible candidates being, in fact, relational. This, it seems clear, was Russell's position in 1899. Because of it he was quite unable to follow Leibniz's treatment of relations as the work of the mind – to do so would have completely undermined Russell's direct realist view of propositions and his method of propositional analysis and returned him to the psychologism from which he and Moore thought they had just escaped. In this way the two main themes of the present paper – the impact of the containment principle on the possibility of contingent and/or synthetic propositions and the status of relations as internal or external – come together. The former was of immense concern to Leibniz and is discussed in detail in Russell's book. The latter was not discussed by Leibniz, who seems to have simply accepted without dispute that all relations were what Russell would subsequently call internal, but it was of enormous importance to Russell. But Leibniz, though he was not prepared to doubt the doctrine, was a good enough philosopher to state it and to do so in a way which would make it easy for Russell to see why it need not be accepted. If this is what actually happened then Russell's study of Leibniz was not only an event of major importance in Leibniz scholarship but for the development of Russell's philosophy as well.

Acknowledgment

This paper owes much to the generosity of Richard Arthur, for severely criticizing an early, halting draft; Jolen Galaugher, for sharing with me her knowledge of the Russell-Couturat correspondence and allowing me to make use of her translations; and Massimo Mugnai, for letting me see a copy of his forthcoming paper on Leibniz and relations and for sharing with me his unrivalled knowledge of the subject. I am grateful also to Clinton Tolley for letting me see his unpublished paper on Leibniz and Kant and to an anonymous referee whose gentle complaint has led to a much improved ending.

References

1. *Works by Russell and Leibniz*

(see also the general list of abbreviations on p. 267 of this volume)

[AMR] Russell, 'An Analysis of Mathematical Reasoning' (1898), in *CPBR2*, pp. 163–242.

[AOG] Russell, 'On the Axioms of Geometry' (1899), in *CPBR2*, pp. 394–415.

[BRLC] Russell, *Correspondance sur la philosophie, la logique et la politique avec Louis Couturat* (1897–1913), ed. Anne-Françoise Schmidt (Paris: Kimé, 2001), 2 vols.

[COR] Russell, 'The Classification of Relations', in *CPBR2*, pp. 138–46.

[CPBR2] *The Collected Papers of Bertrand Russell*, vol. 2, *Philosophical Papers, 1896–99*, ed. by Nicholas Griffin and Albert C. Lewis (London: Routledge, 1989).

[CPBR3] *The Collected Papers of Bertrand Russell*, vol. 3, *Towards "The Principles of Mathematics"*, 1900–02, ed. by Gregory H. Moore (London: Routledge, 1993).

[CPBR4] *The Collected Papers of Bertrand Russell*, vol. 4, *Foundations of Logic, 1903–05*, ed. by Alasdair Urquhart (London: Routledge, 1994).

[CPBR6] *The Collected Papers of Bertrand Russell*, vol. 6, *Logical and Philosophical Papers*, 1909–13, ed. by John G. Slater (London: Routledge, 1992).

[D] *The Philosophical Works of Leibnitz*, ed. by G.M. Duncan (New Haven: Tuttle, Morehouse and Taylor, 1890).

[EFG] Russell, *An Essay on the Foundations of Geometry* (New York: Dover, 1956; 1st edn. 1897).

[FIAM] Russell, 'The Fundamental Ideas and Axioms of Mathematics' (1899), in *CPBR2*, pp. 265–305.

[KAKD] , 'Knowledge by Acquaintance and Knowledge by Description', in *CPBR6*, pp. 148–61.

[L] Leibniz, *Philosophical Papers and Letters*, ed. L.E. Loemker (Chicago: University of Chicago Press, 1956), 2 vols.

[MPD] Russell, *My Philosophical Development* (London: Allen and Unwin, 1959).

[N&P] Russell, 'Necessity and Possibility', in *CPBR4*, pp. 508–20.

[NE] Leibniz, *New Essays Concerning Human Understanding*, transl. A.G. Langley (London: Macmillan, 1896).

[NT] Russell, 'The Nature of Truth', *Mind*, 15 (1906), pp. 528–33.

[OD] Russell, 'On Denoting', in *CPBR4*, pp. 415–27.

[P] *Leibniz: Logical Papers*, ed. G.H.R. Parkinson (Oxford: Clarendon Press, 1966).

[POL] Russell, *A Critical Exposition of the Philosophy of Leibniz* (London: Allen and Unwin, 1975; 1st edn., 1900).

[POM] Russell, *The Principles of Mathematics* (London: Allen and Unwin, 1964; 1st edn., 1903).

[POM/D] Russell, 'The Principles of Mathematics, 1899–1900 Draft', in *CPBR3*, pp. 13–180.

[POM/DPI] Russell, 'Part I of the *Principles*, Draft of 1901', in *CPBR3*, pp. 184–208.

[RL] Leibniz, *The Monadology and Other Philosophical Writings*, ed. by R. Latta (Oxford: Oxford University Press, 1898).

[RNQ] Russell, 'On the Relations of Number and Quantity' (1897), in *CPBR2*, pp. 70–82.

[RWPL] Russell, 'Recent Work on the Philosophy of Leibniz' (1903), in *CPBR4*, pp. 537–66.

[SLBR1] *The Selected Letters of Bertrand Russell*, vol. I, *The Private Years*, 1884–1914 (London: Allen Lane, 1992).

2. Works by Other Authors

[Adams 2010] Adams, R.M., 'The Reception of Leibniz's Philosophy in the Twentieth Century', in G.A.J. Rogers, T. Sorell, J. Kraye (eds.), *Insiders and Outsiders in Seventeenth Century Philosophy* (London: Routledge), pp. 309–14.

[Beaney 2003] Beaney, Michael, 'Early Modern Conceptions of Analysis', *Stanford Encyclopedia of Philosophy*, *http://plato.stanford.edu/entries/analysis* (accessed 15 October 2007).

[Bradley 1893] Bradley, F.H., *Appearance and Reality* (Oxford: Oxford University Press, 1969).

[Caspari 1870] Caspari, O., *Leibniz' Philosophie beleuchtet vom Gesichtspunkt der physikalischen Grundbegriffe von Kraft und Stoff* (Leipzig: Voss).

[Class 1874] Class, G., *Die metaphysichen Voraussetzungen des Leibnitzischen Determinismus* (Tübingen: H. Lauppschen Buchhandlung).

[Connelly Rabin 1996] Connelly, James and Rabin, Paul, 'The Correspondence between Bertrand Russell and Harold Joachim', *Bradley Studies*, 2, pp. 131–60.

[Couturat 1901] Couturat, Louis, *La Logique de Leibniz d'après des documents inédit* (Paris: Alcan).

[Couturat 1902] —, 'Sur la métaphysique de Leibniz', *Revue de métaphysique et de morale*, 10, pp. 1–25.

[Cover O'Leary-Hawthorne 1999] Cover, J. and O'Leary-Hawthorne, J., *Substance and Individuation in Leibniz* (Cambridge: Cambridge University Press).

[Curley 1972] Curley, E.M., 'The Root of Contingency' in Frankfurt [1972], pp. 69–97.

[Dillmann 1891] Dillmann, Eduard, *Eine neue Darstellung der Leibnizischen Monadenlehre auf Grund der Quellen* (Leipzig: O.R. Reisland).

[Duncan 1901] Duncan, G.M., Review of *POL*, *Philosophical Review*, 10, pp. 288–97.

[Erdmann 1834–53] Erdmann, Johann Eduard, *Versuch einer wissenschaftlichen Darstellung der Geschichte de neuern Philosophie* (Riga und Dorpat: E. Frantzen, 1834–53), 6 vols. in 3.

[Erdmann 1840] —, Leibniz, *Opera Philosophica* (Berlin: Eichler).

[Fischer 1889] Fischer, Kuno, *Geschichte de neuern Philosophie* (Heidelberg: Battermann, 1889–93), 8 vols.

[Fischer 1889a] —, *G.W. Leibniz* (Heidelberg: C. Winter).

[Foucher de Careil 1854] Foucher de Careil, L.A., *Réfutation inédite de Spinoza par Leibniz* (Paris: L'Institut de France).

[Foucher de Careil 1855] —, *A Refutation Recently Discovered of Spinoza by Leibniz* (Edinburgh: Constable).

[Foucher de Careil 1857] , *Nouvelles Lettres et opuscules inédit de Leibniz* (Paris: Durand).

[Foucher de Careil 1859–75] —, *Oeuvres de Leibniz* (Paris: Didot), 7 vols.

[Frankfurt 1972] Frankfurt, H.G. (ed.), *Leibniz: A Collection of Critical Essays* (Notre Dame: University of Notre Dame Press, 1976).

[Frege 1892] Frege, Gottlob, 'On Concept and Object' in *Translations from the Philosophical Writings of Gottlob Frege*, ed. by P. Geach and M. Black (Oxford: Blackwell, 1977), pp. 42–55.

[Frege 1980] —, *Philosophical and Mathematical Correspondence*, ed. by G. Gabriel *et al.* (Chicago: University of Chicago Press).

[Fried 1988] Fried, Dennis, 'Necessity and Contingency in Leibniz', reprinted in Woolhouse (ed.) [1981], pp. 55–63.

[Griffin 1991] Griffin, Nicholas, *Russell's Idealist Apprenticeship* (Oxford: Clarendon Press).

[Griffin 1993] —, 'Terms, Relations, Complexes', in A.D. Irvine and G.A. Wedeking (eds.), *Russell and Analytic Philosophy* (Toronto: University of Toronto Press), pp. 159–92.

[Griffin 2007a] —, 'Some Remarks on Russell's Decompositional Style of Analysis', in Michael Beaney (ed.), *The Analytic Turn* (London: Routledge, 2007), pp. 75–90.

[Griffin 2007b] —, 'Bertrand Russell and Harold Joachim', *Russell: The Journal of Bertrand Russell Studies*, N.S. 27, pp. 220–44.

[Griffin forthcoming] —, 'Russell and Moore's Rebellion against Neo-Hegelianism', in M. Beaney (ed.), *The Oxford Handbook of the History of Analytic Philosophy* (Oxford: Oxford University Press).

[Ishiguro 1972] Ishiguro, Hidé, *Leibniz's Philosophy of Logic and Language* (London: Duckworth).

[Ishiguro 1972a] —, 'Leibniz's Theory of the Ideality of Relations', in Frankfurt [1972], pp. 191–213.

[Jevons 1874] Jevons, W.S., *The Principles of Science* (London: Macmillan, 2nd edn., 1879).

[Joachim 1906] Joachim, H.H., *The Nature of Truth* (Oxford: Oxford University Press).

[Jolley 1995] Jolley, Nicholas (ed.), *The Cambridge Companion to Leibniz* (Cambridge: Cambridge University Press).

[Jolley 2005] —, *Leibniz* (London: Routledge).

[Kulstad 1980] Kulstad, Mark A., 'A Closer Look at Leibniz's Alleged Reduction of Relations', *Southern Journal of Philosophy*, 18, pp. 417–32.

[Kvêt 1857] Kvêt, F.B., *Leibnitzens Logik* (Prague: F. Tempsky).

[Latta 1901] Latta, R., Critical Notice of *POL*, *Mind*, n.s. 10, pp. 525–33.

[Mates 1972] Mates, Benson, 'Individuals and Modality in the Philosophy of Leibniz', *Studia Leibnitiana*, 4, pp. 81–118.

[McCullough 1977] McCullough, Laurence B., 'Leibniz on the Ideality of Relations', *Southwestern Journal of Philosophy*, 8, pp. 31–41.

[Mertz 1884] Mertz, J.T., *Leibniz* (Edinburgh: Blackwood).

[Mondadori 1975] Mondadori, Fabrizio, 'Leibniz and the Doctrine of Inter-World Identity', 55 *Studia Leibnitiana*, 7, pp. 21–57.

[Moore 1899] Moore, G.E., Review of *EFG*, *Mind*, n.s. 8, pp. 397–405.

[Moore 1899a] —, 'The Nature of Judgment', reprinted in Moore [1986], pp. 59–80.

[Moore 1900] —, 'Necessity', reprinted in Moore [1986], pp. 81–99.

[Moore 1903] —, 'Kant's Idealism', reprinted in Moore [1986], pp. 233–46.

[Moore 1986] —, *The Early Essays*, ed. T. Regan (Philadelphia: Temple University Press).

[Mugnai 1992] Mugnai, Massimo, *Leibniz' Theory of Relations* (Stuttgart: Steiner).

[Mugnai forthcoming] —, 'Leibniz's Ontology of Relations: A Last Word?' in *Oxford Studies in Early Modern Philosophy*.

[O'Briant 1979] O'Briant, Walter H., 'Russell on Leibniz', *Studia Leibniziana*, 11, pp. 159–222.

[Parkinson 1965] Parkinson, G.H.R., *Logic and Reality in Leibniz's Metaphysics* (Oxford; Clarendon Press).

[Parkinson 1995] —, 'Philosophy and Logic', in Jolley (ed.), 1995, pp. 199–223.

[Peña 1991] Peña, Lorenzo, 'De la Logique combinatoire des *Generales Inquisitiones* aux calculs combinatoires contemporains', *Theoria* (San Sebastián), 2nd series, 6, no. 14–15, pp. 129–59.

[Rescher 1967] Rescher, Nicholas, *The Philosophy of Leibniz* (Englewood Cliffs, N.J.: Prentice-Hall).

[Rescher 1979] —, *Leibniz: An Introduction to his Philosophy* (Oxford: Blackwell).

[Selver 1885] Selver, David, *Der Entwickelungsgang der Leibniz'schen Monadenlehre bis* 1695 (Leipzig: Engelmann).

[Sorley 1875] Sorley, W.R., 'Leibnitz, Gottfrid Wilhelm', *Encyclopedia Britannica*, 9th edn, 1875–89.

[Stein 1890] Stein, Ludwig, *Leibniz und Spinoza* (Berlin: G. Reimer).

[Sylvan 2000] Sylvan, Richard, *Sociative Logics and their Applications*, ed. D. Hyde and G. Priest (Aldershot: Ashgate).

[Thilly 1891] Thilly, F., *Leibnizens Streit gegen Locke in Ansehung der angeborenen Ideen* (Heidelberg: J. Hörning).

[Tolley forthcoming] Tolley, Clinton, 'Humanizing Logic: Kant's Departure from Leibniz' (unpublished).

[Tufts 1902] Tufts, James H., Review of *POL*, *The American Journal of Theology*, 6, pp. 104–5.

[Watling 1970] Watling, John, *Bertrand Russell* (Edinburgh: Oliver and Boyd).

[Werckmeister 1899] Werckmeister, W., *Der Leibnizsche Substanzbegriff* (Halle a.S: M. Niemeyer).

[Wernicke 1890] Wernicke, F.G.F., *Leibniz' Lehre von der Freiheit des menschlichen Willens* (Würzburg: F.X. Bucher.

[Whitehead 1898] Whitehead, A.N., *A Treatise on Universal Algebra* (Cambridge: Cambridge University Press).

[Woolhouse 1981] Woolhouse, R.S. (ed.), *Leibniz: Metaphysics and Philosophy of Science* (Oxford: Oxford University Press).

Nicholas Griffin
Bertrand Russell Research Centre
Mcmaster University
1280 Main St. West
Hamilton, Ontario, Canada L8S 4M2
ngriffin@mcmaster.ca

Cassirer, Reader, Publisher, and Interpreter of Leibniz's Philosophy

Jean Seidengart

1. Cohen's legacy in the reading of Leibniz

As early as 1894, Cassirer began an in-depth study of the writings of Kant, Leibniz, and Descartes, and of Hermann Cohen as well. But it wasn't until the spring of 1896 that he began to attend Hermann Cohen's courses in Marburg. For Cohen, as for the young Cassirer, the philosophy of Leibniz constitutes one of the essential links in the chain of the history of idealism[1] (with Plato leading the line-up); it is even a privileged link that elucidates "Kant's relationship to his predecessors" ("*Kants Verhältnis zu seinen Vorgängern*").[2] Hence the particularly flattering terms used by Cohen when he looks briefly back at the Leibnizian epoch:

> Es ist der unschätzbare Wert des Leibnizischen Idealismus, dass er in dem Denken und der Vernunft gegenüber der Empfindung und der Sinnlichkeit schärfer und klarer als Descartes der Urgedanken des Platonismus regeneriert hat: dass die Natur im Bewusstsein entdeckt, die Materie im Denken konstruiert werden müsse.[3]

It is useless to point out what is excessive in Cohen's remarks, although he proved to be much more rigorous in his 1883 book, *Das Prinzip der Infinitesimal-Methode und seine Geschichte*, which contained the seeds of his entire philosophy. Cohen criticizes Gerhardt – whom he nevertheless congratulates for having unearthed Leibniz's philosophical and mathematical writings[4] – for his failure to understand the scope of his mathematical work,

---

[1] Hermann Cohen had planned to write and publish (at Bädecker) a history of idealism, which in fact never came to be. On this point, see Cohen's letter to Natorp dated 10 October 1885 (Holzhey 1986, II, 157–159; in particular 158).
[2] This is the title of a section of the introduction to the third 1918 edition of Hermann Cohen's famous book, *Kants Theorie der Erfahrung* (Cohen, 1918, 4).
[3] Cohen (1918), 57–58.
[4] "Material [. . .], welches Gerhardt's Herausgabe zugänglich gemacht hat" (Cohen 1883, 13).

apart from Leibniz's brilliant idea of "deriving the differential from the law of continuity" ("*Ableitung des Differentials aus dem Gesetz der Continuität*")[5] Cohen also criticizes Gerhardt for not having understood that "the meaning and the legitimacy of the concept of infinitesimal] are clearer in Leibniz than in Newton" ("*Das natürliche Recht und [der einzige] Sinn des [Infinitesimal-]Begriffs [...] werden deutlicher bei Leibniz als bei Newton*").[6] However, this meaning and this legitimacy are not enough to take the place of a foundation. In short, Cohen reproaches Leibniz's genius for not having provided a full foundation for differential calculus in his theory of knowledge, and thus for letting some obscurities remain.

Now it is precisely to be able to examine these obscurities that young Cassirer undertook a vast research project on Leibniz, for Cohen had already been asking him for some time about what subject he was going to choose for his doctoral dissertation. Above all, in the light of Gerhardt's recent publications that presented Leibniz's writings in an incredibly haphazard manner, scattered across a multitude of letters, opuscules, articles and memoirs, Cassirer sensed the need to try to reorganize them into an overall system. Given that his "Doktorvater" (thesis advisor) was writing the first part of his own system of philosophy, *Logik der Reinen Erkenntnis*, Cassirer naturally focused on the scientific foundations of Leibniz's system. This was the starting point of his book *Leibniz' System*, which he published in 1902.

Coming back to Cohen, apart from his historical approach in 1883 in which he shows himself to be quite unfavorable to Leibniz, we are forced to acknowledge that, starting from the second edition of *Kants Theorie der Erfahrung* in 1885, and especially from *Logik der Reinen Erkenntnis* of 1902, Cohen draws profound inspiration from Leibnizian philosophy. As we know, Hermann Cohen presents an entirely new interpretation of *Kritik der reinen Vernunft* in the second edition of *Kants Theorie der Erfahrung*.[7] The Kantian *a priori* has nothing to do with the psychological subjectivity that would necessarily lead back to a relativism, which in turn could only end in a *scepticism* that scientific knowledge would not be able to tolerate. Failing to fully understand the true nature of Kant's *a priori* amounts to condemning oneself to falling back on sterile oppositions between realism and idealism. The Kantian *a priori*, exempt from all empiricity (and thus seen as having a metaphysical status), has a *transcendental* meaning, for it has a *real* meaning to the extent that it is a *method* and a *principle* in the constitution of *experience*. Accordingly, the truth of a concept is constituted in its objectification. However, the various materials in Kant's *a priori* are not separated (if only for expository reasons); they are unified and totalized in the final unity of the principle of knowledge, namely, in the *transcendental apperception* of the "*I think*". It is the "*I think*" that unifies the *methods* of knowledge, and it does so with the help of *intensive magnitudes*. Indeed, *intensive magnitudes* achieve the dual genesis of the object and the internal sense, and thereby enable the transition from *pure consciousness* to *empirical consciousness*.

---

[5]Cohen (1883), 60.
[6]Cohen (1883), 12.
[7]Cohen (1885).

During his research phase for *Das Prinzip der Infinitesimal-Methode*, Cohen came to understand that the principle of intensive magnitudes played an essential role in Kant's first *Kritik*, and this allowed him to analyze the function of Leibniz's infinitesimal method revamped by the Königsberg thinker. Hence Cohen's enthusiastic praise of the infinitesimal analysis for having established the triumph of pure thought, and thus, of *idealism* in the mathematical science of nature:

> [...] die Physik ist den Weg der Mathematik gegangen, der zur *Hypothesis des Unendlichen* führte. Aus der Bewegung sollte das Seiende, Masse und Kraft zur Bestimmung gelangen. So fiel dem Begriff des Unendlichen die Aufgabe zu, das Seiende zu entdecken. Diese Entdeckung ist die wahrhafte, die wissenschaftliche Erzeugung. *Die Infinitesimal-Analysis ist das legitime Instrument der mathematischen Naturwissenschaft.* [...] *Diese mathematische Erzeugung der Bewegung und durch sie der Natur ist der Triumph des reinen Denkens.*[8]

It is in extending this fundamental idea that Cohen makes Leibniz into the "defender and representative of pure thought",[9] this herald (*"der Verkündiger"*) who opposed to Locke his *"intellectus ipse"*. Because of this and independently of Kant, it was Leibniz who led Cohen along the pathway of his "logic of origin" (*"Logik des Ursprungs"*) which, by that token, is the "logic of pure thought" (*"Logik der reinen Erkenntnis"*).[10]

Before leaving Hermann Cohen, let me mention that in spite of the many innovations he was able to derive from Leibniz's infinitesimal analysis, Cohen aligned in the end with the interpretation of Leibniz's philosophy that young Cassirer laid out in his *Leibniz' System* and in his *Erkenntnisproblem*, both of which were explicitly cited by the founder of the Marburg School in 1914.[11] Now let us go on to how Leibniz was received in Cassirer's works.

2. Cassirer's reuse of *Leibniz's system* in his *Erkenntnisproblem*

Cassirer's first great academic work dealt with Leibniz's system and scientific foundations. This was the subject of the thesis he proposed to tackle under the direction of Hermann Cohen at the University of Marburg. In a mere two years, Cassirer succeeded in accomplishing this genuine feat, the first part of which involved the study of Descartes' philosophy of knowledge, the second, the study of Leibniz's system. He obtained his doctorate with the highest distinctions (*opus eximium* and *summa cum laude*). Now it was precisely in December of 1900 that Cassirer sent the Berlin Academy of Science his doctoral work, to enter the competition on Leibnizian philosophy just organized by this institution.[12] Granted, he was awarded only second prize in 1901, but his biographer,

---

[8]Cohen (1902), 33; Cohen's emphasis.

[9]*"Also auch Leibniz ist Verteidiger und Vertreter des reinen Denkens, gerade weil er ein Neubegründer der Erfahrungswissenschaft ist"* (Cohen 1914, V, 61).

[10]Cohen (1902), 36.

[11]See Cohen (1914) V, 24.

[12]For the 1900 competition, the Berlin Academy of Science had put up the topic: The general presentation of Leibniz's system.

Dimitri Gawronsky, reminds us that the Berlin Academy of Science had already made the same evaluation error 140 years earlier when they awarded second prize only to Kant for his *Deutlichkeit*.[13]

A little later in 1902, Cassirer published his book on Leibniz, entitled *Leibniz' System in seinen wissenschaftlichen Grundlagen*.[14] Within the next year, Louis Couturat wrote a thorough review of this work that was both insightful and often quite lauditory, albeit somewhat critical.[15] However, Couturat was apparently mistaken on the meaning that Cassirer intended to give to his own research on Leibniz's philosophy, an error that is clearly excusable on the part of the French philosopher given that the young German philosopher was then only just starting his career in philosophy. Indeed, as Couturat writes in regards to the position Cassirer adopts in his *Leibniz' System*:

> Il [i.e., Cassirer] s'est proposé de définir la place qu'occupe Leibniz dans le développement historique de l'idéalisme, qui part de Platon pour aboutir à Kant ; et il a trouvé que Leibniz est beaucoup plus kantien, ou, si l'on préfère, que Kant est beaucoup plus leibnizien qu'on ne le croit, et qu'il ne le croyait lui-même (parce qu'il connaissait Leibniz surtout à travers la scolastique wolf-fienne). Il est ainsi amené à considérer Leibniz comme un criticiste avant la lettre, et à faire sortir tout son système d'une *critique* des sciences.[16]

The first sentence of this quotation is particularly true since it applies as well to one of Cassirer's greatest enterprises lasting over 36 years: his four-volume *Erkenntnisproblem*, the first volume being published in 1906 and the last not completed until 1942. The rest of this quotation suggests that Cassirer wanted to make Leibniz into a sort of "pre-Kantian", even though – as I shall try to show here – Cassirer's *neo-Kantism* strove instead to correct and rectify Kantism with the help of Leibnizianism, or rather, by means of Leibniz's philosophemes. This approach remains quite close to that taken by Hermann Cohen, from the second edition of his *Kants Theorie der Erfahrung*[17] until his *Logik*[18] of 1914, as we have seen above. Of course, before using Leibniz to "rectify" Kantian philosophy, it was necessary to rigorously study his philosophy, as Cassirer did right after his first academic work on Descartes.[19] It is not possible in this chapter to go into Cassirer's 1902 book on Leibniz. Instead, we shall focus on the long chapter devoted to him four years later in the publication of his first two volumes of *Erkenntnisproblem* (1906[1], 1911[2]). This will allow us to study a text in which Cassirer stood back a little from his first important work on Leibniz by taking into account certain criticisms directed at him, at a time when he was working on developing his first thematic book, *Substanzbegriff und Funktionsbegriff* (1910), where his own philosophy of science was to emerge.

---

[13] Gawronsky (1949), 12.
[14] Cassirer (1902).
[15] Couturat (1903).
[16] Couturat (1903), 83.
[17] Cohen (1885).
[18] Cohen (1902), 2nd edition 1914.
[19] See Cassirer, E., *Descartes' Kritik der Mathematischen und Naturwissenschaftlichen Grundlagen*, in Cassirer (1902).

At the end of the book review of Cassirer's *Leibniz' System*, Couturat was more direct in his critiques, condensing them into this concise and very enlightening statement:

> En résumé, M. Cassirer a eu le tort, selon nous, de chercher les principes du système, non pas dans la Logique, mais dans une *critique des sciences* dont Leibniz semble n'avoir jamais eu l'idée. [...] Il a toujours pensé, en vrai dogmatique qu'il était, que les lois de la raison sont identiques aux lois de la nature, et que les exigences de l'esprit concordent avec les conditions de la réalité, sans pour cela les fonder. [20]

It is clear that Cassirer read Couturat's review of his book very carefully. He answered him with a lengthy footnote in his chapter on Leibniz, saying that he still held the view he had adopted in his *Leibniz' System*, while nevertheless justifying himself in the face of the criticism:

> So sehr ich mit Couturat darin übereinstimme, daß die Logik des formalen *Grundriß* gebildet hat, nach welchem der Aufbau des Systems unternommen wurde, so sehr ist andererseits zu betonen, daß das *Material* für diesen Aufbau aus der Betrachtung der "realen" Wissenschaften, insbesondere aus den Problemen, die die neue *Analysis* darbot, gewonnen worden ist. Erst aus der Wechselwirkung dieser beiden Motive erklärt sich die Entstehung der Leibnizischen Philosophie. [21]

2.1. Leibnizian generalization of the concept of function

Cassirer purposely put the Leibnizian concept of *truth* at the center of Leibniz's system. He takes a very firm stand on this point – and is perfectly sincere in this – because, as we shall see, he will reuse the Leibnizian conception of truth in his own philosophy throughout his entire intellectual life. Cassirer states:

> Der Gehalt der Leibnizischen Philosophie wurzelt in den formalen Eigentümlichkeiten ihres Erkenntnisbegriffs und empfängt von hier aus erst sein volles Licht. [22]

Cassirer outwardly acknowledges that in Leibniz, logic has taken on an immense task since it is not limited to describing the formal connections of thought (*"die 'formalen' Verknüpfungen des Denkens zu beschreiben"*, but must also deal with the concrete contents of the knowledge itself (*"sie geht auf den sachlichen Gehalt des Wissens selbst"*)[23] by examining the logical grounds of necessary and contingent truths. Granted, Leibniz himself agreed that the human mind cannot analyze contingent truths all the way to the end because they are only approachable, but the path that allows one to approach them is itself precisely determined by general *rational* methods. In other words, one must start

---

[20]Couturat (1903), 99.

[21]Cassirer (1911), II, footnote 1, 141. Cassirer holds firmly to his positions with respect to Couturat, and does not fail to reassert this on several occasions. See for example Cassirer (1911), 164: *"Die Ergebnisse der besonderen Wissenschaften greifen hier bedeutungsvoll in die Entwicklung der allgemeinen [leibnizischen] Methodenlehre ein."*

[22]Cassirer (1911), II, 132.

[23]Cassirer (1911), II, 138.

from an *alphabet of thoughts* to build knowledge from a small number of simple elements.[24] Here is where Cassirer's work on logic and his work on combinatorics overlap. However, Cassirer deviates considerably from Couturat in stressing that scientific discoveries lead to rectifications or total revisions of the basic elements of knowledge, from which one can deduce new knowledge. It is as if scientific discoveries send feedback to the logical connections linking the ideas from which they arise, after which they must be reconstructed all over again.

More specifically, Leibniz is thought to have evolved profoundly in his theory of knowledge, going from a combinatory model (borrowed from arithmetic) to a "functional" model, that is, one transposed directly from the mathematical concept of *function* or *functional equation*. On this point, Cassirer writes:

> Allgemein ist es nunmehr der *Funktionsbegriff*, der sich an Stelle des *Zahlbegriffs* als der eigentliche Grund und Inhalt der Mathematik erweist. Der Gesamtplan der Universalwissenschaft erfährt damit eine charakteristische Umbildung. Wenn das Interesse bisher wesentlich an der Bestimmung der *Elemente* haftete, aus denen die zusammengesetzten Inhalte sich erzeugen sollten, so wendet es sich jetzt vor allem den *Formen der Verknüpfung* zu. Die verschiedenen Arten, wie wir in unserem Denken Inhalte wechselseitig durch einander bedingen, müssen an und für sich und ohne daß wir den materialen Gehalt der einzelnen Inhalte selbst ins Auge zu fassen brauchen, zum Gegenstand der Untersuchung gemacht werden.[25]

As a result, it is the very notion of *number* that is transformed, for unlike Aristotle and Euclid, a number is no longer seen as a "collection of units" nor as a sum of units, but as a *ratio*, i.e., a relation. This relegates Aristotle's logic to the rank of a mere elementary starting point that had to be surpassed in order to fall in line with the functional relations being brought to bear in the new sciences.[26] Accordingly, Leibniz goes on to say, among the new relations to be taken into account are those of the *Analysis Situs*, which are beyond the scope of geometric intuition. Of course Cassirer attached great importance to the *Analysis Situs*, which deals with genuine "relations of situation", missing in simple analytic geometry. Granted, by moving up to the *functional equation* for defining the properties of a figure, analytic geometry had already acquired a means for going beyond sensible figuration, but it continued to draw its elements from intuition, thereby retaining some conceptually non-elucidated theoretical elements. This is where *Analysis Situs* helps remedy the insufficiencies of analytic geometry:

---

[24]Cassirer (1911), II, 141.

[25]Cassirer (1911), II, 109. The concept of function is at the core of Cassirer's theory of knowledge: it is this concept that appeared in the title of his great book of 1910, *Substanzbegriff und Funktionsbegriff*. But he continued to restate its centrality until his last book on the philosophy of science, *Determinismus und Indeterminismus* (1936), 48: "Seit der Renaissance, seit Kepler und Galilei, Descartes und Leibniz hat sich das wissenschaftliche und philosophische Denken im mathematischen Funktionsbegriff das ideale Mittel geschaffen, um dieser Forderung zu genügen. Mit ihm ist eine universelle Form gegeben, in die ständig neuer Inhalt einströmen kann, ohne sie zu sprengen – ja ohne sie auch nur in ihren wesentlichen Zügen zu verändern." Not to forget Cassirer (1929), 335.

[26]Cassirer (1911), II, 110.

[Die analytische Geometrie operiert] statt mit der unübersehbaren Mannigfaltigkeit der Gestalten selbst, mit abgekürzten Zeichen für sie [...], die alle ihre Beziehungen enthalten und getreu wiedergeben. So gelangen wir hier zuletzt zu einer adäquaten, symbolischen Erkenntnis, die das Höchste ist, was wir innerhalb der Grenzen menschlicher Wissenschaft fordern oder erstreben können.[27].

Cassirer skims rather quickly over the new *analysis of the infinite*, which Leibniz had established by showing that it allowed him to resolve most of the difficulties surrounding the question of the "composition of the continuum". Cassirer points out, however, that *infinitesimal calculus* and the general theory of *functions* supplied the basis for taking an important step toward going beyond the *universal science* level, which planned to reduce everything to *numerical ratios:*

[...] so ist jetzt die reine Theorie und der allgemeine Kalkul der Funktionen als das eigentliche und tiefste Instrument für die Zahl- und Größenbestimmung erkannt. Jetzt erst hat die Frage nach der "Zusammensetzung" des Stetigen die scharfe und prägnante Fassung erhalten, die die Voraussetzung ihrer Lösung ist.[28]

All these inventions of Leibniz required philosophical elaboration of the "principle of continuity", which is a fundamental requirement of reason (and thus a methodological principle) for gaining access to the intelligibility of reality and providing a foundation for all forms of knowledge. On this subject, Cassirer quotes Leibniz's well-known letter to Varignon:

So kann man allgemein sagen, daß die gesamte Kontinuität etwas *Ideales* ist, daß aber nichtsdestoweniger das Reale vollkommen vom Idealen und Abstrakten beherrscht wird, so daß die Regeln des Endlichen im Unendlichen ... und umgekehrt die Regeln des Unendlichen im Endlichen ihre Geltung behalten. Denn alles untersteht der Herrschaft der Vernunft; anderenfalls gäbe es weder Wissenschaft, noch Regel, was der Natur des obersten Prinzips widerstreiten würde.[29]

Although the concepts of function and continuity were brought into the picture to complete Leibniz's theory of knowledge, thanks to the contribution of various scientific discoveries, Cassirer's condensed presentation of the philosophy of Leibniz insists that

---

[27]Cassirer (1911), II, 150. On this point, Cassirer quotes (in German) a passage from *De Analysi Situs* (GM V, 183): "Imaginationis ergo supplementum, et ut ita dicam perfectio in hoc, quem proposui, calculo situs continetur, neque tantum ad Geometriam, sed etiam ad machinarum inventiones, ipsasque machinarum naturæ descriptiones usum hactenus incognitum habebit."

[28]Cassirer (1911), II, 155.

[29]Cassirer (1911), II, 160. Note that Cassirer truncated this quotation in several places without indicating that he had done so. The letter to Varignon was written by Leibniz on the 2nd of February, 1702 (GM IV, p. 92). The untouched passage reads: "Toute la continuité est une chose idéale et il n'y a jamais rien dans la Nature qui ait des parties parfaitement uniformes, mais en récompense le réel ne laisse pas de se gouverner parfaitement par l'idéal et par l'abstrait, et il se trouve que les règles du fini réussissent dans l'infini." Cassirer quotes this letter again in *Freiheit und Form* (1916), footnote 1, 33.

the general truth concept ("*allgemeiner Begriff der Wahrheit*") is at the heart of this philosophy because it sheds light on the principal connection between form and matter of
knowledge ("*prinzipieller Zusammenhang zwischen Form und Materie des Wissens*").[30]
Now this view of *truth* is not only a salient feature of Leibniz's philosophy, it is also the
starting point and a central theme of Cassirerian theory of knowledge that marks off all
of his writings, from 1902 up to *An Essay on Man*, his last book (Cassirer, 1944). The
express admiration that Cassirer feels for the Leibnizian concept of function allows us
to assume that it was in Leibniz – or at least in a Leibniz interpreted by Cohen[31]– that
Cassirer discovered the fundamental thesis of *Substanzbegriff und Funktionsbegriff*:

> Leibniz geht von dem *Funktionsbegriff* der neuen Mathematik aus, den er als
> Erster in seiner vollen Allgemeinheit faßt und den er schon in der ersten Kon-
> zeption von aller Einschränkung auf das Gebiet der Zahl und der Größe be-
> freit. Mit diesem neuen Instrument der Erkenntnis ausgerüstet, tritt er an die
> Grundfragen der Philosophie heran.[32]

2.2. The Leibnizian concepts of truth and "expression"

Cassirer begins his presentation of the Leibnizian conception of truth (quoting many passages from Leibniz's famous letter to Arnauld on 9 October 1687) in the same terms as
those he had used to present his own view of truth at the beginning of the introduction
to his *Erkenntnisproblem*,[33] that is, as a means of going beyond and rejecting what one
might call the "copy" theory of truth:

> Die Ansicht, daß alle Erkenntnis das getreue *Abbild* einer für sich bestehenden
> Wirklichkeit sein muß, wird von Leibniz von Anfang an verworfen. Zwischen
> unseren Ideen und dem Inhalte, den sie "ausdrücken" wollen, braucht keinerlei
> Verhältnis der *Ähnlichkeit* zu bestehen. Die Ideen sind nicht die *Bilder*, son-
> dern die *Symbole* der Realität; sie ahmen nicht ein bestimmtes objektives Sein
> in all seinen einzelnen Zügen und Merkmalen nach, sondern es genügt, daß
> sie die *Verhältnisse*, die zwischen den einzelnen Elementen dieses Seins ob-
> walten, in sich vollkommen repräsentieren und sie gleichsam in ihre eigenen
> Sprache übersetzen.[34]

Truth is no longer described in terms of adequacy, as originally formulated by Saint
Thomas Aquinas,[35] at least not if we understand this to mean a likeness to a model, in
which case the mind is reduced to being but a sort of *duplicate*, a copy or double of reality.
With his theory of *expression*, Leibniz made it possible to throw away the naive theory
that saw knowledge as a simple reflection of reality, passively received by the mind. As
Cassirer says in regards to the Leibnizian concept of truth founded on expression: "*Der*

[30]Cassirer (1911), II, 166.
[31]Cohen (1902), § 22, 277: "*Seit Leibniz* aber bezeichnet die Funktion das Gesetz der *gegenseitigen Abhängigkeit* zwischen zwei veränderlichen Größen. Die Infinitesimal-Rechnung hat die Funktion in den Mittelpunkt der
mathematischen Methodik versetzt und zum zentralen Begriffe derselben erhoben."
[32]Cassirer (1911), II, 189.
[33]Cassirer (1911), I, 15–16.
[34]Cassirer (1911), II, 166–167.
[35]Saint Thomas Aquinas presented his view in *Quaestiones disputatae De veritate*, 1256–1259, q. I, a. 1).

erste und entscheidende Schritt zur Überwindung der 'Abbildtheorie' ist jetzt getan."[36]
Cassirer follows closely behind Leibniz when he takes up on his theory of expression
(understood as a *functional correspondence* between the projection of a perspective and
its geometric plane) but while stressing the mathematical nature of *expression*, which he
reduces to that of a *function*, whereas Leibniz deemed the latter to be just one possible
illustration among other forms of expression. Moreover, this is the very reason why Cassirer ends up calling his own conception of truth *"Funktionstheorie der Erkenntnis"*.[37] In
his book on Einstein's theory of relativity, Cassirer honors Leibniz for his radical transformation of the concept of *truth*:

> Die "Wahrheit" der Erkenntnis wandelt sich aus einem bloßen Bildausdruck
> zum reinen Funktionsausdruck. In der Geschichte der modernen Philosophie
> und der modernen Logik stellt sich diese Wendung in voller Klarheit zuerst
> bei Leibniz dar, wenngleich hier der neue Grundgedanke noch in der Fassung
> des metaphysischen Systems, in der Sprache des monadologischen Weltbildes,
> erscheint.[38]

However, as we can see in the above quotation, Cassirer regretfully deplores Leibniz's inclusion of this new view of truth in his metaphysical system. Consequently, it
was not until Kant's work that the Leibnizian conception of truth was finally rid of its
metaphysical assumptions and could at last demonstrate its full power. In other words,
Kant played the role of "purifier" by retaining only the epistemological facet of Leibniz's innovations, and in doing so he acted as a relay between Leibnizian metaphysics and
neo-Kantism:

> Diesen Leibnizischen Wahrheitsbegriff hat Kant aufgenommen und weiter-
> verfolgt, indem er ihn zugleich von allen unbewiesenen metaphysischen Vor-
> aussetzungen, die noch in ihm enthalten waren, zu befreien und abzulösen
> versuchte.[39]

One could legitimately wonder what became of Leibniz's theory of expression in
Kant. It seems that Cassirer considers Kant's concept of *object* to be the direct extension
of the Leibnizian theory of expression, in that the latter combines multiple data into a
unified perception or piece of knowledge.[40] Indeed, Kant defined the *object* as such in his
first Critique:

> *Objekt* aber ist das, in dessen Begriff das Mannigfaltige einer gegebenen An-
> schauung *vereinigt* ist. Nun erfordert aber alle Vereinigung der Vorstellungen
> Einheit des Bewußtseins in der Synthesis derselben. Folglich ist die Einheit
> des Bewußtseins dasjenige, was allein die Beziehung der Vorstellungen auf
> einen Gegenstand, mithin ihre objektive Gültigkeit, [...] ausmacht.[41]

---

[36]Cassirer (1911), II, 168.
[37]Cassirer (1921), 55.
[38]Cassirer (1921), 54.
[39]Cassirer (1921), 55.
[40]See Cassirer (1921), 55: "Auf diesem Wege ist er zu seiner eigenen Fassung des kritischen Gegenstandsbe-
griffs gelangt, in welchem nun die Relativität der Erkenntnis [...] behauptet wird."
[41]Kant (1787), § 17, B137; AK III, 111.

In short, the transcendental question of the relationship between a representation and its object is interpreted by Cassirer as the transposition of Leibniz's theory of expression to the framework of *Kantian relativism*. This Kantian "relativism" results firstly from eliminating the metaphysics of preestablished harmony, and secondly, from limiting knowledge to that of Kant's Copernican revolution.

Henceforth, the Leibnizian conception of truth, recast by Cassirer, is transformed into a sort of structural functionalism wherein the truth consists of a "functional structure-to-structure correspondence" (*"funktionale Entsprechung der beiderseitigen Struktur"*).[42] Cassirer sees this functional correspondence as having already explicitly existed in Leibniz, at the same time as he acknowledges that the latter posited an irreducible difference between *de facto* truths and the truths of reason, whereas the Marburg School tended to minimize this difference:

> [...] es bedarf [...] zwischen der Welt der Wahrheiten und der der Wirk-lichkeiten keiner materiellen, sondern lediglich einer 'funktionellen' Entspre-chung. [...] Zwischen dem Gebiet der Tatsachen und dem der reinen ratio-nalen Prinzipien besteht, bei aller Übereinstimmung der Grundstruktur, doch nicht minder eine dauernde Spannung und ein *Abstand*, der auf keiner Stufe wissenschaftlicher Erkenntnis jemals völlig aufgehoben werden kann.[43]

Cassirer also recognizes that Leibniz's philosophy is a metaphysics of individuality, and that single contingent facts can be subjected to an analysis that is "inexhaustible" because it can be pursued *ad infinitum*. Leibniz himself compares the difference that separates truths of reason from *de facto* truths, as Cassirer remarks, to the difference that separates rational numbers from irrational numbers.[44] Granted, Cassirer knows quite well that for Leibniz, every soul is capable of including, inside itself, the infinity of perceptions that encompass the entire universe, but, due to man's limited expression, this knowledge is *confounded*, being distinct only for God.[45] Despite this critical reservation, Cassirer insists on the *rational* conditions for apprehending the *de facto* truths of which Leibniz speaks:

> [...] die empirische Wahrheit einer einzelnen Erscheinung [beruht] einzig auf ihrer *harmonischen Zusammenstimmung* mit den reinen Vernunftregeln und der Gesamtheit aller übrigen Beobachtungen.[46]

The preestablished harmonious accord between monads in Leibnizian metaphysics, i.e., between substantial subjects, is redefined by Cassirer, epistemologically, in terms of *function*.[47] Even though this term is not inaccurate with respect to Leibniz's writings, it must be acknowledged that Cassirer makes an ongoing effort not to "deontologize" the

---

[42] See Cassirer (1910), 404.

[43] Cassirer (1911), II, 179.

[44] Cassirer (1911), II, 133.

[45] See Leibniz, G. W., *Principes de la Nature et de la Grâce*, 1714, § 13, GP VI, 604; edition Frémont, GF, Paris, 1996, p. 231: "Chaque âme connaît l'infini, connaît tout, mais confusément. [...] Dieu seul a une connaissance distincte de tout ; car il en est la source."

[46] Cassirer (1911), II, 185.

[47] Cassirer (1911), II, 185–186: "daß [die verschiedenen vorstellenden Subjekte] in ihrer reinen *Funktion*, in der Kraft der Vorstellungserzeugung, aufeinander abgestimmt sind und miteinander in Zusammenhang stehen."

presentation of his philosophy by consistently using philosophemes taken mainly from the theory of knowledge. This is in fact quite appropriate in a book devoted to the "problem of knowledge"; it is much less so in his 1902 book dealing with the totality of the "*System*", in which he dedicates only three chapters of Part 3 to metaphysics. On the other hand, in the first two volumes of Leibniz's *Hauptschriften*, published in translation by Cassirer in 1904, nearly half of the papers chosen had a metaphysical theme.[48] Should this be seen as reflecting a desire to compensate for the brevity of *Leibniz' System* on this topic? Judging from Cassirer's book on Leibniz, with its emphasis on the concept of relation, one might end up thinking that for Leibniz, *the relation* takes priority over *the being*, although his metaphysics asserts exactly the opposite in that it remains attached to an ontology of the "*inesse*". In Leibniz's ontology of predication, the existence of substances takes precedence over their relations, although his theory of knowledge implies the contrary. Thus, in a neo-Kantian reading of Leibniz's philosophy, there is a sort of tension between his theory of knowledge and his ontology.

3. About the Leibnizian concept of symbol in *Philosophie der symbolischen Formen*

The very first time the expression "symbolic form" was to appear in Cassirer's writings was in a footnote in his publication of Leibniz's *Hauptschriften*, where he annotates Leibniz's fifth letter to Samuel Clarke.[49] Of course, Cassirer doesn't yet use this expression in the sense he will give it in his *Philosophie der symbolischen Formen*, the idea for which, Dimitri Gawronsky tells us, first came to mind on a certain memorable day in 1917.[50]

It is not in his work on Leibniz that Cassirer got the idea for his philosophy of symbolic forms, although he was deeply inspired by the Leibnizian conception of *symbol* and the role he had it play in the cognition process. As early as 1906, Cassirer considers the establishment of the functional theory of knowledge, underlain at the most basic level by a philosophy of the symbol, to be one of the main consequences of the scientific revolution.[51] The symbol is indeed what acts as a substitute for – or a representative of – the real, and the symbol is what *mediates* the relationship between knowledge and its object, even though there is no resemblance between the representative and the entity it represents. It is precisely in Leibniz that Cassirer began to find a general framework for a theory of symbolization:

---

[48] See Cassirer/Buchenau (1904–1924) I, 285–374; II, 3–488) Added to this are the 40 pages that Cassirer devoted to the introduction of the metaphysical texts in his Volume 2, p. 81–122.

[49] See Cassirer/Buchenau (1904–1924), I, footnote 114, (126, footnote 117 in the new 1996 edition): "Die sinnliche Welt der Phänomene ist zwar kein 'Abbild' der einfachen Monaden, dennoch aber finden sich in ihr gewisse Verhältnisse und Beziehungen wieder, die den Grundrelationen, die wir in den einfachen Elementen denken, in bestimmter Weise entsprechen und sie uns gleichsam in symbolischer Form darstellen: 'les composés symbolisent avec les simples'." We owe Massimo Ferrari for having first brought up this point in the studies of Cassirer. See Ferrari (1996), 174–175.

[50] See Gawronsky (1949), 25.

[51] Cassirer (1911) *Einleitung*, 3: "Die Begriffe der Wissenschaft erscheinen jetzt nicht mehr als Nachahmungen dinglicher Existenzen, sondern als Symbole für die Ordnungen und funktionalen Verknüpfungen innerhalb des Wirklichen."

Die echte Wirklichkeit kann nicht auf einmal ergriffen und abgebildet werden, sondern wir können uns ihr nur in immer vollkommeneren *Symbolen* beständig annähern. Noch einmal tritt somit die zentrale Bedeutung dieses Begriffs für das Ganze der Leibnizischen Lehre deutlich heraus. Der Wert, den der Gedanke der allgemeinen Charakteristik für das System besitzen muß, bestimmt sich nunmehr genauer. Es ist kein Zufall, der uns dazu drängt, die Verhältnisse der Begriffe durch Verhältnisse der "Zeichen" zu ersetzen; sind doch die Begriffe selbst ihrem Wesen nach nichts anderes als mehr oder minder vollkommene Zeichen, kraft deren wir in die Struktur des Universums Einblick zu gewinnen suchen.[52]

This highly particular attachment of Cassirer to the Leibnizian theory of signs and symbols was maintained over time, since fourteen years later in *Die Philosophie der symbolischen Formen* he again draws from Leibniz to broaden his definition of the symbol. Moreover, Cassirer does not define his concept of symbol based on work in linguistics, but starts from logic, mathematics, and physics. Although Volume 1 of *Die Philosophie der symbolischen Formen* is devoted to the study of *language*, Cassirer starts from the theory of knowledge in the physical and mathematical sciences (of Galileo and Leibniz) to describe the nature and function of the symbolic:

Die scharfe Erfassung der Grundbegriffe der Galileischen Mechanik gelang erst, als durch den Algorithmus der Differentialrechnung gleichsam der allgemein logische Ort dieser Begriffe bestimmt [...] war. Und von hier aus, von den Problemen, die mit der Entdeckung der Analysis des Unendlichen zusammenhingen, vermochte Leibniz alsbald das allgemeine Problem, das in der Funktion der Zeichengebung enthalten ist, aufs schärfste zu bestimmen, vermochte er den Plan einer universellen "Charakteristik" zu einer wahrhaft philosophischen Bedeutung zu erheben.[53]

Like Leibniz, Cassirer knows quite well that there are radical differences between formal languages and natural languages. In addition, as the first volume of *Die Philosophie der symbolischen Formen* shows, he never thought it was possible to disregard the great diversity of the natural languages, as we can see from the large corpus of ethnolinguistic material he used. However, Cassirer thinks that the mathematical sciences of nature are the best place for grasping the relations between a functional equation and the magnitudes it is capable of coordinating in the physical world. In this respect, he stays totally inside the philosophical orbit of Hermann Cohen. More specifically, Cassirer thinks that the functional laws of classical physics, thanks to their symbolism, are not only *transcendent* to the different numerical values of the physical magnitudes they coordinate (because, as the law of the series, the function $\phi(x)$ *transcends* the values taken on by the variable), but also, by virtue of the fact that the law is the common ratio of the infinite series, it is also *immanent* in them. It is true that the various numerical values of the variable can be given (at least partially) by careful observation and accurate measurement: this is the *empirical* moment of knowledge. Concerning the form of the function ϕ, it possesses

---

[52]Cassirer (1911) II, 187.
[53]Cassirer (1923), 17.

its own consistency by virtue of its being the *rational* moment. Cassirer's full originality lies in how he transforms this clear-cut opposition between the empirical and the rational, into a transcendental correlation. This is the true meaning of this symbolic transcendence:

> Es bleibt freilich dabei, daß die Funktion ϕ und die *Werte* der Variablen einem ganz verschiedenen Denk*typus* angehören, und daß sie sich somit niemals aufeinander reduzieren lassen: aber diese Irreduzibilität besagt andererseits keineswegs, daß sie voneinander *trennbar* wären. [...] Die Funktion 'gilt' für die Einzelwerte, eben weil sie kein Einzelwert 'ist' – und andererseits 'sind' die Einzelwerte nur, sofern sie zueinander in der durch die Funktion ausgedrückten Verknüpfung stehen.[54]

Cassirer's idealism consists of putting the *form* of the functional relation before the numerical values it integrates and coordinates, because, by virtue of its legality, it has a *universality* and a *necessity* that an isolated empirical case never has. Yet it is precisely this necessity and universality that, for Cassirer, are the distinctive features of objective reality, not the existence of contingent, singular cases.

In conclusion, despite Cassirer's strong intellectual attachment to the philosophy of Leibniz, particularly his work on the "universal characteristic", the thing that all the members of the Marburg School considered unacceptable when it came to accounting for the applicability of mathematics to experience, was Leibniz's reliance on a "preestablished harmony". This is the point at which Cassirer abandons Leibnizian metaphysics to turn to the Kantian approach:

> [...] das Problem der *Anwendbarkeit* der Mathematik hat im Aufbau von Leibniz' System keine Stelle. Eben dieses Problem aber ist es, das Kant, schärfer als je zuvor, stellt, und aus welchem ihm die endgültige Gestalt seiner 'kritischen' Lehre wächst. Er verwirft die dogmatische Entscheidung der 'prästabilierten Harmonie'; er fragt nach dem Grund der Möglichkeit der Übereinstimmung zwischen apriorischen Begriffen und empirischen Tatsachen.[55]

While Cassirer moves away from preestablished harmony and Leibnizian metaphysics to search in transcendental philosophy for non-dogmatic answers to the problem of knowledge, he is well aware that modern mathematics, as he says, "followed the pathway laid out by Leibniz, not by Kant" (*"nicht auf dem von Kant, sondern auf dem von Leibniz gewiesenen Wege weitergeschritten ist"*).[56] This is the most glowing tribute that Cassirer could pay to Leibniz.

References

[Cassirer 1902] Cassirer, E., *Leibniz' System in seinen wissenschaftlichen Grundlagen*, Marburg, Elwert 1902.

[Cassirer 1910] Cassirer, E., *Substanzbegriff und Funktionsbegriff: Untersuchungen über die Grundfragen der Erkenntniskritik*. Berlin, Bruno Cassirer 1910.

---

[54]Cassirer (1929), 381.
[55]Cassirer (1923), 423.
[56]Cassirer (1929), 402.

[Cassirer 1911] Cassirer, E., *Das Erkenntnisproblem in der Philosophie und Wissenschaft der neueren Zeit*, Berlin, Bruno Cassirer ¹1906, ²1911.

[Cassirer 1916] Cassirer, E., *Freiheit und Form*, 1916.

[Cassirer 1921] Cassirer, E., *Zur Einsteinschen Relativitätstheorie*, Berlin, 1921. Reprinted in Darmstadt.

[Cassirer 1923] Cassirer, E., *Die Philosophie der symbolischen Formen. I. Die Sprache*, 1923.

[Cassirer 1929] Cassirer, E., *Die Philosophie der symbolischen Formen. III. Phänomenologie der Erkenntnis*, 1929.

[Cassirer 1936] Cassirer, E., *Determinismus und Indeterminismus*, 1936.

[Cassirer 1944] Cassirer, E., *An Essay on Man. An Introduction to a Philosophy of Human Culture*, 1944.

[Cassirer/Buchenau 1904–1924] Cassirer, E. and Buchenau, A., *Leibniz: Hauptschriften zur Grundlegung der Philosophie*. 2 Vol. Leipzig 1904–1924.

[Cohen 1871] Cohen, H., *Kants Theorie der Erfahrung*. Berlin 1871.

[Cohen 1883] Cohen, H., *Das Prinzip der Infinitesimal-Methode und seine Geschichte*, Berlin, 1883.

[Cohen 1885] Cohen, H., *Kants Theorie der Erfahrung*. 2nd edition Marburg 1885.

[Cohen 1902] Cohen, H., *Logik der Reinen Erkenntnis*, 1902. 2nd edition Berlin: Bruno Cassirer 1914.

[Cohen 1914] Cohen, H., *Einleitung mit kritischem Nachtrag zur "Geschichte des Materialismus" von F. A. Lange*, ⁵1914. Reedited by H. Holzhey, Hildesheim, Olms, 1984.

[Cohen 1918] Cohen, H., *Kants Theorie der Erfahrung*. 3rd edition Berlin: Bruno Cassirer 1918. Werke I,1 Olms Hildesheim etc. 1987

[Couturat 1903] Couturat, L., *"Le système de Leibniz d'après M. Cassirer"* in: *Revue de Métaphysique et de Morale* XI (1903), 83–99.

[Ferrari 1996] Ferrari, M., *Ernst Cassirer, Dalla scuola di Marburgo alla filosofia della cultura*, Florence, Olschki 1996.

[Gawronsky 1949] Gawronsky, D., "Ernst Cassirer: His Life and His Work", in: *The Philosophy of Ernst Cassirer*, collective volume edited by P.A. Schilpp, Open Court, La Salle, Illinois 1949. Reprinted 1973.

[Holzhey 1986] Holzhey, H., *Cohen & Natorp*. Basel: Schwabe & Co. 1986.

[Kant 1787] Kant, I., *Kritik der reinen Vernunft* (B), 1787.

Jean Seidengart
Institut de Recherches Philosophiques (EA 373):
"Les dynamiques de l'invention philosophique,
scientifique & artistique"
Université Paris-Ouest-Nanterre-La Défense
jean.seidengart@sfr.fr

Leibniz on Relativity. The Debate between Hans Reichenbach and Dietrich Mahnke on Leibniz's Theory of Motion and Time

Vincenzo De Risi

1. Introduction

One of the most significant episodes in the early-twentieth-century reception of Leibniz was the reading of his works on natural philosophy as an anticipation of Einstein's Theory of Relativity. In a way, such an interpretation might have been meant to uplift the fame of Leibniz as a scientist, which had been, it was said not without a touch of chauvinism, overshadowed for too long a time by the evershining Newtonian star; as well as to give evidence of the extraordinary forerunning talent of the philosopher, which had, as it was just then being discovered and believed, most clearly manifested itself not only in the edition of his logical work but also in many other fields of science; and even, perhaps, to burnish Einstein's new-born theory through the authority of a highly reputed metaphysician of the past that could inaugurate it. Certainly, it was not the first time that Leibniz's fortune happened to gain so much from misunderstandings and false trails, nor was it probably the last one. Be that as it may, in the 1920s the name of Einstein, at least in philosophical circles in Germany, was often mentioned in connection with that of Leibniz. At times, the forerunning attitude of Leibniz took on the grotesque persona of a great thinker who had done nothing but hint to of speak almost metaphorically about a lot of brilliant ideas that others would later profit from. And the Theory of Relativity was one of the issues in which, I think, such an attitude proved particularly harmful. In fact, instead of producing, as fortuitously but felicitously happened elsewhere, fruitful insights of philosophers and scientists who resumed or even crafted anew Leibniz's conjectures, here the comparison with Einstein mostly produced a worse understanding of the extraordinary innovation that Relativity brought to physics, and an almost total concealment of the historical and epistemological meaning of Leibniz's dynamics. Relativists were too busy developing such a promising science to devote themselves to a thorough historical investigation, while good historians got entangled in the difficulties of tensor calculus and had to leave their theoretical endeavor unaccomplished. This all resulted in more confusion than clarity, and in several misunderstandings that lasted for a long time among physicists and philosophers as well.

In this picture, however, there were a few remarkable exceptions concerning authors who succeeded in mastering both the technical difficulty of the new physical theory and the Leibnizian historical sources. One of them was Cassirer, who, coming from Leibnizian studies and being almost neo-Leibnizian himself, managed to write an all-important book on the philosophical meaning of Einsteinian Relativity– in which actually, and wisely so, he did not straightforwardly see any necessary development of Leibniz's dynamic theory.[1] However, the best by far contribution in this direction was undeniably the one offered by Hans Reichenbach. In the 1924 issue of *Kant-Studien*, he published an important and well-known essay on the theory of motion in Huygens, Leibniz, and Newton, in which, among other things, he compared Leibniz's natural philosophy with Relativity Theory.[2] At the time, Reichenbach was a distinguished, very qualified interpreter of the philosophical implications of the new physics, and he had already dealt with Relativity in two important monographs, *Relativitätstheorie und Erkenntnis Apriori* in 1920, and *Axiomatik der relativistischen Raum-Zeit-Lehre* in 1924, as well as half a dozen specialized articles; his historical penetration of Leibniz's texts, however imperfect, also turned out to be far better than that of many of his contemporaries.[3]

The 1924 article called the attention, among others, of Dietrich Mahnke, who knew little about Einstein's theory, but much about Leibniz, and had already acquired a wide and well-deserved reputation as one of the best Leibniz scholars of his generation. In 1917 he had written a daring book, much more theoretical than historical, entitled *Eine neue Monadologie*, in which he attempted, not without a few naïve claims, to put Husserl and Leibniz together; and near the end of 1924 he was reviewing the proofs of his historical book on Leibniz, *Leibnizens Synthese von Universalmathematik und Individualmeta-*

---

[1][Cassirer 1920]. Cassirer was, moreover, too learned a Leibniz expert to be unaware of the "non-relativistic" loci in Leibniz's physical writings, which in fact he hastens to mention in his essay on Relativity in order to show how the concept of relativity of space and motion had not yet reached full maturity in the eighteenth century. Cassirer will possibly detect some general affinity between Leibniz and Einstein in their methodological views, but not their strictly physical doctrines; cf. [Cassirer 1943].

[2][Reichenbach 1924a]. Here Reichenbach complains (p. 417) that no one had realized until then how much Einstein's theory was indebted to Leibniz's natural philosophy, not even Einstein himself, who had always disputed and fought Newton's theory, nor "the first relativist of the new era", Mach, who had devoted to Leibniz little more than a contemptuous remark on his superabundant metaphysics. Reichenbach exaggerates, because Cassirer had already written his essay four years before, nor was he the only one to speak about it (see for example [Enriques 1922]); it is however true that it was Reichenbach's article that ignited a "plenary" discussion in the academic world about Leibniz and Relativity. His criticism of Mach also is a little ungenerous; in fact, in all likelihood, Reichenbach himself might have been prompted to write on Leibniz and Huygens's theory of motion just because in the *Mechanik* ([Mach 1883]) Newton's theory was juxtaposed to that of these other two scientists.

[3]Besides the two books [Reichenbach 1920b] and [Reichenbach 1924c], see at least [Reichenbach 1920a, 1921, 1922a, 1922b, 1922c, 1924b]. For a historical framing of Reichenbach's contributions on Relativity, see [Hentschel 1990], pp. 358–64.

physik, which was to be published the next year.[4] In the very same weeks, and precisely on Christmas Eve of 1924, Mahnke read Reichenbach's essay in *Kant-Studien* and immediately wrote to him for remarks and questions. Reichenbach answered with a polite letter, and Mahnke replied by expanding at a certain length on his own interpretation of Leibniz's philosophy of nature. These three letters, which do not seem to have been followed by others, are kept today in the *Archives for Scientific Philosophy* in Pittsburgh, and the reader can find a transcription of them in the Appendix.[5]

Taking my cue from the brief but animated exchange between Reichenbach and Mahnke concerning Leibniz and Relativity Theory, I will here attempt to highlight a few important elements of contact, but mainly of distance, between Reichenbach's reading of Leibniz, which is somehow halfway between neo-Kantianism and empiricism, and Mahnke's, which verges, as it were, towards an imperfect phenomenology. Under many respects, in fact, these two interpretations are but the opposite poles between which all other exegeses of Leibniz's scientific and metaphysical work can be placed, while the very narrowness and technicality of this issue in Leibniz's natural philosophy makes it easier to focalise differences and similarities. It is also evident that both authors, Mahnke and Reichenbach, owe to the comparison with Leibniz the full formulation of their own philosophical views and even the parabola of their ideas on the background of the philosophical world of the time. Therefore, we can hope that the cross-reading of their two interpretations will possibly shed some light both onto the theories of Leibniz and those of two of the main philosophical schools of the early twentieth century.

It will be worth briefly introducing the two protagonists of the correspondence (§§ 2 and 3), and then approaching the main interpretive differences between Reichenbach and Mahnke. Two major issues are dealt with in their letter exchange: relativity of motion and causal theory of time, both of them strongly linked in turn to different philosophical interpretations of Relativity Theory. We are dealing with the first in §§ 4–9, and with the second in §§ 10–16.

2. Reichenbach

Hans Reichenbach (1891–1953) is too well known a thinker to need here any profile of life and doctrine. However, it may be interesting briefly to recall his intellectual orientation in 1924. At the time, he was relinquishing the (nearly) neo-Kantian stance espoused in his young years. His 1920 book on Relativity and *a priori* knowledge was vaguely reminding of neo-Kantianism and the Marburg School, and it attempted to reconsider Kantian transcendentalism in the light of Einstein's theory. The attempt, actually, was only half successful, so that on the one hand Reichenbach kept something from Kant's lexical and stylistic features but on the other hand he went explicitly beyond all reasonable orthodoxy

---

[4][Mahnke 1917] and [Mahnke 1925].
[5]The signature of the correspondence is HR-016-37-08, 09, 10. These papers are published here with the permission of the University of Pittsburgh (all rights reserved). I wish also to thank Ralf Krömer for helping me with the transcription of the letters.

by boldly reinterpreting the meaning of apriorism. Which at the time was not enough, apparently, for one to cease being Kantian, as in fact also Cohen, Natorp and Cassirer had no qualms about recognizing that in Kant there was still something alive, but undeniably much was dead; and that the spirit of the *Critique*, not its letter, should be preserved. This attitude particularly shows in the neat refusal of all psychologism in the Marburg reading of Kant's work. The *Critique of Pure Reason* is not about consciousness, but about science: sensibility, imagination, and understanding are but ill-assigned names for logical functions; phenomena and transcendental illusions are the objects of an epistemological theory, by no means of a psychology (be it transcendental or not); space and time have nothing to do with the physiology of perception, but only with the general conditions of natural science; and so on. In Cohen's words, the purpose of the *Critique* is not to explain what occurs in my mind, but what is to be found in a handbook of physics.[6] Reichenbach was more cautious, as he did not think that one could dispense with all the elements of phenomenology of consciousness and yet preserve Kant. In his 1920 book, he also indulges quite often in psychological or even physiological considerations; it is however already evident that the only dimension he is really interested in is the strictly epistemological one. It is also clear, on the other hand, that he must have misunderstood the sense of Kant's transcendental endeavor from the outset, by sacrificing it to the principles of empiricism.

In any case, it was indeed the year 1920 that gave Reichenbach the occasion to reconsider his interpretation. His book elicited in fact the interest of Cassirer, who wrote him a letter to praise it, however remarking that his reading of Kant was definitely too much on the psychologistic side.[7] And his book also elicited the attention of Moritz Schlick, who wrote him many important letters in which he (successfully) attempted to convince Reichenbach that his "relativized *a priori*" was absolutely incompatible with Kant's original project in every possible interpretation of it, and that the constitutive principles according to which Reichenbach reread Kant's transcendental principles were, if truth be spoken, nothing but *conventions*.[8] It so happened, therefore, that, following the suggestions of Cassirer and his own demon, Reichenbach completely abandoned psychologism but along with psychologism, following Schlick's opinion, also Kant. In his second book on Relativity, the one from 1924, Reichenbach already explicitly spoke of conventions instead of *a priori* principles, and exhorted the reader to keep logical conditions of science, the only interesting ones, well apart from merely psychological ones. In his writings from those very same years, in effect, with a neophyte's heat, rarely does Reichenbach miss the opportunity to speak disparagingly of Kant and the many problems

---

[6]Naturally, Natorp's stance was much more nuanced than Cohen's, and more open to a phenomenological reading of the *Critique*. On this point, however, Cassirer seems more to follow Cohen than Natorp. On some neo-Kantian readings of Relativity Theory, see [Ferrari 1995].

[7]Cassirer's letter is also in Pittsburgh, signature HR-015-50-09. It is discussed in [Ryckman 2003].

[8]HR-015-63-22. On this exchange, see [Coffa 1991] and [Friedman 1994]. In the same years, Schlick reasserted also publicly his stance in [Schlick 1921].

of Criticism, all tied as they were to a psychological and phenomenological dimension of the theory rather than to a merely logical-epistemological one.[9]

It is in this period that Reichenbach's encounter with Leibniz takes place. As had been the case with other Marburg neo-Kantians and Cassirer himself, Reichenbach finds in Leibniz a Kant perfectly divested of psychologistic elements. He recognizes in him a spirit similar to Kant, a similar attention to the theory of science, a kind of idealism not incompatible with empiricism but somehow allied with it, and even a number of other very Kantian theories. Here, however, everything appears in the light of pure logic, while there, in Kant, everything was messed up by psychology, the Ego and the faculties of the mind, and apperception, intuition, will, sensibility. Leibniz, thus, such a dephenomenologized Kant, will possibly be the forefather of the philosophical endeavor of Berliner neo-Empiricism at its beginnings.

Reichenbach, of course, could not ignore that a very great quantity of Leibniz's philosophical writings were devoted to a theory of consciousness and sensibility, let us say, thus, to phenomenology; nor that the author of the *Specimen Dynamicum* and inventor of the Calculus had also written the *Nouveaux Essais*. The main point, however, according to Reichenbach, was that Leibniz's phenomenology did not intrude on his philosophy of nature. There was nothing, he thought, in Leibniz's research on dynamics, that might refer to his theory of sensibility; there might have been an excessive amount of metaphysics, and some overly present mentions of God (not so many, actually, and even those anyway occurring more out of pretence than substance) but nothing at all that referred to the psychological structure of consciousness. As a natural philosopher, Leibniz went about things just as a genuine physicist and fashioned a markedly physicalistic dynamics. The opposite could be said of Kant, whose *Metaphysische Anfangsgründe der Naturwissenschaft* seemed to get you into hopeless intricacies unless you had already digested the whole of the *Critique* and the system of the faculties of the mind. In Reichenbach's view, Kant's endeavor of natural philosophy is structurally phenomenological, not physicalist, and therefore should be decidedly rejected.

It is in this way, that is, looking at Leibniz as if he were a physicalist scientist and an enemy of any phenomenology in the doctrine of nature, that Reichenbach approaches the problem of the relationship between seventeenth-century dynamics and General Relativity in his 1924 essay on the theory of motion.

3. Mahnke

At this point, it will be readily understood that the interpretation of Leibniz given by the phenomenologist Dietrich Mahnke (1884–1939) must be radically opposite to that of Reichenbach.[10] Mahnke studied with Husserl in Göttingen in the first years of the century and always regarded himself as a faithful disciple of his; between the two, there was a long

---

[9]See for example [Reichenbach 1924a], p. 426. Reichenbach's disapproval of psychologism was addressed against the Critical perspective of the Kantians of his time, but also against Mach's Empirio-Criticism (and, say, physiologizing Positivism).

[10]For a biography of Mahnke, see [Woltmann 1957] and [Mancosu 2005]. For a brief account of his monadological doctrine, see [Poser 1986].

and important correspondence.[11] As early as 1912, Mahnke had started publishing both historical and theoretical essays on Leibniz, who was soon to become the center of his philosophical interests.[12]

Mahnke's interpretation of Leibniz developed, at least until the mid-1920s, along the lines that Husserl himself had here and there suggested in his works, but the teacher also intellectually profited back from the historical studies of his disciple. Just to sketch it in a simple way, Husserl's interpretation of Leibniz seems to have passed through three stages.[13] In the last years of the nineteenth century, Husserl, engrossed as he was in preparing for the *Logische Untersuchungen*, must have mainly regarded Leibniz as an antipsychologist logician and the author of a project of *mathesis universalis* that he found close to his own research.[14] Mahnke's early essays on Leibniz mostly deal with such topics of logic and combinatorics. However, starting with the transcendentalist turn of Husserl's thought, that is, with the years in which phenomenology expanded itself into a general theory of consciousness, Leibniz's role also partly changed. Now, Leibniz began to be seen as a great forerunner of the theory of intentionality, and along with Leibniz's logical works Husserl studied and quoted his writings on epistemology and the theory of sensibility. In many places the *Ideen* of 1913 show Husserl's keen interest in Leibniz's theory of knowledge.[15] This seems to me to be the period of the greatest ever contiguity between Husserl and Leibniz, and in any case the most creative period of Mahnke, who comes to share his teacher's views and particularly discusses the Leibnizian concepts of expression and idea. Mahnke's most important theoretical work, the above mentioned *Neue Monadologie* from 1917, is mostly an ample rereading of Leibniz's *Monadologie* in the light of Husserl's theory of intentionality:

> Das Erleben einer Monade kann am besten mit einem scholastischen Termi-
> nus, den Franz Brentano wieder zu Ehren gebracht hat, charakterisiert werden
> als eine intentio, die über den augenblicklich gegebenen Erlebnisinhalt hinaus
> gerichtet ist auf einen dem Gewühl des Gegebenen Sinn verleihenden Erleb-
> nisgegenstand. Er gehört zum Wesen jeder bewussten Sinneswahrnehmung,
> wie die Phänomenologie in Sinne Husserls feststellt, dass sie ein Ding wahr-
> nimmt, von dem nur einige Eigenschaften durch sinnliche Qualitäten präsen-
> tiert sind.[16]

In a letter to Mahnke from the same year, Husserl admires his critical and historical investigation on Leibniz's thought; he goes as far as to say that he feels himself a monadologist, «*Ich bin eigentlich Monadologe*», and besides:

---

[11] It is to be found in the third volume of Husserl's *Briefwechsel*, in [Husserl 1994], pp. 391–520.

[12] Besides his two already mentioned important works, see [Mahnke 1912a, 1912b, 1913, 1921, 1924, 1927].

[13] An analogous periodization can also be found in [Van Breda 1966]. A useful collection of essays on Leibniz and phenomenology is [Cristin Sakai 2000].

[14] The classical locus is § 60 of the *Prolegomena* to the *Investigations* (i.e., [Husserl 1900/13]).

[15] In the course of our discussion at this conference, David Rabouin noted that in any case Husserl also went on working on his projects of *mathesis universalis* in the following years and thus mentioning Leibniz in that context. It was therefore much more an enlargement of perspective than a shift of interest.

[16] [Mahnke 1917], § 16, p. 14.

Ich fühle mich heute noch Leibniz nahe, und Ihre Interpretation der Monadologie, so wie Sie dieselbe in wenigen Sätzen andeuten, ist mir ganz und gar
verständlich, ja sie ist, wenn ich nach diesen Sätzen mich orientiere, ganz und
gar auch die meine.[17]

However, towards the end of the 1920s, in Husserl's reception of Leibniz's thought
there will be a third, more explicit but also more critical stage. When he devotes his efforts
to building the world of intersubjectivity, in his *Cartesianische Meditationen*, Husserl will
make frequent references to Leibniz, and he will overtly speak of monads and monadic
community. However, in those years, Husserl will mostly care for marking the difference between his own search for an intersubjective dimension and Leibniz's metaphysical
solipsism. On the contrary, in the same years, Mahnke was addressing his interests elsewhere, so much so that Leibniz ceased being the center of his production. Above all,
however, it was Husserl himself who was no longer the focus of Mahnke's theoretical
attention. In sum, such a critical turn concerning monadology did not directly resonate
through the works of Husserl's Leibnizian disciple. Although he did write to his teacher
a letter praising the *Cartesian Meditations*, Mahnke was by now too far from the studies
of his young years to be able to grasp the difference between the phenomenological and
the monadological method.[18] The early thirties, thus, marked the final divorce between
Husserl and Mahnke through a theoretical crisis that was all consummated about very
important and detailed aspects of the interpretation of Leibniz. Apparently, such an interpretive divide might have been caused by Mahnke's acquaintance with other readings
of Leibnizian philosophy, which led him progressively to abandon his former view. Reichenbach himself might have contributed to that and anyway, as early as his 1924 letters
on Leibniz's theory of motion, with which we are going to deal here, Mahnke already
seemed sometimes to oscillate in and out of the phenomenological horizon.

---

[17] Husserl to Mahnke, 5 January 1917, in [Husserl 1994], pp. 407–408. In the same letter, Husserl also recalls
how much in his young years he loved Leibniz's philosophy: «Ihre Liebe zu Leibniz kann ich sehr wohl nachver-
stehen. In jungen Jahren habe ich mich offenen Augen öfters in der Erdmann-Ausgabe von Leibniz gelesen und
zweifellos hat das auf mich, so sehr anders ich damals noch eingestellt war, stark gewirkt», and he also insists
on the intentional character of Leibniz's theory of knowledge: «Leibniz übersieht nicht ganz die Intentionalität,
das Vorstellen besagt wohl für ihn etwas sehr Weites, ziemlich so weit wie intentionales Erlebnis reicht. Aber
die immanente Motivation, in der sich alle Innerlichkeit entfaltet und die etwas anderes ist als Kausalität, möchte
ich nicht in mathematische Notwendigkeiten auflösen».

[18] Here is the very severe commentary Eugen Fink wrote to Husserl about Mahnke's opinion on the *Cartesian
Meditations*: «Mahnkes Fehlinterpretation der "transcendentalen Implikation der Monaden" ist bedingt durch
das Verfehlen des prinzipiellen Problemhorizontes, in welchem allein diese phänomenologische These zu Recht
besteht: er verkennt der Sinn der "phänomenologischen Reduktion". Das zeigt sich: 1.) am Begriff der Monade, der bei ihm keineswegs ein transcendentaler Begriff, sondern ein **ontologischer** ist. Monade wird mit der
"mikrokosmisch" geschlossenen, individuell-subjektiven Innenwelt (der Immanenz) gleichgesetzt. 2.) Daran,
dass das Problem der Konstitution verstanden wird als die Frage nach dem Zusammenhang von "Innenwelt"
und "Aussenwelt"...» (Fink to Husserl, 30 November 1933; [Husserl 1994], p. 519). Plainly, to Fink's eyes,
Mahnke was a metaphysician and not a phenomenologist anymore. He no longer cared for the problem of a
phenomenological *Konstitution* of the world, but rather for a nearly ontological *Aufbau* of it (if not in the Carnap
meaning, at least along the Helmholtz–Hertz–Schlick line).

4. Leibniz's theory of motion

On the very first rising of General Relativity, one of greatest confusions that originated in the minds of philosophers and scientists was precisely concerned with relativity of motion. Einstein's theory seemed to some, actually to many, to put an end to a metaphysical monster, that is, Newton's absolute space, that had been so much criticized in recent years by Mach and his followers. Therefore, the obvious thing to do could well seem that of going back to the anti-Newtonian "relativists" who had rejected absolute space and absolute motion, and affirming as a result that Descartes, Huygens, Leibniz had all been vindicated according to most recent physics. It was indeed such a historical naiveté that produced an avalanche of misunderstandings, for, even though Einstein's space was no longer Newton's, it was by no means more similar to Leibniz's. And so happened that, amidst quite serious confusions, some of them bearing an exquisitely mathematical nature, between covariance and invariance, between transformations of coordinates and motions of space, between intrinsic and extrinsic tensor calculus, between gravity field and *Strukturfeld*, very few were those who could find their way, and the return path to any pre-Newtonian relativity of motion was a mirage of simplicity and elegance. It is inconvenient here, however, to let us digress into the tangle of historical motives that persuaded some, though not all, of the main thinkers of the relativistic revolution to maintain the incompatibility between new physics and absolute motion, first because such motives are very complex and, second, they have already been splendidly accounted for in other works; nor is it convenient here to discuss the issue in theoretical terms and attempt to specify how many relativistic stances and how many non-relativistic arguments there may be in Relativity Theory, and thus to what extent those scientists may have been right or wrong: in fact, under many respects, the issue is such a complicated one that it is still sharply debated in today's philosophy of science. Suffice it to say that in the various attempts to make out the nature and foundations of relativistic physics, the name of Leibniz represented more a conceptual shortcut, and even, I dare say, a blind alley, than a source of clarity. Reichenbach himself was not completely immune from this in-between charm of history and theory.[19]

---

[19]For example, in 1924 Reichenbach seemed to think that the General Theory of Relativity effectively implemented Mach's Principle in its very strong form affirming that all inertial forces are produced (not just modified) by the distribution of matter in space. For instance, speaking of the Leibniz-Clarke debate, he writes that Clarke (in his *Fifth Answer*, §§ 26–32; GP vii, p. 425) «folgert aus Leibniz' Standpunkt, daß die Teile der rotierenden Sonne ihre Zentrifugalkraft verlieren müßten, wenn alle äußere Materie um sie herum vernichtet würde. Diese Folgerung ist in der Tat richtig, und die moderne Physik gibt sie zu» ([Reichenbach 1924a], p. 429). Which seems to be false because, even though all other matter were nullified, yet the Minkowski (or better the Kerr-Schwarzschild) structure of space-time (given some reasonable boundary conditions) would remain, and thus the rotation of the Sun would produce a centrifugal force. Such incompatibility between Einstein's theory and Mach's strong relativism was not overlooked by the best physicists of the period, such as Weyl and Eddington but, admittedly, it took quite a long time for it to become common knowledge; the discussions on Einstein's cosmological constant and De Sitter's model blurred further the picture. In the second half of the twentieth century, on the contrary, this example of centrifugal forces in Minkowski space was often employed in favor of Newton's absolute space and against Leibniz's doctrine. It is not by chance, I believe, that Reichenbach allows himself to be mistaken on such a salient point just in an essay of historical critique. This may indeed be, I am afraid, one of those high-rank instances of the confusion about the foundations of physics that the many comparisons with

In any case, the first question the shrewdest readers of Leibniz, and certainly Reichenbach among them, had to answer was whether the theory of absolute (i.e., dynamic) relativity of motion, which they made out in General Relativity, was or was not actually to be recognizable in Leibniz's natural philosophy. Contrary to most textual evidence, in fact, less shrewd readers seemed to have no difficulty recognizing it, and thus they got into the worst of troubles: by erroneously likening first Einstein to Leibniz, and then Leibniz to Einstein, they had in fact both thinkers allegedly maintain theories that ensued from neither of them but, at the very best, were a form, possibly even a simplified one, of Machian mechanical philosophy.

In fact, it is well known that, along with a great harvest of Leibnizian statements all intended to reject Newton's absolute space and claim full relativity of motion,[20] there is a little smaller number of them in which Leibniz maintains unequivocally that one needs carefully to distinguish between apparent and true motion. For, although it is apparently impossible to perceive any difference whatsoever if a body A moves towards a body B, or the body B towards the body A, so that motion is relative or, better, it envelops something relative (*involvat aliquid respectivum*),[21] nevertheless a body will be in (true) motion if the *cause* of motion itself is to be found in the body, while it will be at rest if such a cause is not there:

Motus est mutatio situs.
    Movetur, in quo est mutatio situs, et simul ratio mutationis.
    . . .

    Moveri dicitur situm habens in quo causa est mutati situs, seu ex quo mutati ejus cum alio situs ratio redditur. Quod si sufficiens ex ipso ratio redditur, hoc unum movetur, caeteris quiescentibus; sin minus, plura simul moventur.[22]

Leibniz's work have made greater instead of smaller. Reichenbach returns to the subject, and still expresses the same opinion as before, in his *Raum-Zeit-Lehre*, see [Reichenbach 1928], § 34, pp. 246–52. For an expanded treatment of the issue, see below, notes 35 and 36.

[20] Among the very many passages that can be quoted here, I take the following one from a note, left unpublished, in the *Specimen Dynamicum*, in which Leibniz's stance is also most clearly expressed with reference to circular motion: «Motus quoque naturam vidi. Adeo prehendi etiam spatium non esse absolutum quiddam aut reale, adeoque nec mutationem pati nec motum absolutum posse concipi, sed omnem motus naturam ita esse respectivam, ut ex phaenomenis mathematico rigore non debeat determinari posse, quidnam quiescat, aut quanto motu quodnam corpus moveatur, ne circulari quidem motu excepto, quanquam aliter visum sit Isaaco Newtono, insigni viro (quo nescio an majus ornamentum habuerit Anglia erudita), qui cum multa praeclara circa motum dixerit, hujus ope ex ipsa vi centrifuga discerni posse putavit, in quo subjecto sit motus, in quo non potui assentiri. Interim etsi Mathematica determinatio Hypotheseos verae nulla sit, motus tamen verus ei subjecto cum ratione a nobis tribuitur, in quo simplicissimam hypothesin, et ad phaenomena explicanda aptissimam facit; de caetero non tam de subjecto motus, quam de respectivis rerum inter se mutationibis quaeri ad usum sat est; cum nullum sit punctum fixum in universo» ([Leibniz 1982], pp. 22–23; the note was not in Gerhardt, and thus it was not known to Reichenbach, who however could rely on a similar passage in Leibniz's letter to Huygens from 14 September 1694: see below, note 31).

[21] I follow Daniel Garber's translation of *involvere* as *enveloping*; see [Garber 2009].

[22] The first quotation comes from the *Initia rerum mathematicarum metaphysica*, GM vii, p. 20. The second, from Leibniz's letter to Des Bosses, 12 December 1712, in GP ii, p. 473.

It is easy enough to understand the historical reasons that drew Leibniz to such a theory. The most important of them is probably to be found in his logical theory of relations, according to which there cannot exist a "purely extrinsic" relation, i.e., a relation not founded in its own terms.[23] Thus, the force causing motion (which in Leibniz's young years was the quantity of motion, and in his mature years the *vis viva*) cannot be a relational property of the mechanical system as a whole, but it must necessarily inhere (in absolute terms) in either one of the two bodies; and therefore either this or that body will be the real subject of the motion.[24]

On the contrary, it is not as easy to understand in what connection Leibniz could hold this opinion and still call himself a relativist as far as motion was concerned; and in fact his position was not at all understood by his contemporaries. Leibniz went so far as to maintain this theory of "true motion" in a most delicate place of his dispute against Newton, that is, in § 53 of his *Fifth Letter* to Samuel Clarke:

> Cependant j'accorde qu'il y a de la différence entre un mouvement absolu et veritable d'un corps, et un simple changement relatif de la situation par rapport à un autre corps. Car lorsque la cause immediate du changement est dans le corps, il est veritablement en mouvement; et alors la situations des autres par rapport à luy, sera changée par consequence, quoyque la cause de ce changement ne soit point en eux.[25]

On reading such a statement, Clarke saw in it nothing but the capitulation of his adversary, and in some way Leibniz's unconditional surrender, after a year of ferocious disputes, to the absolutist party:

> Whether this learned Author's being forced here to acknowledge the **difference** between **absolute real Motion** and **relative Motion**, does not necessarily infer that **Space** is really a quite different Thing from the **Situation** or **Order** of Bodies; I leave to the Judgement of those who shall be pleased to compare what this learned Writer here alleges, with what Sir **Isaac Newton** has said in his **Principia** Lib. 1 Defin. 8.[26]

Leibniz died before he could reply, and thus modern interpreters have also been left with no suggestions on how he meant possibly to reconcile the theory of *mouvement absolu et veritable* with that of *motus respectivus*.

Today, however, we can hope for more clues to this theoretical quandary. We can count, in fact, on a much greater quantity of writings, letters, as well as private or even very private reflections of the great philosopher than those Clarke and other contemporaries had access to. In the papers now available there are, for example, some particularly

---

[23]On this issue, see especially [Mugnai 1992].

[24]For an interpretation of Leibniz's physics along these lines see [Sklar 1976], pp. 229–32, who takes acceleration as a monadic property; and [Roberts 2003], who (more correctly) takes mv^2 as such a property. Both of them improve on the very anti-historical reconstruction of a "Leibnizian Space-Time" offered by [Earman 1989] (at the price of shadowing some very neat theoretical remarks), but their purely physicalistic reading seems to me still very far from Leibniz's own (phenomenological) solution. For still another approach see [Arthur 1994]; for a historical introduction see, among others, [Bernstein 1982].

[25]GP VII, p. 404.

[26]GP VII, p. 428.

relevant contributions of Leibniz to the discussion on the two chief world systems, that is, the Ptolemaic and Copernican ones. In these studies, Leibniz enunciates another celebrated theory, that of the equivalence of hypotheses, according to which any physical phenomenon, and motion more than anything else, allows itself to be accounted for through a plurality of different principles; rotation of the Sun or rotation of the Earth, thus, would only be two different theories each of them explaining equally well one and the same appearance. There is however, Leibniz adds, a difference between them, and not an irrelevant one: the Copernican hypothesis is *simpler* than the Ptolemaic one.[27] Through it, the very same phenomena are explained more directly, clearly, and easily. There is no doubt that a modern, seventeenth-century scientist ought to prefer the heliocentric system. In a sense therefore (but we will need to understand what this sense may be), the Sun is *really* still, and the Earth *really* in motion: science can, and indeed *must*, assume this hypothesis as the only true one.

That Leibniz considers the theory of simplicity of hypothesis in the same connection of thoughts as he does that of the determination of true motion through a force, is clearly witnessed in passages like the following:

> Ut vero res intelligatur exactius, sciendum est Motum ita sumi, ut involvat aliquid respectivum et non posse dari phaenomena ex quibus absolute determinetur motus aut quies; consistit enim motus in mutatione situs seu loci. Et ipse locus rursus aliquid relativum involvit, etiam ex Aristotelis sententia, qui definivit superficie ambientis. Hinc in rigore omne systema defendi potest, ita ut ne ab angelo quidem Metaphysica certitudine aliquid absoluti determinari inde queat, quoniam ipsa conditio est legum motus, ut omnia eodem modo in phaenomenis eveniant, nec dijudicari possit utrum et quatenus corpus aliquod datum quiescat vel moveatur, nisi rationem majoris explicabilitatis habendo, idque adeo verum est ut ne vis quidem agendi verum sit motus absoluti iudicium.[28]

However, though numerous throughout Leibniz's writings, such statements seem to contribute little to clarify our problem or the compatibility between true and apparent motion. As a matter of fact, the very first reason for disagreement between Reichenbach and Mahnke in their 1924 exchange was on how to interpret the "non-relativistic" oscillations Leibniz incurs in as a forerunner of the Theory of Relativity.

---

[27] In the course of our discussion at this Conference, Ralf Krömer and Jean Seidengart have rightly observed that it is not true that Copernicus's account of the Solar system is (remarkably) simpler than Ptolemy's; for a basic account on the topic, see [Bernard Cohen 1985], pp. 119–22. In all evidence, under the concept of the Copernican system, Leibniz intended the heliocentric theory such as it evolved in the period of a century and a half separating Leibniz himself from the *De Revolutionibus*, that is after the discovery of Kepler's laws and the new mechanics.

[28] *Tentamen de Motuum celestium Causis* (1689), GM VI, p. 146. For some penetrating historical insights of Leibniz's involvement in the debate on Copernicanism, see [Heilbron 1999], and, more to the point, [Bertoloni Meli 1991].

5. The missing argument

Reichenbach's solution, actually a very easy one, was to contend that, although he had
caught a glimpse of dynamic relativity of motion, Leibniz proved nevertheless unable
to defend it against Newton's attacks; in this way, he remained a forerunner, but only a
forerunner, of Einsteinian physics.[29] Leibniz, in other words, faced with overwhelming
experimental evidence, such as the famous bucket experiment, was unable to account for
the existence of centrifugal forces and thus, with tight lips, he admitted the existence of
true motion. It was therefore a contradiction on his part, nothing more than that, and
Clarke was perfectly right in rejoicing and regarding Leibniz's attempt to found a relativistic dynamics as a failure. What was missing in Leibniz's theory was Mach's argument
against absolute space, and this also was the reason why Newton could triumph, at least
until Mach devised his celebrated argument and Einstein in turn implemented Mach's
philosophical intuition in physical terms.[30]

Reichenbach's solution for the possibly conceptual snags in Leibniz's stance, however, may not be very convincing. In fact, it seems to suggest that it was indeed the
correspondence with Clarke, maneuvered backstage, as is well known, by Newton himself, first to present Leibniz with the inertial force problem, and that, between the month
of July 1716 in which he got Clarke's *Fourth Answer* and November of the same year,
in which he died, Leibniz did not succeed in disentangling his theory from such arduous
complexities and making it consistent. The truth is, however, not only had Leibniz been
very well aware of the contradiction brought about by circular motion ever since his trip to
Paris (1672–76)[31], but his above-mentioned definitions of true motion recurred through-

---

[29]This is very clearly reasserted, for example, against Mahnke's quick enthusiasm, in the letter written by
Reichenbach and published in our Appendix. Here, we can just cursorily mention Hermann Weyl's subtler
interpretation, which is just a few years posterior to Reichenbach's essay in *Kant-Studien*. Weyl does cite Huygens, Leibniz, and Mach as supporters of the theory of dynamic relativity of motion but, unlike Reichenbach,
he also accurately shows how Einstein's Relativity corroborates Newton's theory rather than theirs. However,
Weyl adds, Leibniz, unlike Huygens and Mach, also held a concept of inertia as a dynamical principle, and
thus as something determined by the reciprocal influence of bodies: and this concept is right (except, inertia
is actually a property belonging to the *Strukturfeld*, and not to the bodies, which Leibniz could not have the
faintest inkling of). In other words, the same argument we have above carried on by means of living forces, by
showing how according to Leibniz true motion is determined through the presence of such forces, is employed
by Weyl by means of inertial forces – which is well possible because they too are taken into consideration in
Leibniz's dynamics; and, in so doing, Weyl also retrieves some features of Leibniz as a relativist *malgré lui*. In
short: according to Reichenbach, Leibniz is too little of a relativist to be Einsteinian, according to Weyl, on the
contrary, he is too little of an absolutist, but on the right track to become one. This interpretation of Leibniz
seems to me to be a *unicum* in the philosophical landscape of the first half of the twentieth century: cf. [Weyl
1927], § 16. Weyl's ideas have been recently resuscitated by [Lariviere 1987], but Leibniz's space structure
(and inertial structure) surely does not depend on the disposition (*situs*) of bodies, so that, albeit interesting, the
whole operation appears necessarily faulted.
[30]«Damit tritt allerdings ein ernster Riß in die Leibnizsche Bewegungslehre ein, denn mit dieser Unterschei-
dung ist ja die Newtonsche Auffassung zugegeben. Hier fehlt Leibniz das Machsche Argument, das allein die
Relativität der Bewegung auch **dynamisch** verteidigen kann» ([Reichenbach 1924a], p. 428). For a discussion
of Mach's principle in Einstein's physics, see [Barbour Pfister 1995] and here especially [Norton 1995].
[31]Here is a statement in Leibniz's own words: «Comme je vous disois un jour à Paris qu'on avoit de la peine à
connoistre le veritable sujet du mouvement, vous me répondîtes que cela se pouvoit par le moyen du mouvement
circulaire, cela m'arresta ; et je m'en souvins en lisant à peu près la même chose dans le livre de Mr. Newton ;

out his scientific production, and certainly through the many years preceding not only the correspondence with Clarke, but even the publication (1687) of Newton's *Principia*. It was therefore neither Clarke nor Newton, nor any experimental evidence or bucket, that drew Leibniz towards the doctrine of true motion, nor did he lack the time to ponder the consistency of his solution. If there is an actual inconsequence between his assuming both relative motion and true motion, such inconsequence is therefore as structural to Leibniz's natural philosophy as to have been born, so to speak, with it. Which would be a very serious problem, to be sure, but just because of that, and because such inconsistency would have been evident to all thinkers including Leibniz himself, one should doubt that he did not have a more than adequate answer to such an obvious question.

In any case, one is impressed at how steadily Reichenbach's solution to Leibniz's difficulty, beginning as it does with Descartes's and Huygens's theories of motion and ending with Newton's mechanics, all draws on physicalistic principles and never departs from them. Mahnke's answer, on the contrary, which attempts to defend Leibniz from the charge of contradiction, introduces a radically phenomenal element into Leibniz's natural philosophy.

6. Noumenal motions

One needs to distinguish, Mahnke says, between monads and phenomena, as well as between movements of the mind and the body. Monads do change, and the force Leibniz refers to is indeed the force producing such a noumenal change (*"Bewegung"*), which is in the mind (*"psychische oder wenigstens seelenähnliche"*) and certainly absolute (true). However, since this realm of minds, i.e., the set of monads, is expressed and represented through phenomena, and each mental feature of them is embodied in a certain corporeal appearance, we can very well say that each spiritual change must be represented in a phenomenon, and it is actually represented as a phenomenal change, i.e., as a corporeal motion. However, Mahnke says, through the representation process such an absolute spiritual movement is *radikal relativiert*, so that it is expressed through a purely relative motion. Thus, according to Mahnke's reconstruction, any contradiction would be avoided, with a typically Kantian move, by a transition to phenomenalism: one of the horns of the antinomy (true motion) falls into the noumenon, and the other one (relative motion) into the phenomenon.

It is not surprising that Reichenbach did not like the "Kantian" solution. More surprising, under many respects, that Mahnke liked it.

From the strictly textual point of view, such an interpretation seems quite untenable. In fact, in none of the passages in which he distinguishes between relative motion and true

---

mais ce fut lorsque je croyois déjà voir que le mouvement circulaire n'a point de privilege en cela. Et je voy que vous estes dans le meme sentiment. Je tiens donc que toutes les hypotheses sont equivalentes et lorsque j'assigne certains mouvemens à certains corps, je n'en ay, ny puis avoir d'autre raison que la simplicité de l'hypothese, croyant qu'on peut tenir la plus simple (tous consideré) pour la veritable» (Leibniz to Huygens, 4/14 September 1694; GM II, p. 199). Throughout the Parisian unpublished papers, there are in fact innumerable definitions of motion in which the body in motion is said to be the one in which resides the cause of the change. See for instance the *Pacidius Philalethi* from November 1676, in A VI, 3, n. 78, p. 535; C 599.

motion, or just deals with true motion in itself, does Leibniz ever mention monads or any kind of supersensible movement or change. Thus, not only does Leibniz nowhere explain what the meaning would be of such a "radical relativization" of will and perception in a reciprocal motion of bodies, but also he never hints that those should be absolute and this one be relative. Here, I believe, Reichenbach defeats Mahnke easily enough: the controversial passages from the correspondence with Clarke or other writings on absoluteness of motion are definitely concerned with phenomenal forces and true motions of bodies in space, but in no way do they seem to describe metaphorically a simple change in the appetites of a bunch of souls. In those passages, Leibniz writes as a physicalist, not as a phenomenalist.

Admittedly, however, there is something right and important in Mahnke's interpretation. If we drop the discussion on the mental origin of the relative change of bodies, we will simply see (from an almost physicalistic perspective) that Leibniz would be stating that there is a true motion of bodies lying at the base of the relative one that we perceive, but such a true motion is not perceived in a phenomenon. In other words, Leibniz's metaphysics (as well as his theory of relations, as we saw earlier) demands that, if A and B move according to an apparent relative motion, one of them must be in true motion, because it is endowed with force, and the other one (for example) must be not, because it has no force. By looking at the relative motion of A and B, however, we are unable to say exactly where the force may be. True motion is there (owing to metaphysical reasons, it must be there), but it cannot be seen. The question is thus reducible to the distinction between ontological and epistemic scope. This seems to be Mahnke's argument, once it has been divested of its spiritualistic superstructure (a distinction between phenomena and noumena still remains, in the sense that force is regarded as a thing in itself, i.e., it is not perceivable as such). And this also is the reconstruction which the finest and most attentive contemporary interpreters of Leibniz's natural philosophy have offered as far as his theory of motion is concerned.[32]

Now, even putting aside any judgment on this (important) reconstruction, we cannot but be astonished that it was Dietrich Mahnke to suggest it. We can very well see, in fact, that this interpretation of Leibniz's philosophy is strongly grounded in the possibility that there may exist noumenal objects and properties not faithfully expressed in a phenomenon. A change occurs within a noumenon that will be expressed as a corporeal motion: all right. But in the same noumenon also lies the cause for such a motion, and it will by no means be expressed in the phenomenon: nor can we know anything about "true motion". Mahnke's radical relativization of change, therefore, is founded on a limit of phenomenal expression. Surely, we can discuss whether Leibniz would ever concede that there are noumenal properties that are not expressed and cannot be expressed in a phenomenon; many places in his writings seem to incline the reader towards a negative answer. Leibniz's very well-known sentence on observability, so often cited by phenomenologists and idealists, «quand il n'y a point de changement observable, il n'y a point de change-

[32] See above all [Garber 1995], pp. 306–309, an ample and brilliant discussion of the issue. Recently [Jauernig 2008] proposed some variations on this interpretation.

ment du tout»[33], seems to apply easily to forces and, since we do observe a change, but do not observe any force, we may *tout court* be led to conclude that there are no forces. Mahnke's interpretation, in short, has to recognize that, being a metaphysician much more than a physicist, and a Kantian much more than a phenomenologist, Leibniz admitted to the existence of things-in-themselves of which there is no trace in the appearances. But this reasoning strongly contrasts the purely phenomenological interpretation of Leibniz's philosophy that had been offered by Husserl and, chiefly, Mahnke himself in 1917. The *Neue Monadologie*, in fact, was entirely meant to demonstrate that Leibniz excludes the Kantian, unknowable things-in-themselves from his own philosophy, and that monads, as well as the entire noumenal world, are faithfully represented through their phenomenal expressions.[34] Apparently, thus, by 1924 he had changed his mind.

7. Conventions and truth

Mahnke does however resort, in his correspondence with Reichenbach, to yet another strategic argument through which he would like to show that there is no contradiction in Leibniz's theory of motion. This second strategy, instead of drawing on the distinction between phenomena and noumena, draws upon those Leibnizian places in which, as we have seen, the philosopher speaks of a plurality of hypotheses all explaining one and the same thing, and of the need to choose the simplest of them.

Here too Mahnke's stance sounds rather surprising, because he seems to regard such assertions of Leibniz's as an old (and possibly forerunning) form of conventionalism. There is no contradiction between true motion and relative motion, in other words, because strictly speaking motion is neither true nor relative, but conventional. We can only speak of true motion insofar as we have chosen a definite theoretical frame to describe our mechanical system, and our choice has been determined by the convenience of the hypothesis, that is, its greater simplicity. Indeed, Reichenbach is in error in maintaining that it was because he lacked Mach's argument that Leibniz did not get to perfect relativism. In fact, if we read into Leibniz's writings on dynamics carefully enough, we

---

[33] «On replique maintenant, que la verité du mouvement est indipendente de l'observation, et qu'un vaisseau peut avancer sans que celuy qui est dedans s'en apperçoive. Je reponds que le mouvement est indipendant de l'observation, mais qu'il n'est point indipendant de l'observabilité. Il n'y a point du mouvement, quand il n'y a point de changement observable. Et même quand il n'y a point de changement observable, il n'y a point de changement du tout» (*Fifth Paper to Clarke*, § 52; GP vii, pp. 403–404).

[34] See, for instance, passages of the following tenor: «Das transzendentale Bewusstsein, das in allen Einzelseelen lebendig ist und ihre übersubjektive Geisteseinheit ermöglicht, ferner seine objektiv gültigen Begriffe, die den subjektiven Phänomenen überindividuelle und ewige Bedeutung geben, würden am ehesten die Bezeichnung als "Ding an sich" oder nooÚmena verdienen, weil sie die Welt nicht relativ zu einem bestimmten subjektiven Erleben, sondern nur relativ zum Bewusstsein überhaupt, also in absoluter Allgemeingültigkeit wiedergeben. Wir ziehen es aber vor, hier lieber von den "Dingen für alle" oder den "objektiven Dinge" zu sprechen. Denn es gehört zum Wesen jedes "objektiven Dinges", das wirklich existiert, sich auch subjektiv darzustellen, nämlich einem bestimmten Individuum immanent als "Ding für sich" und sämtlichen übrigen transzendent als "Ding für andere" oder sinnliche "Dingerscheinung". Und man könnte nun auch mit gutem Rechte, wie es Schopenhauer, Beneke, Lotze u. a. tun, statt den bloss ideellen "Dingen für alle" vielmehr den real existierenden, ihren eigenen Begriff in voller, evidenter Anschaulichkeit erlebenden "Dingen für sich" das "Sein-an-sich" zuschreiben. Wir wählen daher, um Missverständnisse zu vermeiden, unsere eigene Terminologie» ([Mahnke 1917], p. 98).

can easily realize that the theory of the equivalence of hypotheses also perfectly fits with a relativistic account of circular motion, and that Leibniz knew very well how to reply to the Newtonian example of the bucket.[35] Perhaps, he did not pursue his theory of rotation in detail, but his epistemological considerations are more than enough to show that, if a theory of rotation is at least possible that does not require absolute space, it should certainly be preferred to one requiring it, as in fact we have other, metaphysical and logical, reasons for rejecting such a dogmatic idol. Therefore, it is once again conventionalism, or at least conventionalism along with some metaphysical principles, that rescues Leibniz from contradiction.[36]

---

[35] In this case, Mahnke succeeded in convincing Reichenbach. In the third letter in the Appendix, in fact, Mahnke underlines a passage from Leibniz's *Dynamica* of 1690, in GM VI, p. 197, that Reichenbach had not taken into consideration in [Reichenbach 1924a], when he affirmed that Leibniz lacked Mach's argument (see earlier on, note 30). In his *Raum-Zeit-Lehre* of 1928 Reichenbach adjusts fire and says that, although it was not clearly expounded by Leibniz, the Machian argument, "may legitimately be extrapolated" from the passage suggested to him by Mahnke (§ 34, p. 246 note 1). More recently, Howard Stein (knowing nothing about Mahnke's contribution to the issue) has contended that Reichenbach's concession to Leibniz is excessive, and that from GM VI, p. 197 hardly can we draw anything that may resemble Mach's Principle (see [Stein 1977], p. 7 and note 5 in pp. 31–32). I would think that Stein is right: from Leibniz's argument in the *Dynamica* we can at the most deduce that, if ethereal matter were destroyed in the Leibnizian universe, there would be no gravitation, but by no means that there would be no centrifugal force for a rotating body. Finally, [Earman 1989], p. 65, also states against Reichenbach (he ignores Reichenbach's amendment) that Leibniz considered but explicitly discarded Mach's argument. From a historical point of view this is hard to maintain, but one cannot but agree with Earman (against Reichenbach and Mahnke as well) that Leibniz could not possibly take an instrumentalist position and discard absolute space simply because it may be possible (in abstract terms) to construct a relational theory of motion: he had to explicitly create one.

[36] The issue here is a delicate one. The principle of simplicity alone might not have satisfied Leibniz without calling in other metaphysical principles. Let us go back to the example of Mach's Principle in its strong form (i.e., that matter and energy fully determine the inertial structure of space): from this point of view, Einstein's Relativity, that rejects such a principle, may be termed as much more Newtonian than Leibnizian. Now, let us admit that this aspect of the Einsteinian theory could be fixed, by producing a theory in which in an empty universe inertial forces vanish; let us admit, for example, that this is the case with the Brans-Dicke theory. Let us admit, furthermore, that, because of some experimental inductive indetermination, we are not able to devise any *experimentum crucis* that may discriminate between Einstein's theory and Jordan, Brans, and Dicke's one (according to Mach's words, universe is given to us once, and we just cannot have a bucket rotate without fixed stars in order to check what happens with centrifugal force). Well, which theory should we embrace, according to Leibniz and Mach? The Einstein one is simpler, but involves absolute space (or something like that); the Brans-Dicke eliminates absolute space, but introduces a scalar field that makes equations more complicated (if we want to get to the same experimental results). Mach would possibly choose the Einstein theory, because he has repeatedly said that simplicity and practicality of a hypothesis must be the only guidelines; he never dealt with the problem, actually, because the definition of inertia he suggested in his *Mechanik* must have been as simple as Newton's one. Leibniz on the contrary would possibly choose the Brans-Dicke theory, because he thinks he holds *metaphysical arguments* against absolute space. Not only does absolute space create a problem for the economy of ontology (so much so that our Mach, in order to make ideology thinner, may even be willing to expand ontology through this monster of absolute space), but it also creates a problem for his own concept, which is contradictory or, at least, violates the *principium rationis* and that of the identity of indiscernibles. In conclusion, the search for physical simplicity needs also to be consistent with more general *meta*physical principles. For a Machian theory which is more classical (i.e., "Newtonian") than the Brans-Dicke one, see [Lynden-Bell 1995].

This view of Mahnke's will sound unbelievable to anyone who is aware of the sharp disagreement that just a few years before had caused Husserl and Schlick to split on the issue of conventionalism, and precisely in reference to some views expressed by Husserl in the *Ideen*.[37] While, a little earlier, phenomenologist Mahnke inclined towards Kantianism, now he makes a few steps into neo-Positivism.

Meanwhile, recently converted to conventionalism by Schlick himself, and yet a fervent admirer of Leibniz, Reichenbach rejects Mahnke's interpretation. He says, on the contrary, that Leibniz did not use the simplicity of hypothesis as a choice criterion of the, say, pragmatic kind, but rather as evidence of true motion: which true motion, therefore, according to Leibniz, exists objectively and not only conventionally. It all, however, in Reichenbach's opinion, contradicts both Mahnke's overly generous and modernizing interpretation, and Leibniz's affirmation of relativity of motion.[38]

Clearly enough, in interpretive terms Reichenbach is right, nor has apparently Leibniz ever regarded simplicity as an epistemological criterion of the convenient: as a matter of fact, he has regarded it as an ontological criterion for the creation of the world, in the sense that God creates the world according to the simplest possible laws that are compatible with the greatest possible variety and richness of beings (*simplex est sigillum veri*). It is of little interest here that Leibniz has always left the detail and exact meaning of such a criterion of simplicity in the vague: what matters now is that it can by no means be regarded as a subjective one. Consider, for example, the following passage, that is contemporary and so similar to the one cited above from the *Tentamen*:

> Cum verò nihilominus homines motum et quietem assignent corporibus, eiam illis, quae neque ab intelligentia, neque ab interno instinctu moveri censent, videndum est quo sensu faciant ne falsa dixisse judicentur. Et respondendum est eam Hypothesin eligendam esse, quae est intelligibilior; neque aliud esse veritatem Hypotheseos, quàm ejus intelligibilitatem. Et cum diverso respectu non tam hominum et opinionum, quàm potius rerum ipsarum quae tradendae sunt una Hypothesis aliâ sit intelligibilior et scopo proposito convenientior;

---

[37] The dispute between Schlick and Husserl dated back a long time, having begun with [Schlick 1910] and [Schlick 1913]. However, Schlick's most conspicuous attack was delivered to Husserl in § 18 of the first edition (1918) of the *Erkenntnislehre* ([Schlick 1918/25]), to which Husserl vigorously replied in the 1920 Preface to the second edition of his Sixth *Logical Investigation* ([Husserl 1901/22]). Schlick in part adjusted fire and in part reasserted his criticism in the second edition (1925) of his book. Here the controversy was mostly concerned with Husserl's concept of *ideation*, but Schlick's criticism fell into a wider discussion of conventionalism. Anyway, Schlick will be still attacking phenomenology during the thirties.

[38] See [Reichenbach 1924a], p. 433: «In seltsamer Weise gehen hier moderne und alte Ansichten durcheinander. Leibniz ist sich klar darüber, daß die Auszeichnung der wahren Bewegung nicht auf unmittelbaren Beobachtungen beruht. Auch die Dynamik liefert nicht einen direkten Nachweis der Kraft, denn für jede der gleichberechtigten Auffassungen der Bewegung läßt sich auch eine dynamische Hypothese ersinnen. Nur die Einfachheit der Erklärung ("la semplicité de l'hypothése") ist der Vorzug der ausgezeichneten Bewegungsannahme, also etwa des Kopernikanischen gegenüber dem Ptolemäischen System. Dieses ist nur das für die Erklärung der Phänomene geeignetste (ad explicanda phaenomena aptissima). Aber er meint doch, daß die einfachere Erklärung eben deshalb die wahre wäre (qu'on peut tenir la plus simple, tout consideré, pour la veritable), und so führt ihn die Untersuchung der logischen Äquivalenz der dynamischen Erklärungen doch wieder auf die metaphysische Auszeichnung der einen Bewegung zurück».

etiam diverso respectu una erit vera, altera falsa. Ut proinde veram esse Hy-
poyhesin nil aliud sit, quam recte adhiberi.[39]

Evidently, what Leibniz is attempting here is neither to replace the concept of truth
with that of the convenience of a theory (which Leibniz may possibly term "supernomi-
nalism"), nor to juxtapose the two concepts as if there were the truth on the one side and
then more and less convenient (simpler or less simple) ways of expressing it on the other
side. On the contrary, Leibniz's attempt is that of making the two concepts intrinsic to
one another, so that the simplest theory is true and the others are not, even though they
may express the same facts. Truth coincides with intelligibility; the world is a perfectly
rational thing, produced by reason for the reason, by simplicity for the simplicity of the
mind; so that the principle of simplicity is both epistemological and ontological.[40] Nor
would I be able to say how much Leibniz's "classical" theory of truth (as inherence of a
predicate in the subject, for example) was ready to incorporate this architectonic concept,
simplicity, that was to break into it; to be sure, such an attempt at definition was explicitly made by Leibniz in his mature years, and it was then that the import of Leibniz's
"rationalism" became incompatible, as its seems, with conventionalism.

Reichenbach's latter statement, thus, according to which the simplicity of a hypothesis is not the ground for building a convention but rather an indication for discovering
absolute motion, seems to me very convincing. It seems also to suggest a way to understand Leibniz's view on the whole issue of motion, which, if I am not mistaken, is to be
found somewhere midway between Mahnke's interpretation and Reichenbach's.

(There may be something deeper about this specific disagreement between Reichenbach and Mahnke. Reichenbach recognizes that Leibniz's solution, that of holding what is
maximally simple as true, contains something modern, although it needs to be interpreted
in its epistemically strong and indeed ontological sense. Only, Reichenbach adds, the
principle of simplicity here is misused: in fact, Einsteinian Relativity has shown the radical relativity of motion and thus it is senseless to speak, even though conventionally, of
absolute motion.[41] In sum, Reichenbach makes use of a metatheoretical conventionalism,
according to which it is not the coordinate system that can properly be seen as conventional (it will just be relative), but instead the theory of motion itself (General Relativity,
for example, versus Minkowskian physics with universal forces). On the contrary, when

---

[39] This is a text erroneously published by Couturat (C 590–91) as a Preface to the *Phoranomus*. Although it
has nothing to do with the *Phoranomus*, it was indeed written in the same period, that is, during Leibniz's trip
through Italy between 1689 and 1690. Reichenbach does not seem to know it, but he does mention ([Reichen-
bach 1924a], p. 43 note 2) a text next to it, which had been partially edited by Gerhardt in GM vi, pp. 145–47,
and which is now to be found in the collection of essays *De Praestantia Systematis Copernicani* in A vi, 4c,
n. 377, pp. 2065–75. No doubt because of a mishap, the text published by Couturat has not been included in
the Academy edition of Leibniz's works. For a masterful reconstruction of the circumstances in which Leibniz
composed these texts, see [Bertoloni Meli 1988].

[40] See for instance this most perspicuous passage of Leibniz's from 1700: «Realitas porro aestimanda est et
multitudine et varietate et ordine rerum, et ut verbo dicam quantitate intelligibilitatis, quod etiam indicat omnia
esse propter intelligentes» ([De Risi 2006], p. 58).

[41] We have already repeatedly mentioned Reichenbach's view according to which Einsteinian Relativity would
actually lead to the radical relativity of motion (and space). The present reasoning on Leibniz's making incorrect
use of the principle of simplicity is to be found in [Reichenbach 1924a], p. 433, note 3.

speaking of a plurality of hypotheses, Mahnke does not refer to a plurality of theories but only to a plurality of systems of non-inertial coordinates, which is a completely different thing. The reason for it may be found in Mahnke's understanding of his phenomenologist Leibniz as if admitting to a sort of eidetic intuition of physical laws: so much so that they cannot be conventional, but are simply given in an intellectual intuition;[42] conventional will possibly be the object submitted to a force, but by no means the law of a force.)

8. Obscure motions

Now, let us try and orient ourselves in the complex intricacies of Leibnizian epistemology. In order to do this, we will attempt an interpretation of Leibniz's natural philosophy that can somehow be regarded as a radicalization of Mahnke's phenomenological perspective, which retrieves several features of Reichenbach's physicalistic reading.

We need, at this point, to introduce into our context of natural philosophy the phenomenological notion of different degrees of clarity and distinctness of perceptions and ideas. By this means, we will see that Leibniz deals with the indeterminacy of a phenomenon, that is, with the fact that a phenomenon is subject to a plurality of possible explanations, just because each phenomenon as such is the product of a confused apprehension of reality. A phenomenon is for Leibniz the object of sensible perception, and sensibility, in Leibniz's theory of knowledge, is the organ of confused knowledge. Because of such confusion, there are aspects of a phenomenon not adequately expressive of the thing in itself (be it a monad or an aggregate of monads), and thus the apprehension of things through phenomena is a structurally incomplete one. Thus, the indeterminacy of the physical theory claiming to explain the phenomenal world is nothing but a consequence of the phenomenalness of the world itself.

On the other hand, however, this structural indeterminacy of a phenomenon as such with respect to theory is not of the kind imagined by Mahnke. It is not, I mean, fashioned in such a way that there may be objective contents of the noumenal world that cannot be phenomenally expressed, and thus realms of things-in-themselves (or of properties of the things-in-themselves) lying beyond phenomena and irrepresentable through them. On the contrary, each content is, *qua* content, liable to be phenomenally expressed, and there is no natural mystery impossible to unveil. Now, thus, we can understand those passages on the determination of motion in the solar system according to the non-conventionalist reading that, more faithfully to the texts and the spirit of Leibniz's philosophy, Reichenbach had offered. The architectonic principle of the simplicity of hypotheses is but a means we own

---

[42] See for example: «**Kant** ist aktiver Rationalist, weil er Willensmensch und "Idealist der Freiheit" ist; Leibniz dagegen ist zunächst intuitiver Noëtiker, weil er in erster Linie theoretischer Mensch ist. [...] [Pichler] hält Leibniz also für einen **intuitiven Noëmatiker**, dem Noëma und Hyle wie Form und Inhalt zu einer einheitlichen Gegenstandsanschauung verschmelzen. In Wahrheit dagegen steht Leibniz, wie wir wiederholt gesehen haben, **Husserls Phänomenologie** viel näher. Denn auch er hält die wesensverschiedenen Regionen der noëmatischen Objektivität und hyletischen Subjektivität deutlich auseinander und schaltet zur Vermittlung zwischen ihnen an zentraler Stelle einer dritte, **die noëtische Region der intentionalen Geistesakte**, ein; und auch er gründet wie **Husserl** die Noëmatik auf die Noëtik, die Vernunftobjektivität auf die objektivierende Vernunft...» ([Mahnke 1925], pp. 486 and 513; many more examples can be found in [Mahnke 1917]).

in order to make our ideas on Nature clearer and more distinct. According to it, we are able to affirm that the Earth rotates, and the Sun stays still, and therefore to determine exactly where the active force and the quantity of motion are to be found (in the Earth, that is, and not in the Sun); which would have been impossible according to Mahnke's interpretation, as it rested on properties that were unknowable in principle and on conventionalism as a way to make up for such unknowability.

Admittedly, this true motion of the Earth is itself a very hypothetical one, and thus new observations, or clearer perceptions, or simpler architectonic hypotheses (such as a different theory of motion, or a different theory on the circulation of celestial bodies through the ether) will possibly allow us to maintain that the true motion is neither the Ptolemaic nor the Copernican one, but yet another one. For us, a phenomenon always remains undetermined, just because it is a phenomenon, and true motion can only be grasped hypothetically. Finally, it is interesting to see how, in this context, Quine's dichotomy can be dropped in which he distinguishes between inductive and theoretical indeterminacy of a theory, the first one caused by the insufficiency of experimental data, and the second, more radical one, by the fact that, no matter how much experimental data we may collect, there still remains a great number of possible theories to explain them.[43] Leibniz's indeterminacy belongs to the inductive type, insofar as it lacks data, and only data; but it also belongs to the theoretical type insofar as it is impossible, not just difficult but impossible, to collect all data.

9. False ideals

We need better to clarify these considerations and possibly dispel an easy misunderstanding, which actually seems to me the main point at stake here. One could be tempted, that is, to interpret Leibniz's theory of the indeterminacy of a phenomenon and true motion as a regulative theory. In other words, we may say: true motion exists, but it cannot be determined exactly; however, through progressively better experimental apprehensions and theoretical skills, we will get closer to it, even though we never reach it (because our experiments are finally over, and even our best skills prove hopeless with Nature's blind impenetrableness). Natural as it sounds, such a reconstruction would not very well distinguish between the theory of Leibniz and that of Newton, who besides was perfectly aware that his own absolute space (and absolute motion) was not there to be immediately grasped through experiments but needed to be built through successive approximations of inertial systems. Above all, furthermore, such a theory would more or less be the same as Kant's, as expounded in the section devoted to Phenomenology of the *Metaphysische Anfangsgründe der Naturwissenschaft*: in which in fact absolute space (and thus true motion) is a regulative ideal to which one gets indefinitely closer and closer.[44]

---

[43] See [Quine 1970].

[44] Kant's text in point can be read in KgS IV, pp. 558–65. Kant's stance is actually a little more complex than the one here shortly expounded, because, as regulative ideal, Kant does not assume absolute motion (which is according to him a contradictory concept), but absolute space as a condition of possibility of relative motion. Kant's distinction thus finds itself in between true motion and apparent motion, both of them being relative. It may not be convenient here, however, to go on looking into these further specifications, as they add very little

For Leibniz however things must be different. The crux of the matter lies indeed in the notion of a phenomenon. It, according to Leibniz, more or less faithfully expresses a world in itself, which is a universe of timeless, spaceless, and motionless spirits. In fact, space, time, and motion are but expressive (phenomenal, indeed) correlates resulting from the imperfect, veiled, and confused representation of the noumenal world. If it is so, however, we can easily see that such a presumptive regulative ideal, true motion, which our more and more detailed investigations of the phenomenal world have meant to reconstruct, is in fact a fictitious ideal. Should we ever possess, *ad absurdum*, a perfect (*viz.*, clear and distinct) knowledge of the whole phenomenal world, which would finally enable us to determine true motion and true force without any doubts, oscillations, or slips, we would no longer be holding a phenomenon, but a noumenon. The phenomenon in fact existed because of its very confusion and obscurity. But the noumenon we are holding now is a supersensible, spaceless, timeless, and (as far as we are concerned here) motionless object. All questions concerning true motion and true forces would therefore subside. Such absolute motion thus, which in order to be motion must be phenomenal, and in order to be absolute must not be phenomenal, is nothing but a *contradictio in adjecto*. And it cannot be a regulative ideal.

It is easy to see that such a difference between Leibniz's and Kant's views ultimately rests on their different doctrines of the relationship between sensibility and the understanding. Kant, who has regarded sensibility as a faculty independent from thought, finds no contradiction with a perfect knowledge of a phenomenon qua phenomenon, nor, thus, with assuming this kind of knowledge as a regulative ideal (an ideal that cannot be attained just because of the limited structure of our sensibility); while Leibniz, who has made sensibility an imperfect degree of the understanding, can admit to no perfect knowledge of a phenomenon as such.

More in general, and from the viewpoint of the history of ideas, I may add that Kantian and Enlightenment moral ideal of a continuous progress towards a better knowledge, which *can* and thus *must* be pursued, is partially alien to Leibniz, who on the few occasions he speaks of progress is always very cautious. He seems to me to have on this subject an idea more similar to Husserl's, according to which gaining one more piece of evidence is one and the same thing with producing elsewhere a shading and oblivion. Thus, for Leibniz, even absolute motion (just as absolute space, as a matter of fact) is not an ideal we must or simply can pursue. What we are left with is only partial and local theories that, here and there, from time to time, give us, *per speculum et in aenigmate*, across and beyond phenomena, an intimation or a glimpse of true motion.

---

to the general perspective we are interested in, that is, a regulative construction of an inertial system. In any case, Reichenbach was very well aware of how detailed Kant's stance was, as we can read in [Reichenbach 1924a], pp. 429–30, note 2, where Reichenbach reasserts that Kant's distinction was anyway "anticipated" and even improved on by Leibniz. For an almost contemporary reading on these Kantian topics see also [Schneider 1921]. For an excellent, both historical and theoretical, exposition of Kant's doctrine of absolute space as regulative ideal, see [Friedman 1992], pp. 136–64.

10. Leibniz's theory of time

This reconstruction of Leibniz's theory of motion may at least be a starting point for explaining those apparently contradictory statements on relative and real motion. Before we venture any conclusion, however, let us deal with the other main subject of the correspondence between Reichenbach and Mahnke, which also is the second philosophical core of the interpretation of Leibniz as a forerunner of Relativity: his theory of time.

In fact, we can see that here and there in his writings Leibniz carries on a few considerations in favor of a causal theory of time, a theory, that is, according to which time would be but the perceptual manifestation, almost the epiphenomenon, of deeper and more genuine causal relations.[45] Leibniz however never bothered to write a treatise or even a brief essay on such a theory: thus, all we can rely on today is a small number of very private, short, obscure, and elliptical notes that the great philosopher penned down on a few occasions. His thought on the issue does not in the least appear neat or consistent, and there is every reason to doubt that he ever possessed a full theory of temporality. One may even suppose that not only did he lack such a theory, but he found it impossible to build one as well, because the other parts of his metaphysics left no room for a philosophy of time to establish itself.[46]

Be it as it may, the fact remains that in an extremely mature work of Leibniz's, the *Initia rerum mathematicarum metaphysica* of 1715, the philosopher ventures a bit further into this domain, and begins to lay the foundations of that which actually looks like a complex causal theory of time.

These kinds of theories, in their various articulations, enjoyed some fortune in the course of the nineteenth century (to be sure, not because of any recognition of Leibniz's work, but thanks to Kant's *Critique*), and at the beginning of the new century there was a number of thinkers who wanted to see in Einstein's Theory of Relativity a confirmation of those elder interpretations of the temporal flow.[47] As is also the case with relativity of motion, or Mach's Principle, it is to be doubted that Einsteinian physics may actually require such a philosophical reading, and in fact, in the second half of the twentieth century, many voices were raised to deny the necessity, opportunity, or even possibility of a causal theory of time in the framework of Relativity.[48] In the twenties, however, a

---

[45] On the reductionist or non-reductionist character of this theory of Leibniz's, see [Arthur 1985] and [Cover 1997]. Later on, I will however speak of Leibniz's *causal* theory of time, even though it may be doubted (see § 16) that Leibnizian *ratio* may be so easily equaled to cause in the strict sense.

[46] For a better articulation of this point, see [De Risi 2007].

[47] Nearly all the authors who, around the end of the nineteenth century, grappled with the building of causal theories of time were either Kantian or anyway strongly influenced by the *Analogies of Experience*. See, one example among many, [Lechalas 1895]. At the beginning of the twentieth century, chapter 8 of [Russell 1914], which also discussed a causal theory of time (here a non-Kantian one), could well influence some neo-Positivists (and certainly Carnap) on this subject. The first systematization of Relativity as a causal theory of time is probably [Robb 1914]; the closest to Reichenbach's work is Lewin [1923]; soon after, [Carnap 1925] also appeared. As for Reichenbach himself, he was indeed well aware that he was only giving an *interpretation* of Relativity.

[48] The main point here is that, in the second half of the last century, Relativity became more and more geometrized and therefore, once scientists and philosophers became acquainted with Minkowski's and Lorentz's spaces, very few of them thought that the notion of cause could ever prove clearer than that of time: thus, there

causal theory was very appealing, and it seemed to account for many aspects of the new physics. Reichenbach stepped forward as one of its most lucid and prolific interpreters. In fact, his *Axiomatik* of 1924 essentially is an axiomatics of time.[49] It is by no means surprising, therefore, that in his essay on Leibniz from the same year he attempted– and it was the first time ever anyone bothered with it – to interpret Leibniz's *Initia* by giving a rational reconstruction of it that might be able to fill in its many gaps and show how and to what extent Leibniz had or had not been also a forerunner of such an essential part of Einsteinian physics. There can be no doubt, on the other hand, that for Reichenbach, just as for his contemporaries, the drive to interpret Relativity Theory as a causal theory of time came from Kant, and while he was busy axiomatizing it, he must have had the volume of the *Analytic of Principles* wide open in front of him. Apparently, it also was in the course of this endeavor that Reichenbach became aware of the insufficiencies of Kantian philosophy face to the new physics. So that, once he had accomplished the theoretical work, and he devoted himself instead to the Leibniz he had been discovering, not only did Reichenbach find in him the anticipation of both his own and Einstein's work, but he also once again deemed Leibniz to be much better than Kant.[50]

Once again, the issue at stake was psychologism. For Reichenbach, all the faults of the Kantian position were deeply rooted, to begin with, in Kant's admission that time was a form of intuition, and a pure intuition itself, and that causality as a logical principle would only structure but not produce such an intuitive content, originally acquired by consciousness through other sources. Reichenbach never tired of repeating that Einstein's determination of time, though (unlike Newton's) heavily depending on the observer, nevertheless "does not query the perceptions of the observer, but rather the knower's schema for ordering his knowledge" because it "pertains only to the *logical* conditions of knowledge, not to the *psychological*".[51] Reichenbach knew very well that Cassirer had adroitly

---

were almost no epistemological reasons left for attempting such a reduction; and as for its metaphysical reasons, they were definitely looked down on. See for instance the debate between Bas Van Fraassen and John Earman ([Van Fraassen 1970], [Earman 1973], [Van Fraassen 1973]). Recently, purely geometrical interpretations of Relativity have begun to be questioned in favor of other readings, possibly more attentive to the reasons of dynamics (see [Brown 2005]), which may succeed in reinvigorating causal approaches to the doctrine of time. For a more general criticism of the assumption that Relativity Theory should necessarily require a causal theory of time, and particularly of a few aspects of Reichenbach's axiomatics, see [Friedman 1977]. For a defense of Reichenbach's perspective, though not in the detail of his axiomatization, see finally the classical (and in some respects outdated) work, [Grünbaum 1973], pp. 179–208.

[49] [Reichenbach 1924c]; see also the following [Reichenbach 1925]. Reichenbach will go on dealing with this problem throughout his lifetime; his later studies were published posthumously in [Reichenbach 1956], where, in § 3, he still recognized Leibniz as the first ever theorist of a causal theory of time.

[50] [Reichenbach 1924a], p. 421 note 1, reveals that he was not yet aware of any Leibnizian text concerning the causal theory of time when, just a few months earlier, he wrote his axiomatization (i.e., [Reichenbach 1924c]). A few years later, an only historical discussion of Reichenbach's interpretation of Leibniz's theory of time also appeared in *Kant-Studien*; see [Gent 1926]. The following year, [Weyl 1927] was also dealing (§ 16) with the Leibnizian theory in the *Initia* in reference to Einsteinian Relativity.

[51] [Reichenbach 1924b], § 1 (trans. Maria Reichenbach). As we can read in the correspondence here published in the Appendix, Reichenbach sent this violently antipsychologist essay of his to Mahnke. See also [Reichenbach 1924a], p. 422: «Auch die Kantische Zeitlehre (wie sie etwa in der 2. Analogie der Erfahrung formuliert ist), reicht an die Leibnizsche Erkenntnis nicht heran. Der unglücklichen Kantischen Bestimmung der Zeit als anschaulicher Bedingung der Kausalität ist die Leibnizsche Formulierung als allgemeines Ordnungsschema der

used the dependence of simultaneity from the observer in Special Relativity in order to make the Transcendental Ego seep into the new physics, and had however insisted on an antipsychologist interpretation of it and of Kantianism generally. But we have already seen how Reichenbach had become no longer convinced that this could be a viable reading of Kant, and believed on the contrary that the Kantian route was at least liable to a "*gefährliche Irreführung*".[52]

Of course, not even Leibniz had fully attained the purely causal conception of time of Einsteinian Relativity,

> aber seine Raum-Zeit-Lehre führt direkt zu diesem Resultat hin, während die Kantische ihm im Wege stand, und auch durch die Doppeltheit von transzendentaler Idealität und empirischer Realität den objektiven Charakter von Raum und Zeit nur sehr ungenügend formulierte.[53]

But then, what exactly are the faults with Leibniz's theory? What exactly did Leibniz fail to do as a forerunner of more complete theories? Such are the main issues of the exchange between Reichenbach and Mahnke.

11. Simultaneity lost

Let us, therefore, try and understand something of the causal theory of time so fragmentarily offered by Leibniz. The main locus taken into consideration by all interpreters ever since Reichenbach is the following passage from the already mentioned *Initia rerum mathematicarum metaphysica* of 1715:

> Si plures ponantur existere rerum status, nihil oppositum involventes, dicentur existere **simul**.
>
> . . .
>
> Si eorum quae non sunt simul unum rationem alterius involvat, illud **prius**, hoc **posterius** habetur.

Kausalreihen überlegen. Dies tritt allerdings erst vom Standpunkt einer auf Axiomatik begründeten Erkenntnistheorie klar zutage».

[52] Cassirer's analysis concerning transcendental subjectivity in Relativity is to be found in [Cassirer 1920]. The opinions expressed by [Weyl 1917] (II, § 6, p. 72), who at the time decidedly inclined towards phenomenology, were not very different (but certainly less anxious about being charged with psychologism). Reichenbach's opinion on Cassirer's undertaking, who under this respect tended in fact to assimilate Leibniz's theory of time to Kant's one and thus to overlook any antipsychologist distinctions, is the following: «Cassirer sucht dort [in Cassirer 1902] die nahe Verwandtschaft beider Lehren nachzuweisen, während es mir unter dem Eindruck der inzwischen erfolgten relativistischen Lösung des Problems notwendig scheint, die Leibnizsche Raumlehre der Kantischen noch überzuordnen» ([Reichenbach 1924a], pp. 425–26 note 3; in the same page 426 Reichenbach's words on the "dangerous misunderstandings" can be read). On the psychologism implied in Kant's concept of time, see also [Reichenbach 1924c], § 3. Posterior to Reichenbach's, the most important work on causal theories of time is Henryk Mehlberg's great essay of 1935, in which Leibniz is also said to have anticipated Kant's and Einstein's causal theories, but to have been, unlike Kant, an antipsychologist; which, however, in Mehlberg's opinion is a fault and not a merit because, according to this author, it is indispensable to have a theory that may link one's interior experience of time to the objectified time of physics (cf. [Mehlberg 1935]). On this topic, see also [Putnam 1991].

[53] [Reichenbach 1924a], p. 426.

...

Et ideo quicquid existit alteri existenti aut simul est aut prius aut posterius.[54]

As is quite evident from the *involvere rationem alterius*, here Leibniz is dealing with something like a causal theory of succession, even though, actually, Leibnizian reason does not always let itself be smoothly reduced to the notion of cause. It is evident as well, on the other hand, that the *nihil oppositum involvere* of the first line cannot but refer to the principle of contradiction, that is, to Leibniz's definition of a possible thing (or world) as a non-contradictory whole. At this point, however, our problem also sticks out concerning the legitimacy of the conclusion drawn in those last lines. It is still to be seen, in other words, how it may ever be that two (existing) states of affairs opposed to one another, and which therefore cannot coexist, must be causally connected and therefore ordered into a temporal succession. Or also, vice versa, why two (existing) logically compatible and simultaneous states of affairs should necessarily be causally independent of one another.

The problem is therefore, from Leibniz's point of view, nothing less than the relationship existing between the *principium contradictionis*, which regulates and governs simultaneity, and the *principium rationis*, which on the contrary determines a succession. And, as is well-known to all interpreters, the relationship between these two principles hardly allows itself to be made out in Leibniz's writings; so much so that it will become one of the most challenging issues in the philosophical debates of eighteenth-century Germany, where, in fact, the party of those who maintained the possibility to reduce one principle to the other fought against the irreducible as well party of the anti-reductionists; both of them, finally, being most confused as to the exact relationships occurring between these two maximum principles.[55]

Above all, when we shift the focus from the general terms of the question to the application suggested by Leibniz in his theory of time, and we move from the eighteenth to the twentieth century, we will most clearly see that it is indeed Einstein's Relativity that, by accepting the existence of neither simultaneous nor causally connected events, denies Leibniz's conclusion and confers on the problem its most conspicuous and almost dramatic character.[56]

What we have here is the problem, much debated in Reichenbach's times, of the conventionality of simultaneity in Einstein's theory. In Relativity, in fact, let events O, P, and Q, be given in such a way that Q is in the future lightcone of P but neither of them is in the lightcone of O; then, we will be able objectively to determine that P precedes Q, but by no means to determine whether O is simultaneous with the one or the other of them (because it is not causally connected to either). Thus, we will need to resort to a convention.[57]

---

[54] GM VII, p. 18.

[55] [Mahnke 1917], § 32, p. 32, on the contrary, offers a rather simplistic reading of the issue.

[56] The opposite occurrence, that of both causally connected and simultaneous events, is on the contrary denied by Relativity Theory. In § 13 we will briefly return to the simultaneous causality in Leibniz's system.

[57] [Friedman 1983], pp. 177–78, accurately notes that Reichenbach systematically confuses between conventionality of simultaneity (which concerns the determination of e in the formula concerning the propagation time of a signal) and relativity of simultaneity (which concerns on the contrary the observer's state of motion), and

By the same token, however, Leibniz's theory does not seem to be able to rule out that a state of affairs O may be compatible (i.e., envelop no opposition) with a state of affairs P as well as a state of affairs Q, and yet that P and Q may be incompatible between them and causally connected. Thus, we would again obtain that P and Q are temporally determined between them, without however being able to tell to which of the two O is actually simultaneous. What is missing or problematic in Leibniz's relation of compatibility, in sum, is transitivity. In fact, if O and P, and O and Q are compatible in pairs, this does not necessarily mean that P and Q are compatible as well. On the other hand, it seems highly improbable that Leibniz might have been able or ever wished either to resort to a conventionalist solution, which would be totally foreign to the eighteenth-century way of thinking, or to regard simultaneity as depending on the observer's speed. Thus, he found himself once again, and nearly more by mistake than on purpose, midway between a Newtonian absolutist theory and an Einsteinian relativistic theory, without knowing for sure how to get out of there.

This is the way in which Reichenbach read Leibniz's causal theory of time. It is hardly surprising that, in those years of fervent debates concerning the relativistic notion of simultaneity, he might have seen in the *Initia*, just as before, in Leibniz's theory of motion, both an anticipation and a missed opportunity.[58]

12. Simultaneity regained

Thus, Reichenbach complains that Leibniz's doctrine of time does not completely determine the horizon of simultaneity (because the lack of causal connection is not sufficient for it). Mahnke, as had been the case with relative motion, attempts again to defend Leibniz. In the *Initia*, Mahnke remarks, Leibniz does offer an argument for completely determining the temporal position of a state of affairs whatsoever; it goes as follows:

> Status meus prior rationem involvit, ut posterior existat. Et cum status meus prior, ob omnium rerum connexionem, etiam statum aliarum rerum priorem involvat, hinc status meus prior etiam rationem involvit status posterioris aliarum rerum atque adeo et aliarum rerum statu est prior. Et ideo quicquid existit...[59]

A state of affairs, therefore, *involvit rationem* of the states of affairs subsequent to it, but, owing to the connection among all things, it also *involvit* (let us say: *simpliciter*) those states of affairs simultaneous with it. Such a state of affairs, thus, determines a temporal slice of simultaneity and, through such a slice, it is also temporally determined with respect to those other states of affairs that, directly, are not causally connected with it.[60]

---

that such confusion has produced a number of erroneous interpretations of the meaning of Special Relativity. Perhaps, Reichenbach's reading of Leibniz has not emerged unscathed from such ambiguity either.

[58] See [Reichenbach 1924a], p. 421 (in the text and note 1), and the correspondence with Mahnke in the Appendix.

[59] GM vii, p. 18. The place is mentioned in Mahnke's second letter, but it was also known to [Reichenbach 1924a], p. 421.

[60] On a thorough scrutiny, this account may also appear insufficient, without any additional hypotheses, to solve the problem of a total determination of time, because it does not seem to eliminate the possibility of a branched

As soon as one gives it some thought, this will appear a brilliant solution. A few of today's interpreters, evidently motivated by Reichenbach's pioneer research, have actually attempted to reconstruct a formal axiomatization of Leibniz's theory of time (under so many respects similar to Reichenbach's relativistic axiomatics), and they have found out that some accurate interpretations of the above-mentioned Leibnizian passage allow a very complex but also very consistent theory to be formulated.[61]

However, the exact meaning of Leibniz's phrase *involvere* (*simpliciter*), or, in Reichenbach and Mahnke's words, the *Verknüpfung aller Dinge*, remains still to be investigated.

13. Physical Influx

A first possibility is that of giving a *physical* meaning to this notion. At times, Leibniz himself seems inclined to think that his own cosmology, the account of a filled-up universe in which matter pushes other matter, may imply that sÚmpnoia p£nta and that each part of this physical universe is immediately connected with everything else and bears the footprints of it.[62] As a result, there would be a sort of interaction, a *Community* (*Gemeinschaft*) of all states of affairs that could very well be the physical ground in which to found the relation of simultaneity.

Such a Kantian solution is not however likely to be the one really provided by Leibniz, nor is it likely to succeed in defining simultaneity either. First of all, it is far from clear how, in Leibniz's natural philosophy, such a dependence of each thing from the whole should be physically realized. Even admitting, in fact, a causal effect of the whole universe on each part of matter, it would by no means naturally follow that the universe

---

time, that the General Theory of Relativity should also take into consideration. However, additional hypotheses may actually be found in Leibnizian metaphysics, such as, for example, that the cause for a state of affairs must be a unique other state of affairs (given that a cause equals the effect, or that causality actually is an internal relation, etc.). In any case, here the story has gone too far from Leibniz's real issues and thus it gets inexorably lost in the fogs of interpretive vagueness. It is however true that the relativistic axiomatization of [Reichenbach 1924c] has held to the present day a few valuable results concerning Special Relativity (or even General Relativity, if one admits the existence of a global Cauchy surface), while Gödel's models, and Carter's, Penrose's, and others' subsequent studies have made it difficult to accept Reichenbach's axiomatics of time for the General Theory of Relativity in its amplest formulation. Anyway, see the reserves expressed in [Winnie 1977].

[61] See [Winnie 1977], as well as [Arthur 1985], no doubt the best ever contributions to the interpretation of Leibnizian philosophy of time. Richard Arthur has recently returned to the subject in his forthcoming book on Leibniz, furtherly improving on axiomatization and convincingly replying to the remarks that had been variously addressed to his reconstruction. A very comprehensive discussion of Leibniz's philosophy of time is also to be found in the recent [Futch 2008].

[62] Cf. *Monadology*, § 61, in GP vi, p. 617. The idea of a physical connection of the world can already be found in Leibniz's earliest writings, and thus it possibly precedes his more metaphysical one, linked with the abstract concept of expression and the notion of a complete concept. Such a more abstract idea of connection can also be found in the *Monadology* (§ 56, GP vi, p. 616), and Leibniz seems to think of the two as being equivalent. In my opinion, they are not equivalent at all, and on the contrary their strength is very different, as if the physical notion were but a sort of fossil, so to speak, surviving through Leibniz's mature system or at least his most general expositions of it.

could be reconstructed starting from one of its parts, since a plurality of different causes could well have equal effects. Most of all, such a causal effect of the whole universe on a single part of it, which has been sometimes maintained from the viewpoint of metaphysics, seems to be denied from the viewpoint of physics, because Descartes's theory of vortices, which Leibniz adopts in all his cosmological systems, has been built indeed with the purpose of justifying local motion and a non-global causal interaction (a vortex itself is the closure domain of physical causality).

Second, even admitting its existence, we should ask ourselves whether a physical interaction might be suitable for determining not only succession but also simultaneity. It is certainly true, in fact, that Leibniz's theory, just as all pre-relativistic physics, does not accept the notion of a maximum speed of propagation of signals and thus of causal effects; however, unlike Kant, Leibniz decidedly denies the possibility of an action at a distance, *viz.*, the possibility for such a speed to be actually infinite. This consideration seems to suffice to exclude that a relation of simultaneity may ever be grounded on an instantaneous, causal *Wechselwirkung*.[63]

Third, finally, we are confronted with the problem that Leibniz's philosophy lacks, if metaphysical rigor be applied, any *real* physical influx whatsoever. As a result, the very empirical causality on which one wanted to ground temporality, and thus the condition of possibility of a phenomenon, would only be derivative and phenomenal.[64]

In sum, no matter how many different approaches we take to this argument, we can hardly think that Leibniz could have ever accepted the solution suggested in the *Third Analogy*. On the contrary, he might have come close to Einsteinian Relativity at least in the following sense: simultaneity must not be grounded in any form of reciprocal causality (as if *A involvit simpliciter B* were equal to *A involvit rationem B et B involvit rationem A*), but on the absence of any (physical) causal connection. Leibniz's theory thus presented at least the advantage of being disengaged from the formidable obstacles affecting the theories of those who accept a simultaneous causality or a temporal distance between cause and effect that "*verschwindend ist*".[65]

---

[63]For a discussion of the importance that the highest propagation speed of a signal had, possibly wrongly so, in [Reichenbach 1924c], see the above-mentioned locus of [Friedman 1983]. As is well known, [Reichenbach 1920] thought that the finiteness of the highest speed of a signal was an *a priori* synthetic principle. Naturally, the main requirement here is for the action to be instantaneous, not "at a distance"; cf. note 2 to Mahnke's second letter, in the Appendix.

[64][Mehlberg 1935] maintains on the contrary, but I think erroneously, that Leibniz's theory of a pre-established harmony "implicitly" contains Kant's theory of physical influx.

[65]As for Kant, see *KrV*, A 202–203, B 247–49; KgS III, pp. 175–76. More in general on this topic, cf. for example the above-mentioned debate between [Van Fraassen 1970] and [Earman 1973]. It is however true that, by resorting indeed to an extra-logical structure, i.e., pure intuition, Kant would perhaps be allowed a few steps towards solving the problem (for example, if the topological structure of time did not depend on any causal structure but were given in an *a priori* intuition); this solution, however, would definitely be precluded in Leibniz's case. Note, finally, that (at least in the 1680s) Leibniz seems to accept a simultaneous causality (cf. A VI, 4A, n. 132, pp. 563–64; A VI, 4A, n. 133, p. 568; A VI, 4A, n. 147, pp. 628–29; discussed in [Futch 2008], pp. 118–25), but this was probably meant as a very abstract and metaphysical one, and not as a form of physical influx (see the next section on the expressive relation); moreover, Leibniz may well have changed his mind on these topics after the correspondence with Arnauld, see below note 72.

14. Metaphysical influx

In any case, such a Kantian and physicalist solution – interpreting Leibniz's word *involvere* as physical influx– was not propounded by Reichenbach, who on the contrary, regarding it as totally inadequate to the new physics, thought it most ungenerous to force it on Leibniz. Following Leibniz's statements on the absence of any real physical influx, it is however still possible to interpret the interconnection between all things according to a much more metaphysical sense. Then, we can resort to the abstract notion of expression, and note that, according to Leibniz, each monad more or less obscurely perceives the entire world, and thus in representational terms it contains all the world within itself, and so, in a sense, each monad also contains the possibility to reconstruct it.[66] If, in my instantaneous perceptual act, I express the whole universe, then evidently I also know (though obscurely) what is simultaneous with such an act. The metaphysical relation of expression instantly "propagates" throughout the universe (just to make use of a physical metaphor for that which is no longer physical), and thus it substitutes for the immediate action at a distance that Leibniz denied in the scope of natural philosophy.

This one, needless to say, is the solution propounded by Mahnke.[67] It could not meet with Reichenbach's approval, as it once again put a whole metaphysical doctrine and rational psychology of cognitive faculties back into the core of Leibniz's physical system. As we can see, also concerning the subject of time the stances of the two contestants widely diverge, the one going towards rigorous physicalism, the other towards phenomenology and metaphysics.

Most of today's interpreters of Leibniz's causal theory of time leave it undetermined what such a Leibnizian connection between all things might ever lead to. However, generally speaking, they incline more towards Mahnke's interpretive optimism than Reichenbach's pessimism; and they definitely regard universal connection as an *immediate* and *total* relation between states of affairs (just as expression seems to be). Starting with such a "Mahnkian" perspective, they succeed somehow in consistently and formally reconstructing the causal theory of time Leibniz offered in the *Initia*; through a special "connection axiom", the indeterminacy of simultaneity which Reichenbach complained about in Leibniz's theory is also corrected; the relation of compatibility between states of affairs is demonstrably transitive; temporal slices prove to be objectively determined, and so does, finally, an *absolute* horizon of simultaneity. Through these proceedings, any relativistic temptation is eliminated from Leibniz's theory, so that Mahnke himself could already conclude that: «Von einer Relativierung der Gleichzeitigkeit kann also bei Leibniz durchaus nicht die Rede sein».[68]

---

[66] That is, here emphasis is given to the above-mentioned § 56, instead of § 61, of *Monadology*.

[67] It is expounded in the third letter in the Appendix.

[68] Mahnke's second letter in the Appendix.

15. Obscure times

This interpretation of Leibniz's thought is convincing, but perhaps a little scant. Here, I would like to attempt by no means to deny but rather to extend it in order to show how it may be possible, within the framework built so far, also to regain some of the critical comments and motives suggested by Reichenbach. The result will once again be achieved by radicalizing the phenomenological character of Mahnke's interpretation.

First of all, let us see what happened with our original problem, the one concerning the relationship between principle of contradiction and principle of reason. As follows from the interpretation suggested by Mahnke, the phrase *nihil oppositum involvere* in Leibniz's definition of simultaneity should be equated to the phrase *involvere simpliciter* on the connection between all things, that is, if A and B do not oppose one another, then they envelop one another. Compelled to clarity by their formulae, modern interpreters simplify even further and posit that the opposite of *nihil oppositum involvere* be indeed *involvere rationem*, so that all three Leibnizian *involvere*'s in the passage from the *Initia* are reduced to one primitive notion, that of causality (or better, of reason). By this means, the principle of contradiction has vanished and left no marks, and our problem has been cut off through a definition that affirms what was to be demonstrated: if two things oppose one another, and thus they are not simultaneous, then they are causally enveloped; and then again, through the connection axiom, if two things, on the contrary, do not oppose one another, then they envelop one another.[69]

However, I wish to suggest that we try to get, first, a more poignant meaning out of that notion of not involving opposition, i.e., a meaning of simple compossibility in one and the same time (resting on the *principium contradictionis* as the ground of possibility), and, second, a weaker meaning for that phrase *involvere simpliciter*, which does not imply that, given a state of affairs (a monadic content in a given instant) then the whole universe is also necessarily (*ex hypothesi*) determined. It seems to me evident enough, in fact, that the strong notion (one of total expressivity) of the Leibnizian *involvere* would lead us straightforwardly to that which some Leibnizian interpreters have called superessentialism or, in short, to affirm that each monad is structurally linked with an entire world. By this reasoning, however, Leibniz's system seems to me to be subject to a peculiar modal collapse, which in the end would make all considerations on freedom, possible worlds, etcetera, utterly useless, and send Leibniz over to abhorred Spinozism instead.[70]

The interpretation we have been considering here, in fact, *trivializes* the notion of simultaneous compossibility (expressed, as I see it, by the *nihil oppositum involvere*), in the sense that any (existing) state of affairs is only compatible with *one* (existing as well) *series* of other states of affairs simultaneous with it.

---

[69] I am here referring again to [Winnie 1977] and [Arthur 1985].

[70] Such has been, of course, the opinion of several interpreters of Leibniz's thought ever since the eighteenth century to the present. Nor is it an opinion easy to deny, as in fact there are many places in which Leibniz sounds as an *obtorto collo* Spinozian. In [De Risi 2007], pp. 463–77, I attempted to offer a "phenomenological", or, if one wishes, "transcendentalist" reading of Leibniz's modal system. What I am expounding here about simultaneity is but a specific instance (and even stripped of some rather important arguments) of that general interpretation.

Most of all, however, it *trivializes* the concept itself of cause, because it is actually a theory that, despite all contrary appearances, does not deal with a plurality of objects and states of affairs, but with one thing (the universe) in its different states. So that, one could say, the fact that now you are reading is the cause for the rising of the sun tomorrow, because your reading necessarily envelops the whole state of the universe in this instant, and this is on its own the cause for the sun to rise tomorrow. There is a sense in which this last conclusion would be surely regarded as true by Leibniz, but not without a number of *distinguo*'s. In any case, such as it is, it makes the concept of cause totally unserviceable for any scientific purposes or accounts of reality. It reduces indeed the complexity of the causal nexus of the world to a total ordering, which is thus easily identified with temporal succession.

It seems to me that a way to avoid such a simplified reasoning and retrieve, on the contrary, a full concept of possibility and contingency of simultaneous things, may be that of considering the full potentialities of the expression concept in earnest. It is true, in other words, that the principle of determination of the horizon of simultaneity must be found in a relation of expressivity (and not, for instance, in a physical influx), and thus that Leibniz's theory of time ineradicably bears such a metaphysical flavor; nevertheless, the concept of expression can and must be phenomenologically articulated according to different degrees of perceptual clarity. The set of my clear (either distinct or not) perceptions has a very limited horizon. Such a horizon, one that, of course, I can extend beyond its physiological limits through architectonic and rational arguments, and plainly through my knowledge of Nature (so wide is indeed the meaning of Leibniz's concept of perception), will however always be a finite one. Beyond it, lies an immense ocean of obscure representations, which for me are nothing. In fact, what I do not perceive now (in this large meaning), I cannot even say it exists now, and beyond my narrow horizon of simultaneity the rest of the universe is for me completely undetermined as to its temporal position. In short, I build a local system of simultaneity, in which all the rules and the axiomatization we have previously seen keep their validity (a neighbourhood of trivial causality), without however being able to determine, and thus trivialize, any globally causal structure.

What we have here may seem a sort of "psychologism", that is, an epistemic conception of modality in general and temporal modality in particular; however it is not, because, as we have already noted with absolute motion, perceptual obscurity is for Leibniz a condition of possibility of perception itself, and by no means a contingent feature of it. Without obscurity there is no sensibility, no phenomenon, and thus *no time* either. Not only is the horizon of absolute simultaneity not determined objectively by any perceptual act, it is also non-determinable in principle; not only does it not exist *really*, it does not exist *ideally* either, i.e., as a regulative ideal (because it is contradictory). There is therefore a strong meaning in which we cannot hold a temporal slice of simultaneity to be more discovered than, on the contrary, *constituted* through a succession of perceptual acts. Certainly, there is a strong analogy between the possibility for a global horizon of absolute simultaneity (and thus a total ordering of time) to be determined, and the possibility for absolute space and true motion to be determined; the local character of Leibniz's constructions of space and relative motion is thus reducible, as I see it, to the same origin as his only local constructions of simultaneity. The two issues are linked to one another, also in the sense that for determining true motion, as we have seen, knowledge of the cause of motion itself is necessary; just as it is necessary, of course, for building a causal theory of time. It is therefore the same "transcendental obscurity" shrouding our knowledge of causes that determines relativity of motion along with relativity of time.[71]

As we can see, we have eventually obtained a negation of the concept of absolute simultaneity, which however has nothing at all to do with Einsteinian Relativity, because it is by no means concerned with the relativity of the reference system, or the conventionality of metrics, or the isotropy of propagation of signals. Here, the negation of simultaneity simply derives from an integrally phenomenological reading of lines of thought lying within Leibniz's metaphysics. Under many respects, such a reading offers an idea of time much more similar to that of, say, Bergson or Husserl than of Einstein. It comes as no surprise that Reichenbach did not venture into such an interpretation of Leibniz, since we have seen how decidedly he rejected any intervention of phenomenology in physics as well as any intervention of the psychology of temporal perception in determining simultaneity. It is also true, on the other hand, that, as was the case with absolute motion, Dietrich Mahnke could well be closer to such a solution, and yet he preferred another one that, for being simpler, ended however by losing the very phenomenological perspective he wanted to start from in the first place. [72]

16. Note on topology

In the margins of this debate on Leibniz's theory of time, I believe another theme may be worth mentioning which Reichenbach dealt with in his 1924 essay on Leibniz and which is of some relevance in the discussion on Relativity. In his *Axiomatik*, Reichenbach had articulated his construction of time succession into two steps: first, one had to build what he called the *topological* structure of a temporal continuum, which, actually, more than a topological structure was a structure of ordering; second, one had to supplement it with a proper *metrical* structure, that is, the measure of the distance between those instants previously ordered. Reichenbach thought that Leibniz himself, in the *Initia*, achieved the construction of a topological, rather than metrical, structure of time; and that such

---

[71] [Arthur 1985] penetratingly remarks that for Leibniz a causal theory of time treated in entirely phenomenal terms is out of the question, because it would presuppose the condition that our knowledge of the causes of motion be complete, which is impossible. Leibniz's theory of time thus necessarily rests on a metaphysical substratum (*viz.*, on a theory of expression and monadic activity), that prevents us from ever calling it a fully physicalist theory and, to some extent, even a causal theory *tout court* (if, that is, we mean cause as a purely physical and phenomenal relation).

[72] As I have already repeatedly advanced strong reserves on the completeness or clearness of Leibniz's doctrine of time, there is no need to reassert here that such interpretation is a very free one, which, starting from general premises of Leibnizian metaphysics, attempts to reconstruct, or better construct for the very first time, a theory on this issue. It aims at possibly providing a point of view alternative to, but not exclusive of, Winnie's and Arthur's reconstructions, as in fact it seems impossible to me to determine what, if any, might have been Leibniz's actual historical stance. Finally, note that there is a place (A VI, 4A, n. 147, p. 629) in which Leibniz explicitly speaks of simultaneous events as being not compossible but conecessary. This is, however, a passage prior to his correspondence with Arnauld and his reflections on possible Adams and general Adams, nor do we know for sure whether he would still maintain such a stance throughout his mature and old years.

an achievement had to be ascribed to his profound studies on geometry, the *analysis situs*, which had once again anticipated the twentieth-century investigations in the field of topology.[73]

Now, there seems to me to be very little topology in Leibniz's studies on *analysis situs*, which mostly and almost exclusively deal, in fact, with metric geometry.[74] There is however something true in Reichenbach's judgment, even though it may better be ascribed to a certain embarrassment of Leibniz's theory than to its foreseeing.

As it happens, in fact, the most important relation in Leibnizian geometry, that of situation (*situs*), is defined through the congruence concept. Congruence is in turn most of the times defined through invariance of quality and quantity of a figure (i.e., as a similarity that does not modify measure). But if we finally ask how exactly Leibniz defines quality and quantity, we find ourselves again pushed back from mathematics to metaphysics or, better, phenomenology. After he attempted various ways (say, since the nineties of the seventeenth century), Leibniz in fact resolved to take a step that I regard as very important in the general mechanics of his thought: to define (and not just characterize) quality and quantity through a coperceptual act.[75] Such a "phenomenological turn" in Leibniz's philosophy, which in the same period would bring about a number of other changes in his metaphysical system, implied however as a consequence that *situs*, i.e., distance, could only be applied to those things coperceivable in principle, that is, to things existing simultaneously. While this view had the advantage of properly characterizing *space* through the situation concept (since space is the order of simultaneity), it also had the fault, as it were, of forcibly excluding any situational consideration from the definition of time: a conclusion that Leibniz does not hesitate to draw to its full coherence. Thus, he says that temporal instants are endowed with *positio* in respect to one another, i.e., with a reciprocal relation of order but by no means with *situs*.[76] This is the (more phenomenological than mathematical or physical) reason why Reichenbach thinks he can find such a peculiar construction of the relation of time ordering in the *Initia*; which in fact, as it implies

---

[73] For the theoretical aspects of the issue, see axioms IV, 1 and IV, 2 in §§ 10 and 11 of [Reichenbach 1924c]. Leibniz's locus is GM vii, p. 25, and Reichenbach's interpretation can be read in [Reichenbach 1924a], pp. 422–23.

[74] For an argumentation of it, here I cannot but refer the reader to Chapter 2 in [De Risi 2007].

[75] An early characterization in these terms already appears in a letter to Gallois from 1677, but not until the *Specimen Geometriae luciferae* (1695) and other writings of the same period will it become a proper definition. The definition sent to Gallois reads as follows: «Apres avoir bien cherché, j'ay trouvé que deux choses sont parfaitement semblables, lorsqu'on ne les sçauroit discerner que **per compraesentiam**, par example, deux cercles inegaux de même matiere ne se sçauroient discerner qu'en les voyant ensemble, car alors on voit bien que l'un est plus grand que l'autre. Vous me direz : je mesureray aujourdhuy l'un, demain l'autre ; et ainsi je les discerneray bien sans les avoir ensemble. Je dis que c'est encor les discerner **non per memoriam, sed per compraesentiam** : parce que vous avez la mesure du premier presente, non pas dans la memoire, car **on ne sçauroit retenir les grandeurs**, mais dans une mesure materielle gravée sur une regle, ou autre chose. Car si toutes les choses du monde qui nous regardent, estoient diminuées en même proportion, il est manifeste, que pas un ne pourroit remarquer le changement» (A iii, 2, n. 79, p. 227–28; A ii, 1, n. 158, p. 380 ; GM i, p. 180).

[76] «An dicere licebit? **Positio** est modus discernendi etiam ea quae per se discerni non possunt, ut duo puncta per se nil habent quo discernantur, at positione discernentur. **Situs** erit positio coexistendi, est ergo positionis species. Etiam instantium datur positio, non situs» (around 1700; in C 540–41).

no situational relations and thus no consideration of distance, Reichenbach mistakes for a topological one.

Leibniz's time is structurally non-metric; it is above all a time of consciousness. And the question even arises whether it can be metricized at all. Leibniz, as we have seen, does not seem to concede much room to conventionalism; and it is by no means clear according to what (non-empirical, of course) principle the *spatialization* of time, i.e., the assignation of a situation to any single instant, should ever be realized.[77]

This problem has not only mathematical but also strictly philosophical reverberations. In fact, over the years, in the framework of Leibniz's system the situation concept ended up by taking on the most general role of a perceptual synthesis. According to a very well-known definition of Leibniz's, a perception is the expression of the Many in the One, and this expression is instantiated by situational relations that produce indeed a spatial representation (a phenomenon). Leibniz however, owing to the reasons we have here briefly suggested, has no analogous synthesis for time, nor (to employ a Kantian terminology) genuinely dynamical categories.[78] Among the many consequences of it, Leibnizian real time (i.e., change) turns out to be disarticulated into instants, just as it is in some recent physical accounts. It is interesting to note how directly some of today's cosmological hypotheses (such as Julian Barbour's, for instance) claim to derive from a Leibnizian model as far as "Machian" relativity of motion is concerned, and how directly as well they link the impossibility of any real synthesis for time with the impossibility of determining absolute space and true motion.[79]

For our purposes, it will be sufficient to have remarked that Leibniz's construction of temporal "topology" did not after all rest on topology but on phenomenology, and that Reichenbach, once again, grasped something profound that had eluded many, or even all other interpreters, however twisting it into a physicalist perspective that little befitted Leibniz's philosophy.

---

[77] In the above-mentioned passage in the *Initia* Leibniz seems to suggest that conditions should be laid down on the existence of empirical objects in time. This is however a most delicate passage, because, as a condition of possibility of an objective (metric) time, it requires a phenomenal condition that should be conditioned in turn by that very time. For a reading along these lines, see [De Risi 2007], p. 359 and note 47. In any case, we must not forget that Leibniz believed some physical laws to be themselves *a priori*, and thus it may be possible to rely on some of such functional laws in order to determine, for instance, some periodical motions. For a grasp of the importance given by seventeenth-century physics to the business of determining uniform time, see for example the early work *Theoria Motus abstracti*, in GM vi, p. 70; GP iv, p. 231; A vi, 2, n. 41, p. 267; and then the text from 1679, *On the universality of Number, and on Time*, in [De Risi 2007], p. 622. For a very naïve but "topologically" flavored solution, see [Mahnke 1917], § 53, p. 52.

[78] *Positio* is not a synthetic modality, as we can see in the manuscript of the *Initia rerum mathematicarum metaphysica* (cf. [De Risi 2007], p. 483 note 31). And neither is the *appetition*, which permits the transition from one representation to another (*Monadologie*, §§ 11–16; GP vi, pp. 608–609), for it is only mental and ideal (a not necessarily fulfilled intentionality).

[79] This is actually Leibniz's well-known doctrine of *transcreatio*, which first appears in his *Pacidius Philalethi* from 1676 (A vi, 3, n. 78, p. 560; C 617), but will recur until the last years of his reflections (Leibniz to De Volder, 11 October 1705; GP ii, p. 279). The *Pacidius* is full of arguments derived from Sextus Empiricus; as for the problem of the relationship between causality and time, see *Adv. Phys.* B, 232–35. For an overview of Barbour's dynamic theories, see [Barbour 1999]; for the connection with Leibniz, see some passing remarks in [Barbour 1989].

17. Conclusions

We have thus compared and contrasted Reichenbach's and Mahnke's interpretations on two very relevant issues of Leibniz's philosophy: motion and time. It seems that a consistent reconstruction of Leibniz's thought should take into consideration a few convincing arguments from one interpretation and a few from the other, but also carefully avoid being caught in a somehow amphibious, half phenomenalist and half positivist, reading of Leibnizian philosophy. On the contrary, it is indeed a certain radicalization of the phenomenological perspective that best seems to allow the many oscillations and complexity of an always in-progress thought to retrieve its compactness. Such a reading also very well marks, in my opinion, both the limits and strong points of those interpretations from the twenties.

In this connection, Reichenbach's purely physicalist perspective proves inadequate, because the theory of sensibility (and indeed the whole phenomenological doctrine) of Leibniz's philosophy must also lie in the core of his natural philosophy, and Leibniz, *homo metaphysicus*,[80] was not, did not wish to be, nor could have ever been, a pure physicist in the twentieth-century and positivist meaning of the word. Just as Kant, he needs a theory of phenomena in order to build a philosophy of nature, and he needs a theory of the obscurity of representations in order to justify his own physical assumptions on relativity of motion or temporal simultaneity. Reichenbach however, who in the early twenties was definitely relinquishing Kantianism, had by then become deaf to such needs.

On the other side, Reichenbach's physicalist perspective has at least the advantage of, so to speak, keeping phenomena and noumena together. In this way, Reichenbach consistently rejects both the conventionalist and the pseudo-Kantian reading suggested by Mahnke. Coming from phenomenological studies, Mahnke had possibly (as far as I can understand) more chances to reconstruct Leibniz's genuine thought, but he let himself be persuaded by neo-Kantian suggestions, or simply a certain *Zeitgeist* in academic German philosophy, to expound Leibniz's phenomenology as if it were mere phenomenalism, and thus to insist on the divorce between things-in-themselves and phenomena, and the infinite progress of knowledge.[81] By doing so, he betrayed a direction of thoughts that

---

[80]We owe this peculiar expression to Leibnizian scholar Bilfinger.

[81]This point is also most clearly witnessed in the subsequent (1933) debate on mathematical epistemology that, within the Husserlian school, divided Mahnke and Oskar Becker, and that is recounted in [Mancosu 2005]. Becker, for the sake of phenomenological method, excluded the possibility of accessing certain non-constructive, transfinite, ideal objectivities. And not only did he exclude it for an embodied mathematician (for a *Dasein*, thus), but also for an ideal mathematician as a pure Ego. A pure Ego, which we may regard as an angelic intelligence of classical metaphysical tradition, Becker said, is anyway well distinguished from the actually infinite divine understanding. Mahnke admits to this last point, reaffirming furthermore that such was also Leibniz's view, and that angelic intelligences and the soul of the world, as far as intellectual power is concerned, are undeniably distinguished from God also in Leibniz's system. And yet, Mahnke goes on, it is well possible that a pure intelligence, which is, as such, uninvolved with time and mortality, can indefinitely *approximate* God's understanding. This reasoning however, which would serve (some 20th-century interpretations of) Kant's transcendental Ego very well, seems to me most erroneous as for Leibniz's doctrine, as Leibniz has in fact always insisted that there is no continuity of sorts between divine and human understanding but, on the contrary, an irreducible leap. So much so that Leibniz carefully distinguishes between the hypercategorematic infinite, which is the true divine Absolute, and a categorematic infinite, which would be the actual infinite attained

was well present, and I would say dominant, in Leibniz's last writings, according to which the expression concept became the core of his entire philosophy, and there was nothing residual in the realm of monads that might not be represented through phenomena. So much so that Leibnizian phenomenalism would come close to phenomenology at least under the following respect: in its maintaining that in a phenomenon, and not beyond it, resides everything that can and must be known. Leibniz's philosophy thus ended, in some way, by resembling more the empiricism of Reichenbach than the Kantianism of the things-in-themselves. Such a feature of Leibniz's philosophy seems to me to have been very well understood by young Mahnke, imbued as he still was with phenomenology, but to have faded away in his later years. Here, I have attempted to show a few stages in his slowly abandoning pure phenomenology in favor of other philosophical stances, such as they appear through Mahnke's statements in his correspondence with Reichenbach, and then, above all, in his subsequently departing from Fink and Becker. By the 1930s, from the point of view of orthodox phenomenology, Mahnke definitely appeared as a metaphysician who was building worlds of noumena and realms of monads beyond phenomena.

It seems to me that Reichenbach's flight from neo-Kantianism and Mahnke's flight from phenomenology, which both took place around 1924, are indeed the most revealing keys to judge of the two thinkers' respective interpretations of Leibniz, and to account for the sometimes very peculiar stances the two contestants happened to defend, often against their own creed or inclination, but always originally and profoundly. In this laborious quest for a possible pathway of their own, pinned against the background of their studies on Leibniz's philosophy, also lies, I think, the most genuine interest of the correspondence between Reichenbach and Mahnke.

Appendix. The Mahnke-Reichenbach correspondence

HR-16-37-08

Greifswald, 25. 12. 24.

Sehr geehrter Herr Doktor,

Ihr Aufsatz über die Bewegungslehre bei Newton, Leibniz und Huyghens [besser Huygens] im neusten Heft der Kantstudien, das ich soeben erhalte, interessiert mich so ausserordentlich, dass ich Ihnen darauf gleich heute – am Weihnachtstage! – schreiben muss. Denn er hilft mir bei einem Ziele, das ich seit Jahren verfolge, den lange verkannten Leibniz in seiner wissensch. Bedeutung zu rehabilitieren und seine aktuelle Gegenwartsbedeutung auf fast allen Gebieten seiner universalen Lebensarbeit nachzuweisen. Dass

---

through a progression, and thus the achieved regulative ideal (that is: God, according to some Kantian or neo-Kantian perspectives as well as in Mahnke's opinion), which categorematic infinite for Leibniz is even contradictory (Leibniz to Des Bosses, 1st September 1706, GP II, pp. 314–15; *Nouveaux Essais*, II, XVII, § 1; A VI, 6, n. 2, p. 157; GP V, p. 144). In 1933, therefore, I think, Mahnke was much more a neo-Kantian than a genuinely Leibnizian or a phenomenologist.

Leibniz der Relativitätstheorie sehr nahe steht, ist mir auch bereits aufgefallen. Ich verweise auf meine "Neue Monadologie", Ergänzungsheft Nr. 39 der Kantstudien, S. 90, und vor allem "Leibniz und Goethe" S. 98 ff. (Ich sende Ihnen mit gleicher Post ein Exemplar dieser vor 5 Monaten erschienenen, vor 13 Monaten verfassten Schrift.) Sie beherrschen allerdings die moderne Relativitätstheorie unvergleichlich besser als ich, dafür glaube ich aber wesentliche Ergänzungen zu Ihrer Leibnizdarstellung geben zu können. Lesen Sie bitte einmal die in Anm. 66 zitierten Stellen nach (ausserdem mir seit der Drucklegung noch aufgefallen sind: Couturat, Opuscules et fragments inédits de Leibniz, Paris 1903, S. 485 f., 590–593 sowie die Ihnen bekannten Briefe an Clarke); Sie werden dort, wie ich glaube, finden, was Sie S. 428, 429, 434 bei Leibniz mit Unrecht vermissen. Doch ich will vorläufig nichts weiter über meine Auffassung dieser Stellen verraten, um Sie ganz "vorurteilslos" an das Studium herangehen zu lassen. Später wäre mir eine Diskussion mit Ihnen sehr erwünscht, da ich in meinem in Vorbereitung befindlichen Leibnizbuche (in der Kafkaschen Sammlung) auch auf diesen Punkt näher eingehen möchte, als ich in "Leibniz und Goethe" beiläufig konnte.

Mit dem Ausdruck vorzüglicher Hochachtung

Ihr ganz ergebener

Dietr. Mahnke.

HR-16-37-09

25. Jan. 1925.

Herrn Oberstudienrat Dr. Mahnke,
Greifswald
Moltkestr. 7.

Sehr geehrter Herr Doktor,

ich danke Ihnen sehr für Ihr freundliches Schreiben zu meinem Leibniz-Aufsatz und auch für die liebenswürdige Uebersendung Ihrer Schrift über Leibniz und Goethe. Leider konnte ich Ihnen nicht eher antworten, da ich erst jetzt von einer mehrwöchentlichen Vortragsreise zurückgekehrt bin. Ich bin mit Ihnen ganz einig darin, daß Leibniz heute lange nicht genug gewürdigt wird, und daß es eine Forderung der historischen Gerechtigkeit ist, auf seine weitschauenden Arbeiten hinzuweisen. Ich bin Ihnen für den Hinweis auf einzelne Stellen bei Leibniz jederzeit sehr dankbar, da es wohl möglich ist, daß mir einiges entgangen ist. Bisher konnte ich leider nur diejenigen Stellen nachschlagen aus Ihren Zitaten, die sich in der Cassirerschen Ausgabe finden. Diese Stellen enthalten allerdings nichts Neues gegenüber den schon von mir zitierten Stellen, denn immer wieder findet sich darin der Gedanke, daß es im metaphysischen Sinne doch eine absolute Bewegung gibt; und über die bloße Versicherung, daß alle Annahmen über die Verteilung der Bewegung gleichberechtigt wären, kommt Leibniz anscheinend nicht hinaus. Es war mir in meiner Arbeit darum zu tun, nachzuweisen, daß trotzdem die Leibnizsche Auffassung schon einen wesentlichen Fortschritt bedeutet; anderseits glaube ich, daß man hier doch nicht zu weit gehen darf und dem alten Philosophen schon Ansichten unterlegen darf, die erst bei dem heutigen Stand der Wissenschaft möglich sind. Das würde seine Lehre geradezu entwurzeln.

Ich werde sehr gern, sowie ich die Zeit finde, die andern von Ihnen zitierten Stellen nachsehen. – Heute schicke ich Ihnen gleichzeitig eine kleine Arbeit von mir über die relativistische Zeitlehre, die Sie vielleicht interessieren wird.

Ich bin mit ergebenem Gruß

Ihr

[Hans Reichenbach]

                                                                    HR-16-37-10
Greifswald, 21.2.25.

Sehr geehrter Herr Doktor!

Nachdem endlich das Manuskript meines im Druck befindlichen Leibnizbuches abgeschlossen und dem "hungrigen" Setzer übersandt ist, beeile ich mich, Ihnen für Ihren freundlichen Brief vom 25.1 und die liebenswürdige Übersendung Ihres Scientia Aufsatzes über die relativistische Zeitlehre bestens zu danken. Ich habe den Aufsatz mit grossem Interesse gelesen. Er räumt mit grosser Klarheit und Überzeugungskraft allerlei Gegenargumente hinweg. Sie haben auch ganz recht – und darum interessiert mich als Leibnizforscher Ihre Arbeit noch besonders, wenn Sie Ihre Axiomatik der relativistischen Raum-Zeit-Lehre als eine (unbewusste) Weiterführung Leibnizscher Gedanken ansehen (Kantstudien, Bd. 29, S. 421). Auch das ist mir natürlich völlig klar, dass es eben eine sehr wesentliche Weiterführung ist. Sie bezeichnen den Unterschied sehr deutlich, wenn Sie sagen, Leibniz habe noch nicht bemerkt, dass die Gleichzeitigkeit keineswegs schon eindeutig definiert sei, wenn man verlange, dass die Ursache der Wirkung nie und nirgends zeitlich nachfolge (dass also, wie ich sagen möchte, eine ursächliches Folgen "aus" auch immer ein zeitliches Folgen "auf" sei). Leibniz ist in der Tat der Meinung, dass erstens in der Erlebniswelt jeder einzelnen Monade die Zeitordnung durch die Begründung jedes Zustandes durch einen andern (den "vorher gehenden") bestimmt festgelegt ist und zweitens die Zustände jeder Monade denen jeder andern durch eine funktionale Entsprechung ein-eindeutig zugeordnet sind, so dass also mit der Zeitordnung in jeder Monade die jeder andern mitgegeben ist. (Diese funktionale Zuordnung, meist prästabilierte Harmonie genannt, ist die "Verknüpfung aller Dinge" in der von Ihnen S. 421 f. zitierten Stelle.) Von einer Relativierung der Gleichzeitigkeit kann also bei Leibniz durchaus nicht die Rede sein. In dieser Hinsicht liegt es mir völlig fern, Leibniz "schon Ansichten unterzulegen, die erst bei dem heutigen Stand der Wissenschaft möglich sind", wie Sie mir schreiben. Leibniz hat gewiss nicht die Einsteinsche Relativitätstheorie vorweg genommen, die auf den neuen empirischen Erkenntnissen beruht. Ja, er begründet seine Relativitätstheorie überhaupt nicht, wenigstens nicht zunächst, auf physikalische Erfahrung, sondern auf metaphysische Spekulation und zeigt erst nachträglich ihre Übereinstimmung mit der Empirie.

Soweit stimme ich mit Ihnen überein. Im Unterschiede von Ihnen aber glaube ich nun zeigen zu können, dass Leibniz wenigstens die Relativität der räumlich-zeitlichen Bewegung ganz konsequent und in jeder Hinsicht behauptet und begründet hat. Die Stellen, die Sie dagegen anführen, handeln gar nicht von der physikalischen, räumlich-zeitlichen Bewegung, sondern von dem, was nach Leibnizens Metaphysik dieser phänomenalen Bewegung in Wahrheit zu Grunde liegt: von der psychischen* "Kraft", die gleichzeitig Energie und Entelechie ist, der Fähigkeit zu geistiger "Bewegung", zu Vorstellungsänderung und Willensstrebung. Solche "Bewegungskraft" aber kommt allerdings "absolut" jeder Monade zu, nur in verschieden hohem Grade. Wenn aber diese absolute innere Wirklichkeit äusserlich, körperlich "erscheint", als räumliche Bewegung, so wird sie dadurch radikal relativiert. Man kann jeden Körper als ruhend oder beliebig bewegt ansehen u.zw. nicht nur in Translationsbewegung, sondern auch in Rotationsbewegung befindlich – alle "Hypothesen" sind "äquivalent", keine kann durch die Erscheinungen widerlegt werden, und es gibt keine andre Entscheidung zwischen ihnen als auf Grund der "Einfachheit". Die Tatsache dieser Relativität benutzt Leibniz gerade, um zu zeigen, dass die wahre Wirklichkeit etwas ganz anderes als räumliche Ausdehnung und Bewegung ist, nämlich innere (letztlich seelenartige) Tätigkeit. (Gerh. Leibnizens math. Schr. VI 251).

Sie vermissen (S. 428 f., 434) im Briefwechsel mit Clarke u. Huygens eine nähere Ausführung seiner Relativitätstheorie der Rotationsbewegung und meinen, wir würden wohl nie erfahren, wie Leibniz sie sich gedacht hatte und was er Clarke erwidert haben würde. Nun, das letztere hat Cassirer in seiner (u. Buchenaus) Ausgabe der Hauptschriften I S. 220 f. schon ganz richtig unter Hinweis auf Gerh. VI 197 festgestellt[†]. Und das erstere finden Sie ausführlich in Leibnizens grosser Dynamica Gerh. VI 484 ff., 500 ff. und vor allem 507–11, auch in der kürzeren Bearbeitung, Specimen dynamicum Gerh. VI 247–253. (Diesen II. Teil haben Buchenau-Cassirer leider I 272 nicht mit angeschlossen; den I. Teil hat Leibniz 1695 in den Acta Eruditorum veröffentlicht, so zu sagen als "Fühler", ob seine grosse Dynamik auf Verständnis würde rechnen können – er fand kein Verständnis, und so hat er die fast druckfertige Dynamik und auch den II. Teil des Spec. dyn. ungedruckt gelassen; auf diese Weise ist es so lange unbekannt geblieben u. z.T. jetzt noch unbekannt, wie weit er die spätere Wissenschaft schon vorausgenommen; vgl. mein "Leibn. u. Goethe" S. 34 ff., Anm. 56, 58, 62, 63, 66.)

Ich glaube, wenn Sie diese Stellen nachlesen, werden Sie Leibniz dasselbe (nicht mehr, aber auch nicht weniger allerdings in anderer Weise!) zubilligen, wie Sie S. 435, 2. Absatz und S. 437 unten Huygens nachrühmen.

Mit dem Ausdruck vorzüglicher Hochachtung

Ihr ganz ergebener

Dietrich Mahnke

Verzeihen Sie, bitte, die nachträglichen Streichungen und Zusätze, die ich im Interesse der Deutlichkeit glaubte machen zu müssen!

---

* oder wenigstens seelenähnlichen, wenn auch unterbewussten "Repräsentationskraft" – repraesentare = mathematisch darstellen, nicht notwendig = vorstellen

[†] Leibniz gibt von seinem Standpunkt der Nahewirkung durch Druck und Stoss eine ganz analoge Lösung wie Mach vom Standpunkt der Newtonschen Fernwirkungstheorie der Gravitation!

References

[Arthur 1985] R.T.W. Arthur, *Leibniz's Theory of Time*, in [Okruhlik Brown 1985], pp. 263–313.

[Arthur 1994] R.T.W. Arthur, Space and Relativity in Newton and Leibniz, "British Journal for Philosophy of Science", 45, 1994, pp. 219–40.

[Barbour 1989] J. Barbour, *Absolute or Relative Motion?*, Cambridge 1989.

[Barbour 1999] J. Barbour, *The End of Time. The next Revolution in Physics*, London 1999.

[Barbour Pfister 1995] *Mach's Principle: From Newton's Bucket to Quantum Gravity*, eds. J. Barbour and H. Pfister, Boston 1995.

[Bernard Cohen 1985] I. Bernard Cohen, *Revolution in Science*, Cambridge 1985.

[Bernstein 1982] H.R. Bernstein, Leibniz and Huygens on the 'Relativity' of Motion, "Studia Leibnitiana Sonderheft", 13, 1982, pp. 85–102.

[Bertoloni Meli 1988] D. Bertoloni Meli, Leibniz on the Censorship of the Copernican System, "Studia Leibnitiana", 20, 1988, pp. 19–42.

[Bertoloni Meli 1991] D. Bertoloni Meli, Public Claims, Private Worries: Newton's Principia and Leibniz's Theory of Planetary Motion, "Studies in History and Philosophy of Science", 22, 1991, pp. 415–49.

[Brown 2005] H.R. Brown, *Physical Relativity: Space-time Structure from a Dynamical Perspective*, Oxford 2005.

[Carnap 1925] R. Carnap, Über die Abhängigkeit der Eigenschaften des Raumes von denen der Zeit, "Kant-Studien", 30, 1925, pp. 331–54.

[Cassirer 1902] E. Cassirer, *Leibniz' System in seinen wissenschaftlichen Grundlagen*, Marburg 1902.

[Cassirer 1920] E. Cassirer, *Zur Einstein'schen Relativitätstheorie*, Berlin 1920.

[Cassirer 1943] E. Cassirer, Newton and Leibniz, "Philosophical Review", 52, 1943, pp. 366–91.

[Coffa 1991] A. Coffa, *The Semantic Tradition from Kant to Carnap. To the Vienna Station*, Cambridge 1991.

[Cover 1997] J.A. Cover, Non-Basic Time and Reductive Strategies: Leibniz's Theory of Time, "Studies in the History and Philosophy of Science", 28, 1997, pp. 289–318.

[Cristin Sakai 2000] *Phänomenologie und Leibniz*, eds. R. Cristin and K. Sakai, Freiburg 2000.

[De Risi 2006] V. De Risi, Leibniz around 1700: Three Texts on Metaphysics, "The Leibniz Review", 16, 2006, pp. 55–69.

[De Risi 2007] V. De Risi, *Geometry and Monadology. Leibniz's Analysis Situs and Philosophy of Space*, Basel 2007.

[Earman 1973] J. Earman, Notes on the Causal Theory of Time, in [Suppes 1973], pp. 72–84.

[Earman 1989] J. Earman, *World Enough and Space-Time. Absolute versus Relational Theories of Space and Time*, Cambridge 1989.

[Earman Glymour Stachel 1977] *Foundations of Space-Time Theories*, eds. J. Earman, C.N. Glymour and J.J. Stachel, Minneapolis 1977.

[Enriques 1922] F. Enriques, *Per la storia della logica. I principi e l'ordine della scienza nel concetto dei pensatori matematici*, Bologna 1922.

[Ferrari 1995] M. Ferrari, Il neocriticismo tedesco e la teoria della relatività, "Rivista di filosofia", 86, 1995, pp. 239–81.

[Friedman 1977] M. Friedman, Simultaneity in Newtonian Mechanics and Special Relativity, in [Earman Glymour Stachel 1977], pp. 403–32.

[Friedman 1983] M. Friedman, *Foundations of Space-Time Theories. Relativistic Physics and Philosophy of Science*, Princeton 1983.

[Friedman 1992] M. Friedman, *Kant and the Exact Sciences*, Cambridge 1992.

[Friedman 1994] M. Friedman, Geometry, Convention and the Relativized A Priori, in [Salmon Wolters 1994]; now also in [Friedman 1999], pp. 59–70.

[Friedman 1999] M. Friedman, *Reconsidering Logical Positivism*, Cambridge 1999.

[Futch 2008] M.J. Futch, *Leibniz's Metaphysics of Time and Space*, Boston 2008.

[Garber 1995] D. Garber, Leibniz: Physics and Philosophy, in [Jolley 1995], pp. 270–352.

[Garber 2009] D. Garber, *Leibniz: Body, Substance, Monad*, Oxford 2009.

[Gent 1926] W. Gent, Leibnizens Philosophie der Zeit und des Raumes, "Kant-Studien", 31, 1926, pp. 61–88.

[Grünbaum 1973] A. Grünbaum, *Philosophical Problems of Space and Time*, 2nd ed., eds. R.S. Cohen and M.W. Wartofsky, Dordrecht 1973.

[Heilbron 1999] J.L. Heilbron, *The Sun and the Church: Cathedrals as Solar Observatories*, Cambridge 1999.

[Hentschel 1990] K. Hentschel, *Interpretationen und Fehlinterpretationen der speziellen und allgemeinen Relativitätstheorie durch Zeitgenossen Albert Einsteins*, Basel 1990.

[Husserl 1900/13] E. Husserl, *Logische Untersuchungen. Erster Teil. Prolegomena zur reinen Logik*, Halle 1900; 2nd ed. 1913.

[Husserl 1901/22] E. Husserl, *Logische Untersuchungen. Zweiter Teil. Untersuchungen zur Phänomenologie und Theorie der Erkenntnis*, Halle 1901; 2nd ed. 1922.

[Husserl 1913] E. Husserl, *Ideen zu einer reinen Phänomenlogie und phänomenologischen Philosophie. Erstes Buch: Allgemeine Einführung in die reine Phänomenologie*, "Jahrbuch für Philosophie und phänomenologische Forschung", 1, 1913.

[Husserl 1994] E. Husserl, *Briefwechsel*, ed. K. Schuhmann, vol. 3, Dordrecht 1994.

[Jauernig 2008] A. Jauernig, Leibniz on Motion and the Equivalence of Hypothesis, "Leibniz Review", 17, 2008, pp. 1–40.

[Jolley 1995] *The Cambridge Companion to Leibniz*, ed. N. Jolley, Cambridge 1995.

[Lariviere 1987] B. Lariviere, Leibnizian Relationism and the Problem of Inertia, "Canadian Journal of Philosophy", 17, 1987, pp. 437–48.

[Lechalas 1895] G. Lechalas, *Étude sur l'espace et le temps*, Paris 1895.

[Leibniz 1982] G.W. Leibniz, *Specimen Dynamicum*, eds. H.G. Dosch, G.W. Most and E. Rudolph, Hamburg 1982.

[Lewin 1923] K. Lewin, Die zeitliche Geneseordnung, "Zeitschrift für Physik", 13, 1923, pp. 62–81.

[Lynden-Bell 1995] D. Lynden-Bell, A Relative Newtonian Mechanics, in [Barbour Pfister 1995], pp. 173–78.

[Mach 1883] E. Mach, *Die Mechanik in ihrer Entwicklung historisch-kritisch dargestellt*, Leipzig 1883.

[Mahnke 1912a] D. Mahnke, *Leibniz als Gegner der Gelehrteneinseitigkeit*, Stade 1912.

[Mahnke 1912b] D. Mahnke, Leibniz auf der Suche nach der allgemeinen Primzahlgleichung, "Bibliotheca mathematica", 13, 1912, pp. 29–61.

[Mahnke 1913] D. Mahnke, Die Indexbezeichnung bei Leibniz als Beispiel seiner kombinatorischen Charakteristik, "Bibliotheca Mathematica", dritte Folge, 13, 1913, pp. 250–260.

[Mahnke 1917] D. Mahnke, Eine neue Monadologie, "Kant-Studien Ergänzunghefte", 39, 1917.

[Mahnke 1921] D. Mahnke, Die Neubelebung der Leibnizschen Weltanschauung, "Logos", 9, 1920–21, pp. 363–79.

[Mahnke 1924] D. Mahnke, Leibniz und Goethe, Erfurt 1924.

[Mahnke 1925] D. Mahnke, Leibnizens Synthese von Universalmathematik und Individualmetaphysik, "Jahrbuch für Philosophie und phänomenologische Forschung", 7, 1925, pp. 305–612.

[Mahnke 1927] D. Mahnke, Leibniz als Begründer der symbolischen Mathematik, "Isis", 1927, pp. 279–93.

[Mancosu 2005] P. Mancosu, "Das Abenteuer der Vernunft": O. Becker and D. Mahnke on the phenomenological foundations of the exact sciences, in [Peckhaus 2005], pp. 229–43.

[Mehlberg 1935] H. Mehlberg, Essai sur la théorie causale du temps, "Studia Philosophica", 1, 1935, pp. 119–260; "Studia Philosophica", 2, 1937, pp. 111–231.

[Mugnai 1992] M. Mugnai, Leibniz's Theory of Relations, "Studia Leibnitiana Supplementa", 28, 1992.

[Norton 1995] J. Norton, Mach's Principle before Einstein, in [Barbour Pfister 1995], pp. 9–57.

[Okruhlik Brown 1985] The Natural Philosophy of Leibniz, eds. K. Okruhlik and J.R. Brown, Dordrecht 1985.

[Parrini Salmon Salmon 2003] Logical Empiricism. Historical and Contemporary Perspectives, eds. P. Parrini, W.C. Salmon and M.H. Salmon, Pittsburgh 2003.

[Peckhaus 2005] Oskar Becker und die Philosophie der Mathematik, ed. V. Peckhaus, München 2005.

[Poser 1986] H. Poser, Monadologien des 20. Jahrhunderts, "Studia Leibnitiana Supplementa", 26, 1986, pp. 338–45.

[Putnam 1991] H. Putnam, Reichenbach's metaphysical Picture, "Erkenntnis", 35, 1991, pp. 61–75.

[Quine 1970] W.V.O. Quine, On the Reasons for Indeterminacy of Translation, "Journal of Philosophy", 67, 1970, pp. 178–83.

[Reichenbach 1920a] H. Reichenbach, Die Einsteinsche Raumlehre, "Die Umschau", 24, 1920, pp. 402–405.

[Reichenbach 1920b] H. Reichenbach, Relativitätstheorie und Erkenntnis Apriori, Berlin 1920.

[Reichenbach 1921] H. Reichenbach, Die Einsteinsche Bewegungslehre, "Die Umschau", 25, 1921, pp. 501–505.

[Reichenbach 1922a] H. Reichenbach, La signification philosophique de la théorie de la relativité, "Revue philosophique de la France et de l'Étranger", 94, 1922, pp. 5–61.

[Reichenbach 1922b] H. Reichenbach, Relativitätstheorie und absolute Transportzeit, "Zeitschrift für Physik", 9, 1922, pp. 111–117.

[Reichenbach 1922c] H. Reichenbach, Der gegenwärtige Stand der Relativitätsdiskussion, "Logos", 10, 1922, pp. 316–78.

[Reichenbach 1924a] H. Reichenbach, Die Bewegungslehre bei Newton, Leibniz und Huyghens, "Kant-Studien", 29, 1924, pp. 416–38.

[Reichenbach 1924b] H. Reichenbach, Die relativistische Zeitlehre, "Scientia", 1924, pp. 361–74.

[Reichenbach 1924c] H. Reichenbach, Axiomatik der relativistischen Raum-Zeit-Lehre, Braunschweig 1924.

[Reichenbach 1925] H. Reichenbach, Die Kausalstruktur der Welt und der Unterschied von Vergangenheit und Zukunft, "Sitzungsberichte der Bayerischen Akademie der Wissenschaften", 1925, pp. 133–175.

[Reichenbach 1928] H. Reichenbach, *Philosophie der Raum-Zeit-Lehre*, Berlin 1928.

[Reichenbach 1956] H. Reichenbach, *The Direction of Time*, ed. M. Reichenbach, Berkeley 1956.

[Robb 1914] A.A. Robb, *A Theory of Time and Space*, Cambridge 1914.

[Roberts 2003] J.T. Roberts, Leibniz on Force and Absolute Motion, "Philosophy of Science", 70, 2003, pp. 553–73.

[Russell 1914] B. Russell, *Our Knowledge of the External World*, Chicago 1914.

[Ryckman 2003] T.A. Ryckman, Two Roads from Kant: Cassirer, Reichenbach and General Relativity, in [Parrini Salmon Salmon 2003], pp. 159–93.

[Salmon Wolters 1994] *Logic, Language, and the Structure of Scientific Theories*, eds. W. Salmon and G. Wolters, Pittsburgh 1994.

[Schlick 1910] M. Schlick, Das Wesen der Wahrheit nach der modernen Logik, "Vierteljahrsschrift für wissenschaftlichen Philosophie und Soziologie", 34, 1910, pp. 387–477.

[Schlick 1913] M. Schlick, Gibt es intuitive Erkenntnis?, "Vierteljahrsschrift für wissenschaftliche Philosophie und Soziologie", 37, 1913, pp. 472–88.

[Schlick 1918/25] M. Schlick, *Allgemeine Erkenntnislehre*, Berlin 1918; 2nd ed. 1925.

[Schlick 1921] M. Schlick, Kritizistische oder empiristische Deutung der neuen Physik?, "Kant-Studien", 26, 1921, pp. 96–111.

[Schneider 1921] I. Schneider, *Das Raum-Zeit-Problem bei Kant und Einstein*, Berlin 1921.

[Sklar 1976] L. Sklar, *Space, Time, and Space-Time*, Berkeley 1976.

[Stein 1977] H. Stein, Some philosophical Prehistory of General Relativity, in [Earman Glymour Stachel 1977], pp. 3–49.

[Suppes 1973] *Space, Time and Geometry*, ed. P. Suppes, Dordrecht 1973.

[Van Breda 1966] H.L. van Breda, Leibniz' Einfluß auf das Denken Husserls, "Studia Leibnitiana Supplementa", 5, 1966, pp. 124–45.

[Van Fraassen 1970] B. van Fraassen, *An Introduction to the Philosophy of Space and Time*, New York 1970.

[Van Fraassen 1973] B. van Fraassen, Earman on the Causal Theory of Time, in [Suppes 1973], pp. 85–93.

[Weyl 1917] H. Weyl, *Das Kontinuum. Kritische Untersuchungen über die Grundlagen der Analysis*, Zürich 1917.

[Weyl 1927] H. Weyl, *Philosophie der Mathematik und Naturwissenschaft*, München 1927.

[Winnie 1977] J.A. Winnie, The Causal Theory of Space-Time, in [Earman Glymour Stachel 1977], pp. 134–205.

[Woltmann 1957] H. Woltmann, Dietrich Mahnke 1884–1939, in *Niedersächsische Lebensbilder*, ed. O.H. May, Hildesheim 1957.

Vincenzo De Risi
Max Planck Institute for the History of Science
Boltzmannstraße 22
D-14195 Berlin, Germany
vderisi@mpiwg-berlin.mpg.de

Interpretations of Leibniz's *Mathesis Universalis* at the Beginning of the XXth Century

David Rabouin

Introduction

In his doctoral dissertation, completed in 1922 under the direction of Edmund Husserl and published in 1925 in the *Jahrbuch für Philosophie und Phänomenologische Forschungen*, Dietrich Mahnke proposed a very valuable overview of the so-called "Leibniz Renaissance". As indicated by the choice of his title: *Leibnizens Synthese von Universalmathematik und Individualmetaphysik*, this renaissance was seen by Mahnke as marked by a tension between two Leibnizian programs: that of a "universal mathematics" and that of a "metaphysics of individuation". His agenda was to propose a way of reconciling these two programs through a point of view inspired by the development of Husserlian phenomenology. In this paper, I will concentrate on the first program, "universal mathematics" or *mathesis universalis*, and see how the interpretation of this Leibnizian theme was indeed a key point in the demarcation between different ways of articulating logic, mathematics and philosophy at the beginning of the XXth century. I will pay particular attention to the way in which commentators carefully selected their texts in the Leibnizian corpus. It will be an occasion to exhibit certain postulates lurking behind classical interpretations of Leibniz in the studies by Russell, Couturat, Cassirer, or Brunschvicg. I will then contrast these readings with another interpretation of Leibniz's *mathesis universalis*, permitted by a better access to the texts and a somewhat calmer discussion around the relationship between logic, mathematics and philosophy.

1. The battlefield

Supporters of "universal mathematics" as forming the core of Leibnizian philosophy – what Manhke describes as a form of "universal panlogism" – did not constitute a homogeneous field. There were major conflicts between them, especially on the front of epistemology. A first camp, represented by authors like Bertrand Russell and Louis Couturat, tended indeed toward a strong "conceptual realism" (*Begriffsrealismus*), as Manhke puts it – whereas the other direction, represented by authors like Cassirer, tended to an idealistic interpretation of Leibniz (*Vernunftidealismus*)– more a *panmethodologismus* than a *panlogismus* stricto sensu.

The diagnostic of this opposition was certainly not new, as can be seen in the famous joint review of Couturat (1901) and Cassirer (1902) by Russell in 1903. But the idea of gathering these different authors around the topic of *mathesis universalis* was indeed quite original and did not correspond to the way they portrayed their respective programs. One of Mahnke's great insights was to point to this theme as a kind of greatest common divisor between the different forms of "panlogism" and the proper place of opposition in their interpretations of Leibniz. My aim in this study is to follow this hypothesis.

In Russell's language, the greatest common divisor was best seen as "logic", i.e., in the thesis according to which Leibniz's metaphysical system derived from his research in logic and the foundation of mathematics: "Dr. Cassirer, like M. Couturat, regards Leibniz's Logic and his investigations of the principles of mathematics as the source of his metaphysical system" (Russell (1903), p. 191). What was then at stake was a proper characterization of what "logic" meant. Since Russell's intention was to secure his own camp by lining up with Couturat against Cassirer, he did not have to give a very precise characterization of these terms and could rely on a brute opposition between "symbolic" and "transcendental" logic:

> Unlike M. Couturat, the present author [*scil.* Cassirer] has not yet grasped the very modern discovery of the importance of Symbolic Logic. In the philosophy of mathematics, his views appear to agree closely with those of Prof. Hermann Cohen, to whom the book is dedicated, and to whom acknowledgments are made in the Preface. We find, accordingly, in spite of occasional references to Dedekind and Cantor, but little realisation of even the arithmetising of mathematics, and none at all of the still more recent "logicising", if such a word be permissible. Mathematics, for Dr. Cassirer, is not synonymous with Symbolic Logic, and Logic is synonymous with theory of knowledge.
>
> In both these respects, the work is Kantian, and supposes Leibniz, at least in a measure, to be also Kantian. The very rare merit of not imputing one's own philosophy to the author one is discussing belongs to M. Couturat's work, but not, I think, to Dr. Cassirer's; and as mathematics have of late conclusively disproved the Kantian doctrines as to their principles, the result is to rob Leibniz of his most extraordinary merit – I mean, the realisation of the supreme importance of Symbolic Logic. [Russell (1903), p. 191]

However, if one pays a little more attention to what *opposes* Russell to Couturat, one sees immediately that the appeal to "symbolic logic" is not enough and that *mathesis universalis* comes to the forefront to settle this divergence:

> Chapter vii. [*scil.* of Couturat's *La Logique de Leibniz*] deals with *Universal Mathematics* – a subject which appears to be precisely identical with what Mr. Whitehead has called *Universal Algebra*. Although M. Couturat deals with this subject in a different chapter from that devoted to the Logical Calculus,

he does not clearly state, any more than Leibniz does, the exact difference between the two. The fact is that the *Ars Combinatoria,* or Universal Mathematics, is more formal than the Logical Calculus; it is concerned with deductions from the assumption of a synthesis obeying such and such laws, but otherwise undefined. We may say that, in this subject, our signs of operation, our + and × and whatever other such signs we may employ, are themselves variables, subject merely to hypotheses as to their formal laws; whereas in every other branch of mathematics, and in the Logical Calculus itself, only the letters are variable, and the signs of operation have constant meanings.

It might seem, from this account, as though Universal Mathematics were the most general of all mathematical subjects, and in a sense this is true. But it is emphatically not the logically first of such subjects, for itself employs deduction and the logical kinds of synthesis, which are explicitly dealt with in the Logical Calculus (...). This science, therefore, is logically subsequent to the Logical Calculus. [Russell (1903), pp. 186–187]

In order to understand this passage properly, one has to keep in mind that in 1903, Russell was already working with A. N. Whitehead on a project which replaced the planned second volume of Whitehead's *Universal Algebra* and which would become the *Principia Mathematica.* The opposition between Couturat and Russell could hence be seen as a consequence of the twofold heritage of the programs of universal algebra at the end of the XIXth century: on the one hand a conception of logical calculus as a branch of a general theory of structures (i.e., different forms of calculus), which was the initial program of Whitehead; on the other hand, a conception of these logical calculi as depending on a general theory of deduction, from which the different mathematical structures would derive by specification of new axioms[1]. According to Russell's interpretation, *Mathesis universalis* had clearly to do with the first line of development. In agreement with Couturat, he considered this theory as "the most general of all mathematical subjects", but added an important *caveat*: "not the logically first of such subjects". This is certainly why he himself did not pay much attention to this topic in his book on Leibniz (Russell (1900)). In fact, as far as I know, the only reference to the theme in this book is to an isolated passage from Leibniz (1704), the only one in which Leibniz seems to identify "universal mathematics" with a form of logical calculus:

Je tiens que l'invention de la forme des syllogismes est une des plus belles de l'esprit humain. C'est une espèce de *Mathématique universelle* dont l'importance n'est pas assez connue; et l'on peut dire qu'un *art d'infaillibilité* y est contenu, pourvu qu'on sache et qu'on puisse s'en bien servir, ce qui n'est pas toujours permis. Or il faut savoir que par les *arguments en forme,* je n'entends pas seulement cette manière scolastique d'argumenter dont on se sert dans les Collèges, mais tout raisonnement qui conclut par la force de la forme, et où l'on n'a besoin de suppléer aucun article, de sorte qu'un *Sorites,* un autre tissu de syllogisme qui évite la répétition, même un compte bien dressé, un calcul

---

[1]On this twofold heritage, see Hintikka (1997), which contains as an appendix the classical Van Heijenoort (1967).

d'algèbre, une analyse des infinitésimales me seront à peu près des arguments en forme, parce que leur forme de raisonner a été prédémontrée, en sorte qu'on est sûr de ne s'y point tromper. (G.W. Leibniz (1704), livre IV, Chap. 17, § 4, [GP V 460–461 ; A VI, 6, 478])[2]

This passage was certainly one of the most famous descriptions of the Leibnizian *mathesis universalis* in the general literature[3]. But, as we will see later, the representativeness of this description is unfortunately inversely proportional to its celebrity.

The link between *mathesis universalis* and the heritage of the "Universal Algebra" program was explicit in Couturat. An article published in the *Revue de Métaphysique et de Morale* in 1904 makes this point particularly clear: after having repeated the "logicist" interpretation according to which Leibniz's great insight was to have foreseen the characterization of mathematics as a purely *formal* science, he goes on by directly identifying "universal mathematics" and "universal algebra"[4]. On this occasion, he sets out very clearly the genealogy of this program:

> It is only little by little, after Möbius' invention of barycentric calculus, Bellavitis' calculus of equipollence, Grassmann's geometrical calculus, Hamilton's quaternions, Staudt's projective geometry, set theory, group and substitution theory, and finally Boole's logical calculus, that it became possible to conceive that Mathematics is not tied to a particular nature of objects, but is a general method of demonstration and invention [My translation][5].

---

[2] In Peter Remnant and Jonathan Bennett's translation (Leibniz (1981), p. 478): "I hold that the invention of the syllogistic form is one of the finest, and indeed one of the most important, to have been made by the human mind. It is a kind of universal mathematics whose importance is too little known. It can be said to include an art of infallibility, provided that one knows how to use it and gets the chance to do so – which sometimes one does not. But it must be grasped that by 'formal arguments' I mean not only the scholastic manner of arguing but also any reasoning in which the conclusion is reached by virtue of the form, with no need for anything to be added. So: a sorites, some other sequence of syllogisms in which repetition is avoided, even well drawn-up statements of accounts, an algebraic calculation, an infinitesimal analysis- I shall count all of these as formal arguments, more or less because in each of them the form of reasoning has been demonstrated in advance so that one is sure of not going wrong with it.".

[3] See, amongst many others, E. Husserl *Logische Untersuchungen* I (1900), § 60; H. Weyl (1926), p. 12. It is worth noting that it appears also on the very first page of Couturat (1901), Book I, chapter 1.

[4] «Kant concevait, avec tous ses contemporains, les mathématiques comme les sciences du nombre et de la grandeur, et même, plus étroitement encore, comme les sciences de l'espace et du temps, et non pas comme une science ou plutôt une méthode purement *formelle*, comme un ensemble de raisonnements déductifs et hypothétiquement nécessaires. Ici encore, on ne saurait lui reprocher de n'avoir pas prévu l'avenir, encore que, sur ce point aussi, Leibniz ait vu plus clair et plus loin que lui, et ait conçu fort nettement la Mathématique universelle, et plus spécialement l'Algèbre universelle (qu'il appelait la Caractéristique) comme applicable à toutes les formes possibles de déduction. Mais ces anticipations géniales étaient encore inconnues ou méconnues, et passaient alors pour des rêves d'utopistes» [Couturat (1904). This article was published as an appendix to Couturat (1905/1980), p. 304.]

[5] «Ce n'est que peu à peu, à la suite de l'invention du calcul barycentrique de Möbius, du calcul des équipollences de Bellavitis, du calcul géométrique de Grassmann, des quaternions de Hamilton, de la Géométrie projective de Staudt, de la théorie des ensembles, de la théorie des substitutions et des groupes, enfin du calcul logique de Boole, qu'on est parvenu à concevoir que la mathématique n'est pas liée à une nature particulière d'objets, mais est une méthode générale de démonstration et d'invention.» (*ibid.*)

I shall come back to the details of this interpretation later. For the time being, I just want to sketch out the different camps of our battlefield and as such will now turn to Cassirer's interpretation. At first sight, the idea of *mathesis universalis* does not play a central role in Cassirer (1902). But one has to keep in mind that *Leibniz's System* incorporates as a first part *Descartes's Kritik der mathematischen und Naturwissenschaftlichen Erkenntnis*, the dissertation which Cassirer defended at Marburg in 1899. Yet, the central thesis of this first study, deeply influenced by the works of Natorp and Cohen, was centred on the idea of *mathesis universalis*.

Like Natorp before him, Cassirer claimed in his thesis that there was a first Cartesian revolution, illustrated by the *Regulae ad directionem ingenii*, where Descartes appeared as a true precursor of Kant. This revolution was characterized by the transition from a model of rationality centred on the notion of substance and the criterion of truth as *adequatio* to a model centred on the knowing subject and the criterion of truth as *certitudo*. According to the Neo-Kantians (and still further, to authors like Husserl and Heidegger) this revolution signaled the beginning of the modern era in philosophy – an era which corresponded with the beginning of "Modern Science" and the destitution of the old Metaphysics in favour of the primacy of the theory of knowledge.

However, according to Cassirer, Descartes was not able to follow this impulsion to its last consequences and fell again quickly into metaphysical considerations. I will not enter into the details of this relapse; the important fact for us is that "*Mathesis universalis*" appearing in Descartes's *Regulae ad directionem ingenii* as the science of "order and measure"[6] was therefore seen as a proper name of the "modern" revolution in philosophy and science, when ontology was brought back to its proper setting: the order imposed by the subject on any rational knowledge[7].

Cassirer's study on Leibniz follows very clearly the same line of interpretation. Its strategy consists indeed in showing that Leibniz overcame some of Descartes' limitations by integrating a dynamical view absent from the static Cartesian conception of knowledge. In this sense, Leibniz was giving the program underlying the concept of *mathesis univeralis* its full power. As I will show later in more details, this reading was also that of Léon Brunschvicg. A key point in this line of interpretation was, of course, the invention of differential calculus, presented as a foundation for a new concept of identity and a basis for many of Leibniz's metaphysical positions (a thesis inherited from Cassirer's teacher: Hermann Cohen). This orientation is very apparent in the structure of Cassirer's book: starting from a reflection on Quantity, core of the old (i.e., Cartesian) "universal mathematics", it proceeds then to a reflection on what Leibniz saw as the "qualitative" part of mathematics, which Descartes failed to take into account and which deals with: place (*situs*), continuity and intensity, core of the new (i.e., Leibnizian) *mathesis universalis*.

My aim is not to enter into the details of these readings, but to follow Mahnke's proposal according to which the interpretation of *mathesis universalis* was a key point in their opposition. Through this particular focus, one sees indeed very clearly the delineation of our battlefield. Not only was Leibniz's thought exploited in a very vivid quarrel over the

---

[6] *Regulae ad directionem ingenii*, Rule IV [AT X, 378, 5–8].
[7] See also E. Cassirer (1906).

relationship between logic, mathematics and philosophy, but it was used in the construction of a certain image of the "modern rationality" and the "scientific revolution"[8]. This point is of particular interest for historians of science since it was of great influence in their discipline (see, for an example in the history of mathematics, Jakob Klein's famous book of 1936[9]).

2. The choice of weapons

Now that we have a clearer idea of our battlefield, I will concentrate more precisely on the way in which these interpretations used Leibniz's texts. For this purpose, I will pay particular attention to Couturat's reading – by far the richest in this regard – and, as a counterpoint, to Brunschvicg's.

As we have seen, Couturat devoted a complete chapter of his 1901 book to "Universal Mathematics". One can certainly express, along with Russell, some surprise at this choice. Many readers at that time (and still now!) did indeed consider *Mathesis Universalis* to be just one amongst many figures of the famous Leibnizian dream of a "Universal Characteristic" (*Caracteristica Universalis*). Since Couturat devoted his chapter IV to this "Universal Characteristic" and chapter VIII to the various logical calculi, Russell could not understand why one should pay particular attention to "Universal Mathematics".

Notwithstanding their disagreement on the nature of "logic", one obvious difference between the two strategies was that Couturat was not simply engaged in a process of interpreting Leibniz's philosophy. His goal was also to give access to unpublished manuscripts. In this new corpus, he would certainly find at least one text in which *Mathesis universalis* was presented, as in Leibniz (1704), as a general logic, encompassing analysis and synthesis, i.e., *ars combinatoria* and algebra[10]. But this text would then define this discipline as a logic *of imagination* (*logica imaginationis*)[11]. This indication oriented Couturat's study in a very specific direction, in which the other authors were not much interested, namely the question of a proper characterization of the objects of mathematics as *imaginabilia*.

Following Leibniz's texts, Couturat came up against another important obstacle. At the beginning of the other main text dedicated to *Mathesis universalis*, and already published by Gerhard, one reads the following *caveat*:

> Many people have tried to illustrate Logic by comparing it to a Computation, and Aristotle himself expressed himself in a mathematical manner in the Analytics ; in the same way Arithmetic and Algebra, but most of all the *Mathesis*

---

[8] See also E. Husserl (1934–1937), Chap. 9, § f.

[9] Klein's study ends with John Wallis, but it aims at giving a genealogy of the transformation of the concept of *mathesis universalis* culminating with Leibniz. According to Klein's teleology, this development had to be seen in terms of a new definition of the concept of «number», considered after Wallis as a purely symbolic entity.

[10] *Haec Elementa Matheseos universalis, multo plus different a Speciosa hactenus cognita, quam ipsa Speciosa Vietae aut Cartesii differt a Symbolica veterum.(. . .). Tradetur et Synthesis et Analysis, sive tam Combinatoria, quam Algebra (Elementa Nova Matheseos Universalis* (1683?), A VI, 4, A, 513).

[11] *Mathesis Universalis tradere debet Methodum aliquid exacte determinandi per ea quae sub imaginationem cadunt, sive ut ita dicam Logicam imaginationis* (Ibid.)

which is truly *universalis* could be treated by logical means, as if they were Mathematical Logic, so that in fact *Mathesis universalis* or Logistic could coincide with the Logic of Mathematicians: hence our Logistic appears from time to time under the name of Analysis of Mathematics[12].

As Leibniz explains in the following development of this text, the goal of his planned treatise (one of the longest fragments on *Mathesis universalis*) was to develop a presentation of elementary mathematics (algebra and differential calculus) following a structure analogous to the distinction existing in Logic between notions, propositions, arguments and methods[13]. As he explicitly states in the preceding quote, Leibniz's move here goes in the *opposite* direction of any algebraization of logic: his originality was presented, on the contrary, as a kind of *logicization of algebra*.

On the ground of these documents, Couturat's strategy consisted therefore of reconstructing a path in the texts allowing a reading of these passages compatible with his "logicist" postulates. He started by taking the analogy between Logic and Mathematics as a *formal* one[14] and tried to characterize the object of mathematics through it. Beginning with the characterization of mathematical objects as magnitudes, inherited from Descartes' *mathesis universalis*, Leibniz's move was hence presented as accomplishing a step further commanded by his claim of a dependence of algebra upon *ars combinatoria*, "general science of forms and formulas ("science générale des formes et des formules"). Paradoxically, this brought Couturat in a very striking proximity to Kant's ideas, which is worth noting. Indeed he characterized first mathematics as dealing with what is susceptible to exact and precise determination in the realm of sensible intuition[15].

Of course such a position asked, in Couturat's eyes, for important qualifications. How could one accept, if mathematics is a purely formal science, that Leibniz's *Mathesis universalis* be so tied to intuition? What is the link between formal logic (or logical calculus) and "logic of imagination"? Couturat's strategy would here consist in leaving as soon as possible the shaky ground of "universal mathematics" for what he considered as a more solid one: "*ars combinatoria*". In this path, an important move was to interpret the link to imagination as a link not to intuition, but to the *form* of intuition, and more generally to the "form of thought". This would justify the progressive identification of *mathesis universalis* (or what he calls the "general and formal part of Mathematics") with

---

[12]GM VII, 54. My translation.

[13]*In Logica autem sunt Notiones, Propositiones, Argumentationes, Methodi. Idem est in Analysi Mathematica, ubi sunt quantitates, veritates de quantitatibus enuntiatae (aequationes, majoritates, minoritates, analogiae, etc.), argumentationes (nempe operationes calculi) et denique methodi seu processus quibus utimur ad quaesitum investigandum* [GM VII, 54].

[14]«Tout d'abord, il [*scil.* Leibniz] constate une analogie formelle entre la Logique et la Mathématique» (p. 282).

[15]«Ainsi la Mathématique universelle de Descartes est dépassée et enveloppée par la Mathématique universelle de Leibniz. Voici, en résumé, comment il comprend et divise celle-ci. La Mathématique n'a pas pour matière seulement le nombre et la grandeur, mais tout ce qui, *dans le domaine de l'intuition sensible*, est susceptible de détermination exacte et précise ; c'est, selon son expression, la *Logique de l'imagination*» (Couturat (1901), p. 290. My emphasis)

ars combinatoria as part of the «science of forms» and, finally, the partial identification with formal logic[16]

Through this identification between *ars combinatoria* and *mathesis universalis*, one could then easily establish the final thesis: "universal mathematics" *is* the "general science of relations", the analogy between logic and mathematics becoming then a "partial identity"[17]:

> En somme, Leibniz a eu le mérite d'apercevoir (bien avant les découvertes et les progrès modernes qui ont rendu cette vérité manifeste) qu'il y a une Mathématique universelle dont toutes les sciences mathématiques relèvent pour leurs principes et leurs théorèmes les plus généraux, et que *cette Mathématique se confond avec la Logique elle-même, ou du moins en est une partie intégrante. Il n'y a plus seulement entre la Logique et la Mathématique une analogie formelle, mais une identité au moins partielle.*C'est que, d'une part, la Mathématique universelle constitue, comme on l'a vu, la science générale des relations (...). D'autre part, la logique formelle s'étend jusqu'à coïncidence avec la Mathématique. En effet, c'est le caractère formel des raisonnements qui garantit la valeur universelle et nécessaire de la déduction (Couturat, 1901, p. 317–318. My emphasis)

I shall now turn to Brunschvicg's analysis in order to make clear that, even if Couturat was more faithful to the texts than authors like Russell, he selected carefully the passages on *mathesis universalis* illustrating his postulates. I chose Brunschvicg rather than Cassirer for several reasons: first, Brunschvicg comes after the battle and offers a panoramic view comparable to that of Russell (1903), with the important difference that he lines up on Cassirer's side. Moreover, due to this position, Brunschvicg's study of 1912 allows a more balanced point of view on the debate and offers the advantage of being more focused than Cassirer's. The last advantage is to bring the debate outside of the neo-Kantian/logicism quarrel and to point to a more general frame[18].

Brunschvicg (1912) does not give a central role to *mathesis universalis* in his reading of Leibniz. But, like Cassirer:

1) he dedicated an entire section of his book to *mathesis universalis* in Descartes' thought (Chap. VII, Section B);
2) he presented this topic as the proper name of a new era in the dialog between Reason and Nature[19];

---

[16]"En définitive, la Mathématique proprement dite, c'est-à-dire la Logistique, est subordonnée à la Combinatoire, et celle-ci à la Logique elle-même. La Combinatoire paraît même faire partie de la Logique. En tout cas, l'une et l'autre réunies composent la *science des formes*. Et par là il faut entendre non seulement les formules mathématiques et les «formes»algébriques, mais toutes les formes de la pensée, c'est-à-dire les lois générales de l'esprit. *La Combinatoire ainsi conçue est la partie générale et formelle des Mathématiques ; elle étudie toutes les relations qui peuvent exister entre des objets quelconques, et leur enchaînement nécessaire et formel. En un mot, c'est la science générale des relations abstraites*" (Couturat (1901), p. 299–300. My emphasis)

[17]This definition had tremendous success in the literature on Leibniz's *mathesis universalis*, see H. Burkhardt (1980), p. 321; W. & M. Kneale (1984), p. 336 sq.; M. Schneider (1988), p. 172.

[18]On Brunschvicg's relation to Kantism see L. Fedi (2001).

[19]See, for example, Section B, p. 113 where "Universal Mathematics" and "Cartesian mechanism" are identified.

3) he considered Leibniz's move as a way to overcome the limitation of the Cartesian program, insisting on the creation of differential calculus, dynamical conception of knowledge and the link of mathematics to physics (Chap. X).

The core of Brunschvicg's response to the logicist interpretation consisted in insisting on the difference between "real" and "ideal" logic. To his eyes, the view defended by authors like Russell and Couturat had systemically confused Leibniz's grandiose *ideals* and the *real* way in which he had worked. This confusion led to a misunderstanding of the way in which Leibniz overstepped the limitations of the Cartesian program. According to Brunschvicg, this step did not find its impulsion in a new dogmatic stance, grounded in a grandiose view of Logic as forming the core of any kind of rationality; its origin was more likely to be found in a series of concrete obstacles presented to this program in its implementation:

> Les embarras qui ont entraîné l'échec, ou tout au moins limité la portée de ces diverses doctrines [*scil.* celle de Descartes et plus généralement tous les dogmatismes scientifiques] décèlent la faiblesse du préjugé dogmatique. La meilleure méthode pour l'intelligence mathématique de l'univers n'est nullement celle qui, dans certains cas élémentaires, présente l'application la plus facile ; car cette facilité même, de nature à séduire le philosophe, paralyse le savant en présence des problèmes complexes que la réalité ne peut manquer de poser. C'est celle qui dans l'apparence du simple sait déjà discerner la complexité, la subtilité, caractéristique du réel ; les principes n'y sont plus des formes déterminées et closes, destinées à opérer la cristallisation du système scientifique ; ce sont des ressorts d'action, des armes pour l'extension illimitée du savoir positif. Descartes, comme les Grecs, se meut dans le domaine du fini ; Leibniz fait intervenir l'infini dans la génération du fini. La science de l'infini sert à trouver les quantités finies : *Itaque Matheseos Universalis pars superior nihil aliud est quam scientia infiniti, quatenus ad inveniendas finitas quantitates prodest.* (Brunschvicg (1912), p. 208)

As is obvious from the final quote, the core of the opposition was the intervention of the infinite in the program of a general analysis. Not that Russell and Couturat ignored this aspect (which is indeed central in their opposition), but Brunschvicg very directly opposed the idea that it could fit with their characterization of mathematics as a purely *formal* science. Of course, one can *transform* Leibniz calculus into formal deductions, but, notwithstanding the fact that this formalization is largely anachronistic, it would lose the very meaning of the introduction of this tool in mathematics as a way of *explaining* natural phenomena:

> Les degrés successifs de différenciation créés par l'algorithme de l'analyse de l'infini sont donc tout autre chose qu'un jeu d'écriture symbolique ; *ils constituent une méthode véritable d'explication, qui, à l'épreuve va se révéler universelle*(...). De fait, la réforme de la mécanique cartésienne, que Leibniz accomplissait dans cette même année 1686, ne pourra s'expliquer complètement sans l'intervention de concepts que seule l'analyse infinitésimale pouvait fournir ((1912), p. 215. My emphasis)

The reference attached by Brunschvicg in the note to this passage is the article: *Considérations sur la différence qu'il y a entre l'analyse ordinaire et le nouveau calcul des transcendantes* (published in the *Journal des Savants* in 1694). There Brunschvicg finds the following quote:

> Notre méthode étant proprement cette partie de la Mathématique générale[20] qui traite de l'infini, c'est ce qui fait qu'on en a fort besoin, en appliquant les Mathématiques à la Physique, parce que le caractère de l'Auteur infini entre ordinairement dans les opérations de la Nature [GM V, 308].

As one can see in these two passages, Brunschvicg did have at his disposal some texts in which *Mathesis universalis* (or "Mathématique générale") was not directly linked by Leibniz to Logic, but to the elaboration of the differential calculus as a tool for the knowledge of Nature ("les operations de la Nature"). In fact, he could have brought many other texts, where Leibniz insisted on the fact that his *mathesis universalis* overstepped Descartes' in the same range that differential calculus overstepped algebra (considered here in the Cartesian sense of an algebraic theory of magnitudes) and in the context of the knowledge of Nature[21].

One can see clearly in this brief overview to which point each of these readings chose very carefully the texts to be used. This indicates, by contrast, how subtle Mahnke's point of view was, since, in a way, *mathesis universalis* was indeed at the origin of these different orientations. But how could we understand this coexistence of apparently contradictory meanings of "universal mathematics" in Leibniz's thought? After all, Leibniz did obviously support *all* of these theses: *mathesis universalis* is a "logic of imagination"; a "logic of Mathematics" (a logic of algebra); its pars superior is the "science of the infinite" and plays a crucial role in the application of mathematics to physics.

3. After the battle

I would like now to take a step back and assess these different interpretative strategies from the point of view of their opposition *in the choice of the texts*. In the corpus which I will now consider, I will limit myself to the published texts, so that only one of my documents would not be known by the commentators mentioned in the above section. But the simple fact of taking into account the fragments *Mathesis generalis* (LH XXXV I, 9 a. Bl. 1–4 and 9–14), on which Couturat mentioned just the title (Couturat (1903), p. 543) and to restore the integrity of the *Elementa Nova Matheseos Universalis*, which he skilfully cut, changes quite deeply the understanding of the topic in Leibniz's work. Before doing so, let me start by quickly giving a general picture of the principal testimonies on *Mathesis universalis*, which we have seen up to this point.

---

[20] It should be noted that Leibniz makes no difference between the terms "general mathematics", "universal mathematics", "mathesis generalis/universalis", "mathematica in universum", etc.

[21] See, for example, *Animadversiones in partem generalem Principiorum Cartesianorum*, on art. II, 36; *Praefatio* to a treatise on *Mathesis universalis* [GM VII, 50]; *De legibus naturae et vera aestimatione virium motricium* [*Acta eruditorum*, 1691; GM VI, 211]; Letter to De Volder (january 1699?), where Leibniz mentions his *Lex aestimandi seu Matheseos vere universalis regula* [GP II, 156].

By looking at the various references, we have encountered so far, we can see that Leibniz's texts on *mathesis universalis* are of three different types.

Encyclopedic Projects	*Initia scientiae generalis. Conspectus speciminum* (summer to fall 1679?) [A VI, 4, A, 362 sq.] *Initia et specimina scientiae novae generalis* (spring 1682?) [A VI, 4, A, 442–443] *Guilielmi Pacidii Plus Ultra* [A VI, 4, A, 673 (april to october 1686 ?)]
Anti-cartesian articles and letters on Physics	*Animadversiones in partem generalem Principiorum Cartesianorum*, on art. II, 36. *Praefatio* to a treatise on *Mathesis universalis* [GM VII, 50] *De legibus naturae et vera aestimatione virium motricium (Acta eruditorum*, 1691 ; GM VI, 211) And many letters…
Projects of Treatises	Mathesis universalis = *scientia de quantitate in universum* *De arte characteristica inventoriaque analytica combinatoriave in mathesi universali* (1679) [A VI, 4, A, 315–331] *Mathesis universalis* (1694–1695) [GM VII, 53–76] *Mathesis generalis* (ca. 1700) (LH XXXV I, 9, Bl. 9–14) *De ortu, progressu et natura algebrae* [GM VII, 203–216.] «New» Mathesis universalis = *logica imaginationis* (*Idea Libri cui titulus erit) Elementa nova matheseos universalis* [1686 ? ; A VI, 4, A, 513–524]

1. A first series is constituted of purely *programmatic* texts. They usually consist of a simple table of contents of an encyclopaedia to come. Here, *Mathesis univeralis* is considered as belonging to the beginnings (*initia*) or the samples (*specimina*) of the project of a "General Science" (*scientia generalis*). In these texts, Leibniz insists on the fact that his own contribution to "universal mathematics" (*mea Mathesis generalis*)[22] is linked to the constitution of new types of calculus and should comprise a new kind of mathematics usually described as an extension from the science of quantity to the "science of quality"[23].

---

[22] A VI, 4, A, 443. *Initia et specimina scientiae novae generalis.*

These texts were central in interpretations like Couturat's and fall directly under Brunschvicg's critiques: they give no hint of the *real content* of what *mathesis universalis* was in Leibniz's mathematical practice, they are merely sketches of an *ideal*, which he moreover never properly succeeded in realizing.

2. The second group consists of polemical texts directed against Descartes and the Cartesians (mainly articles on physics and differential calculus). In this context, Leibniz often criticises the Cartesian program of *mathesis universalis* by claiming that it did not take into account the "true" universal mathematics, grounded on the *aestimatio rationum* (i.e., the new differential calculus as a tool for the understanding of nature).

These texts were central in Cassirer's and Brunschvicg's interpretations: the extension of mathematics to the realm of "quality" is not presented here as an extension to logical "forms", but to the intensive core of the laws of nature, the expression of which needs the use of differential calculus. But they also fall under the same type of critiques: they are all very allusive and don't explain in detail what the content of *mathesis universalis* is supposed to be or why exactly the differential calculus could be part of a *universal* mathematics.

3. The third group of texts consists of treatises, that Leibniz began and never finished, but in which we have many more details on the *content* of the "universal mathematics".

Strangely enough, these texts are the least quoted and written about by commentators. One reason is certainly that in the longest of them (GM VII, 53–76), Leibniz relies on the *traditional* definition of *mathesis universalis* as a "general science of quantity".

A first striking feature that appears in this description is the importance given by the commentators to the *programmatic* and *polemic* texts. This is quite easy to explain: since these texts are very allusive, they give to the reader a nice opportunity to "fill in the blanks" as desired and in this way to support interpretative postulates. This strategy was very obvious in Couturat's reading which jumped from programmatic declarations on *mathesis universalis* to programmatic declarations on *ars combinatoria* allowing, to his eyes, a form of *identification* between the two projects. He then commented in great detail about the *second* program (as a "general science of forms"), to finally arrive at the conclusion that the analogy between logic and mathematics (which is explicitly stated at the beginning of the text entitled *Mathesis universalis*) is in fact a "partial identity" relying on the role of the "general science of relations". This strategy is still very common in modern commentaries, be it grounded on an identification with *ars combinatoria* or with *caracteristica universalis*. It contributed to hiding a very simple and important fact: in the texts just mentioned, *there is no trace of such identifications*.

The second striking feature, symmetric to the first, is the importance given by Leibniz himself to the *traditional* definition of *mathesis universalis* as a "universal science of quantity" – whereas the insistence on the programmatic declarations could have induced the false impression that Leibniz did not pay much attention to it and was more interested

[23]*Initia scientiae generalis. Conspectus speciminum* (1679) [A VI, 4, A, 362 sq.] ; *Guilielmi Pacidii Plus Ultra* [A VI, 4, A, 673 (1686)]. On the situation of *mathesis universalis* in Leibniz's encyclopaedic projects see M. Schneider (1988), pp. 162–169.

in *reforming* the (Cartesian) concept of "universal mathematics". In fact, when Leibniz starts to write on the subject in 1679, he is clearly dealing with the Cartesian meaning of *mathesis universalis* as an algebraic calculus of magnitudes (*De arte characteristica inventoriaque analytica combinatoriave in mathesi universali* (1679) [A VI, 4, A, 315–331]). At that time, he was mainly interested in the privilege of this theory over other mathematical theories as regards the use of symbolism. When he planned to write a big treatise on "universal mathematics" including differential calculus in 1694–1695, he still followed the traditional definition: *scientia de quantitate in universum*, understood as a science *de ratione aestimandi* [GM VII, 53]; and again in 1700, in a fragment devoted to the definition of number: *mathesis generalis est scientia magnitudinis in universum* (LH XXXV, 1, 9). It is highly noticeable that the program of an *aestimatio rationum* is not presented in this context as part of a new science of quality, but as a part of the universal science *of quantity* (the part which deals with the infinite)[24].

Let us now have a quick look at the project of a "new" *mathesis universalis,* the content of which is sketched out in the *Elementa nova matheseos universalis* [A VI, 4, A, 513–524]. It is not my purpose to enter here into the details of this fascinating piece[25], but one can at least notice that the "real" text, now provided by the *Akademie* edition, is quite different from the one presented in Couturat (1903). In fact, Couturat *cut* the key passage of the text, in which Leibniz explains in what sense the new "universal mathematics" should be considered as a "logic of imagination" (and, as we saw, Couturat supplemented in his 1901 study his own interpretation of the meaning of this intriguing characterization). In this passage, Leibniz makes clear that this logic of imagination is grounded on the basic relationship of discernability of co-existing elements and that this analysis is primarily based on an analysis of *spatial* relationships (similitude, *hypallela*, congruence, etc.). This is sufficient to cast serious doubts on the thesis according to which Leibniz had a purely *formal* theory in mind (of which indeed we have no testimony, except the passage of Leibniz (1704)[26]). This impression is reinforced by the fragment *Mathesis generalis* (LH XXXV I, 9), in which Leibniz analyses the concept of numbers through its *spatial* representation[27].

Conclusion

Dietrich Mahnke was certainly right in pointing towards the idea of "universal mathematics" as a key notion in the development of the "Leibniz Renaissance". Not only were the interpretations of this program radically different in authors like Couturat and Cassirer or

---

[24] See for example *Mathesis universalis*, GM VII, 69 : *Itaque* Matheseos universalis pars superior *revera nihil aliud est quam Scientia infiniti, quatenus ad inveniendas finitas quantitates prodest.* Or *Animadversiones ad Weigelium* (1690) in Leibniz (1857), p. 148–149.

[25] I give some indications in Rabouin (2005) and in my dissertation (part VI : "Une reconstitution de la *mathesis universalis* leibnizienne", accessible at http://www.rehseis.cnrs.fr/spip.php?article229).

[26] One should also notice that Leibniz (1704) is not in contradiction with the rest of the corpus either. Leibniz always claimed that there is a formal part of mathematics, consisting in "analysis of truth" as opposed to "analysis of notions".

[27] For an analysis of this piece see Grosholz and Yakira (1998).

Brunschvicg, but it was also central in the opposition between Russell and Couturat. By looking at the textual settings supporting the different interpretations in conflict, one sees very clearly that they rely on a very careful and partial selection of the texts. But one could also go a little bit further and propose to see the very tension between "Universal Mathematics" and "Metaphysics of individuation" as an artefact generated by this use of the Leibnizian texts.

Let us recall one of the postulates of the "Panlogismus" itself, as expressed by Russell: "Leibniz's Logic and his investigations of the principles of mathematics [are] the source of his metaphysical system". As we have seen, this postulate, common to all of our authors, was at the core of the different interpretations of Leibniz's *mathesis universalis*. But this postulate leads to crushing of a very important distinction: for Leibniz, Metaphysics and Mathematics, no matter how close they stand, will always differ in the *type of object* that they can consider. This is precisely a point that Leibniz makes clear in the passage of the *Element Nova Matheseos universalis*, which Couturat cut: mathematics deals with objects defined by a kind of indiscernability (or, in modern terms, equivalence relation). Hence they stop where metaphysics begins[28]: with the real entity, constituting an individual being in and of itself (according to the famous Leibnizan motto: "to be a thing is to be one thing"). In this sense, the "formality" of mathematics, on which Leibniz certainly insists on many occasion, or its "application" to the knowledge of Nature, are no guarantee of its universality (in a metaphysical sense). If there is a place where this was explicitly stated, it is precisely in the introduction of the "new" concept of *mathesis universalis*: *Itaque hic excluduntur Metaphysica*.

References

[Brunschvicg 1912] Brunschvicg, Léon, *Les étapes de la philosophie mathématique*, Paris, Alcan, 1912.
[Burkhardt 1980] Burkhardt, Hans, *Logik und Semiotik in der Philosophie von Leibniz*, München, Philosophia, 1980.
[Cassirer 1998] Cassirer, Ernst, *Leibniz' System in seinen wissenschaftlichen Grundlagen* (1902) [including, as its first part : *Descartes's Kritik der mathematischen und Naturwissenschaftlichen Erkenntnis* (1899)], reed. *Gesammelte Werke*, t. I, Hamburg, F. Meiner, 1998.
[Cassirer 1999] Cassirer, Ernst, *Das Erkenntnisproblem in der Philosophie und Wissenschaft der neueren Zeit*, vol. I, (1906), reed. *Gesammelte Werke* t. II, Hamburg, F. Meiner, 1999.
[Couturat 1901] Couturat, Louis, *La Logique de Leibniz d'après des documents inédits*, Paris, Alcan, 1901.
[Couturat 1988] Couturat, Louis, *Opuscules et fragments inédits de Leibniz, extraits des manuscrits de la Bibliothèque de Hannovre*, Paris, Alcan, 1903, rééd. Hildesheim, New-York, Olms, 1988.
[Couturat 1980] Couturat, Louis, *Les Principes des mathématiques* (1905), reed. Paris, Albert Blanchard, 1980.

---

[28]A VI, 4, A, 514, 19–23: *Numero differunt quae ne quidem comparatione inter se discerni possunt, sed referenda sunt ad externa locum scilicet et tempus, an autem dari res solo numero differente in natura, hoc solo scilicet quod revera non sunt unum, sed plura, non est hujus loci, sed ad Metaphysicam pertinet ; nobis sufficit talia reperiri posse, quae imaginatione, sive sensuum apparentia discerni non possint.*

[Descartes 1964] Descartes, René, *Œuvres de Descartes*, publiées par C. Adam et P. Tannery, 11 vol., nouvelle présentation en coédition avec le CNRS, Paris, Vrin, 1964–1974 **[AT]**.

[Fédi 2001] Fédi, Laurent, "L'esprit en marche contre les codes: philosophie des sciences et dépassement du kantisme chez Léon Brunschvicg", in *Les philosophies françaises et la science: dialogue avec Kant*, sous la dir. de L. Fedi et J.-M. Salanskis, Cahiers d'Histoire et de Philosophie des Sciences n° 50, ENS éditions Lille, 2001, pp. 119–142.

[Hintikka 1997] Hintikka, Jaakko, *Lingua Universalis vs. Calculus Ratiocinator. An ultimate presupposition of Twentieth-century philosophy*, Dordrecht, Kluwer Academic Publishers, 1997.

[Husserl 1975] Husserl, Edmund, *Logische Untersuchungen* I (1900), reed. Husserliana, vol. XVIII, Den Haag, Martinus Nijhoff, 1975.

[Husserl 1954] Husserl, Edmund, *Die Krisis der europäischen Wissenschaften und die transzendantale Phänomenologie* (1934–1937), La Haye, M. Nijhoff, 1954.

[Klein 1936] Klein, Jacob, "Die griechische Logistik und die Entstehung der Algebra", in *Quellen und Studien zur Geschichte der Mathematik, Astronomie und Physik; Abteilung B: Studien*, 3, n. 1 (1934), pp. 18–105 and n. 3 (1936), pp. 122–235.

[Leibniz 1971] Leibniz, Gottfried Wilhelm von, *Nouvelles lettres et opuscules inédits*, éd. Foucher de Careil, 1857, rééd. Olms, Hildesheim-N.Y, 1971.

[Leibniz 1981] Leibniz, Gottfried Wilhelm von, *Nouveaux Essais sur l'entendement humain* (1704), transl. Peter Remnant and Jonathan Bennett: *New Essays on Human Understanding*, Cambridge, Cambridge University Press, 1981.

[Grosholz 1998] Grosholz, Emily and Yakira, Elhanan, *Leibniz's Science of the Rational*, Studia Leibnitiana Sonderheft 26, Steiner Verlag, 1998 (with a transcription and a commentary of LH XXXV, I, 9).

[Kneale 1962] Kneale, William & Martha, *The Development of Logic*, Oxford, Clarendon Press, 1962, rééd. .

[Mahnke 1964] Mahnke, Dietrich, *Leibnizens Synthese von Universalmathematik und Individualmetaphysik*, Max Niemeyer, Halle, 1925, *Jahrbuch für Philosophie und Phänomenologische Forschungen*, pp. 305–612 (reed. fac-simile, Stuttgart, 1964).

[Rabouin 2005] Rabouin, David, "Logique, mathématique et imagination dans la philosophie de Leibniz", *Corpus*, n. 49, "Logiques et philosophies à l'âge classique", hiver 2005, p. 165–198.

[Russell 1900] Russell, Bertrand, *A Critical Exposition of the Philosophy of Leibniz*, London, Unwin, 1900.

[Russell 1903] Russell, Bertrand, " Recent Work on The Philosophy of Leibniz ", *Mind*, vol. 12, n. 46, 1903, pp. 177–201.

[Schneider 1988] Schneider, Martin, "Funktion und Grundlegung der Mathesis Universalis im Leibnizschen Wissenschaftsystem ", *Studia Leibnitiana Sonderheft* 15, 1988.

[Van Heijenoort 1967] Van Heijenoort, Jean, "Logic as Calculus and Logic as Language", *Synthese*, vol. 17 (1967), pp. 324–330.

[Weyl 1926] Weyl, Hermann, *Philosophie der Mathematik und Naturwissenschaft*, Munich, R. Oldenbourg (*Handbuch der Philosophie*), 1926.

David Rabouin
Université Paris 7 – CNRS – Laboratoire SPHERE UMR 7219
Equipe Rehseis – Case 7093, 5 rue Thomas Mann
F-75205 PARIS CEDEX 13, France
David.Rabouin@ens.fr

Leibnizian Traces in H. Weyl's *Philosophie der Mathematik und Naturwissenschaft*

Erhard Scholz

1. Introduction

After a phase of radical mathematical innovations between 1918 and 1925, often with strong repercussions in physics (from foundations of analysis, via general relativity, differential geometry and unified field theory to the representation theory of Lie groups), Hermann Weyl turned toward writing his contribution *Philosophie der Mathematik und Naturwissenschaft* on the philosophy of mathematics and natural sciences, in the sequel abbreviated PMN [Weyl 1927, Weyl 1949], for the *Handbuch der Philosophie* [Baeumler/Schroeter 1927]. It was a time of reorientation for him with regard to foundations of mathematics and to the question of how mathematics may contribute to understanding of the external (natural) world. The phase of his most radical interventions into the foundations of mathematics in a constructivist perspective from 1916 to 1919 and an intuitionist one, 1919 to 1922, lay just behind him. Likewise a period was closed (1918 to 1922), in which he was convinced that one could unify the two most recent pillars of mathematical physics, Einstein's geometric theory of gravity (general relativity) and Hilbert's attempts to formulate a dynamistic field theoretic explanation of matter ("foundations of physics", Mie-Hilbert theory). Weyl proposed his *purely infinitesimal geometry*, a generalization of Riemannian geometry, and used it for formulating a geometrically unified field theory, the first in a series of attempted unified classical field theories which followed [Vizgin 1994, Goenner 2004, Goldstein 2003]. His turn towards the study of "infinitesimal symmetries" during this work brought him into research in the representation of Lie groups (1923 to 1925) which was generally considered to be the most important mathematical research work of his whole career [Hawkins 2000, Eckes 2011].

After this outburst of scientific activity for about eight years, a time that was already deeply permeated by philosophical motivations, Weyl took the task of writing his *Handbuch* article as a chance to rethink much of his earlier philosophical convictions. Later he came to like talking about this kind of reflection by using the good old German word *Besinnung* which denotes a contemplative kind of reflection rather than an analytic one [Weyl 1954]. In this sense, he allowed himself an interlude of holding back his own thoughts and activities with regard to the "new quantum mechanics" of Heisenberg, Jordan, Born and Dirac, although he was well aware of their work and started to have some forward pointing ideas about it [Scholz 2006]. For more than half a year, between summer 1925 and early 1926, he delved into philosophical literature far beyond his earlier interests in this field, which had been concentrated around the philosophies of Kant, Poincaré, Mach, Husserl, Fichte, listed in their time-order, which arose from his close communication with his wife Helene, a Husserl scholar, and Fritz Medicus, a Zürich expert in the philosophy of German idealism and a personal friend [Sieroka 2007, Sieroka 2009, Sieroka 2010a]. Apparently he used the chance to carefully read in Leibniz's works, among others, and to quote them extensively. Leibniz became the by far most frequently quoted author in Weyl's PMN, with 79 entries, against the next ones, Kant: 43, and Newton: 41. On no other occasion, neither earlier nor later, did Weyl ever refer so strongly and explicitly to Leibniz as a philosopher.

It has to be added that Weyl did not aim at a systematic exposition of Leibniz's thought (nor did he with any other philosopher). He rather used the references to philosophers' works interwoven into the development of his own ideas on the philosophy of the mathematical sciences. This did not have merely the function of ornaments by classical text fragments, but rather served the purpose of an examination of Leibnizian thoughts and a debate over them in the light of more recent developments in the object sciences that he had been considering.

This paper does not pretend to give anything like a systematic evaluation of Weyl's way of presenting Leibniz either. That would be a task for a professional philosopher. Coming from a background in the history of mathematics, I just want to present those aspects which apparently made Leibniz so important for Weyl in the middle of the 1920s.

2. General reflections on modern mathematics

Weyl's own position with regard to the foundations of mathematics and the recent developments in mathematical logic, axiomatics and set theory was still shaped by a constructivist perspective with strong intuitionistic sympathies [Feferman 2000, van Dalen 2000, Sieroka 2009]. For a general exposition of the philosophy of mathematics to a broader audience, he had to express himself in a more balanced manner than he used to do the years before [Hesseling 2003]; and in fact he wrote a short first chapter in PMN on mathematical logic and axiomatics, in which Leibniz appeared as a figure of peripheral reference, not much more.[1]

---

[1] For instance, Weyl quoted from Leibniz's letters to Clarke:

> Leibniz speaks of a '... relation between L and M, without consideration as to which member is preceding or succeeding, which is the subject or object. ... One cannot say that both together, L and M, form the subject for an *accidens*; ... It must be said, therefore that the relation ... is something outside of the subjects; but since it is neither substance nor *accidens* it musts be something purely ideal, which is nevertheless worthy of examination.' (Leibniz 5th letter to Clarke, §47) [Weyl 1949, 4f.].

Weyl assigned a more interesting role to Leibniz in his discussion of *modern axiomatics*. After introducing the modern axiomatic method and the role of models for investigating consistency and independence of an axiomatic system, with full acknowledgement of Hilbert's "ingenious construction of suitable arithmetical models" [Weyl 1949, 22], Weyl presented an axiomatically founded mathematical theory as a "logical mould (Leerform) of possible sciences" (ibid. 25), a formulation which he liked and repeated frequently. That gave him the chance to call upon Leibniz as an inspirator of a development in this direction:

> Leibniz takes some decisive steps towards the realization of *mathesis universalis* in the sense here indicated and clearly understood by him. The theory of groups above all, that shining example of 'purely intellectual mathematics' belongs within the framework of his *ars combinatoria*. [Weyl 1949, 27]

This quote shows also that Weyl did not intend a historical reconstruction of Leibnizean thought, but rather read him in a presentist perspective (group theory as part of *ars combinatoria* etc.).

In his discussion of *number* (natural, rational) and *continuum* (real numbers), Weyl of course presented Dedekind cuts and nested interval constructions of the reals, but did not withhold his constructivist sympathies with respect to the ontology of such infinities. The determination of localizations in a continuum stood, for Weyl, in the tension between "the real" and "the ideal" and could be understood as paradigmatic for gaining (ideal) knowledge of (real) things. He insisted that a "real thing" can never be given, but has to be "unfolded" by an infinitely continued process (here he referred to Husserl's "inner horizon").

> For this reason it is impossible to posit the real thing as existing, closed and complete in itself. [Weyl 1949, 41]

In this context, the concept of continuum was pivotal in driving "toward epistemological idealism". Here he could again cite:

> Leibniz, among others, testifies that it was the search for a way out of the 'labyrinth of the continuum' which first suggested him the conception of space and time as orders of the phenomena. 'From the fact that a mathematical solid cannot be resolved into primal elements it follows immediately that it is nothing real but merely an ideal construct designating only a possibility of parts'. (Correspondence Leibniz de Volder, Leibniz Phil. Schr. II, p. 268). [Weyl 1949, 41]

Referring to Leibniz's introduction of *monads* as an attempted path towards giving a metaphysical foundation to the world of phenomena, Weyl continued with a Leibnizian argument on the continuum:

> 'Within the ideal or the continuum the whole precedes the parts ... The parts are here only potential; among the real (i.e., substantial) things, however, the simple precedes the aggregates, and the parts are given actually and prior to the whole ...'(letter to Remond, Phil. Schr. III, 622) [Weyl 1949, 41]

Such a view of the continuum as a whole which had to be stipulated in the intuition, rather than being postulated formally by means of transfinite set theory, was particularly close to his own semi-intuitionist understanding of the continuum and prepared the way to a discussion of the two controversial points of view (set theory versus intuitionism) in the next two subsections of his book.

3. Hilbert's foundational program in the light of symbolic mathematics

In his PMN Weyl discussed both, Brouwer's intuitionist and Hilbert's formalist program, for the foundation of mathematics. Although he still sympathized with the intuitionist perspective, he was quite clear that the loss of the principle of excluded middle was akward for mathematics. He ended this passage by an often quoted remark:

> And the mathematician watches with pain the larger part of his towering edifice which he believed to be built of concrete blocks dissolve into mist before his eyes. [Weyl 1949, 54]

That was a positive motif for turning toward Hilbert's foundational program which Weyl called *symbolic* rather than formalist mathematics. He discussed Hilbert's foundational enterprise for arithmetics with a relatively open mind, including references to recent progress made by J. von Neumann [Weyl 1927, 49], in the later English edition also to P. Bernays and W. Ackermann [Weyl 1949, 60]. As long as only finite sequences of proof derivations were considered, everything worked well and was acceptable also from the intuitionist standpoint. But clearly Hilbert's proof theory aimed at more. For this reason he introduced a specific transfinite logical axiom rule, which should suffice to safeguard the transfinite parts of set theoretic mathematics, so Hilbert hoped, by a purely logical analysis of the deductive structure it allowed for. In 1926 Weyl remarked that von Neumann had recently shown the consistency of those parts of mathematics which treat the "series of all natural numbers as a closed totality of existing objects", comparable to his own point of view in *Das Kontinuum* of 1918 [Weyl 1918], i.e., as long as countable transfinite sets are concerned. Weyl added that the more complicated case of the transfinite dealing with "the totality of all possible sets of numbers", i.e., the uncountable transfinite, was still wide open.

> Only the realization of the consistency proof, or at least the attempts at it, disclose to us the highly sophisticated (verzwickt) logical structure of mathematics, its maze (Gewirr) of circular back references which do not allow to survey whether they might not lead to blatant contradictions. [Weyl 1927, 1st. ed., 49]

In the English translation more than twenty years later, he felt no need to soften his argument; rather to the contrary he reminded the reader of Gödel's results in 1931 which, according to Weyl, "precipitated a catastrophe" for Hilbert's proof theoretic program [Weyl 1949, 61].

But that was not even decisive for Weyl's view of the achievements of Hilbert's approach. Leibniz reentered Weyl's reflections, perhaps to relativize both, Hilbert's achievements and Weyl's own former sceptical reaction to it.

> The described symbolism evidently attacks again, in a refined form, the task which Leibniz had set himself with his 'general characteristics' and *ars combinatoria*. But is it really more than a bloodless ghost of the old analysis that confronts us here? Hilbert's mathematics may be a pretty game with formulas, more amusing even than chess; but what bearing does it have on cognition, since its formulas admittedly have no material meaning by virtue of which they could express intuitive truth? The subject of mathematical investigation, according to Hilbert, is the concrete symbols themselves. [Weyl 1949, 61]

The "old analysis" of Leibniz and others had still developed a symbolic enterprise which aimed at a better understanding of the relationships and laws of nature. For Weyl, the "symbolic" character of mathematics contained more than just its formal side. He adressed the reader:

> ... This last remark reminds us that it is the function of mathematics to be at the service of the natural sciences ... (ibid.).

For Weyl, the "symbolic" consisted of more than the syntactical structure; it aimed at more than a formalist game, and its justification presupposed more than a formal analysis of consistency. It ought, according to Weyl, "furnish" knowledge, and knowledge contained somehow a claim for "truth".

> It seems that we have to differentiate carefully between phenomenal knowledge and insight ... and theoretical construction. Knowledge furnishes truth, its organ is 'seeing' in the widest sense. Though subject to error it is essentially definitive and unalterable. Theoretical construction seems to be bound only to one strictly formulable rational principle, that of concordance (...) which in mathematics (...) reduces to consistency. In connection with physics we have to discuss in greater detail the question. [Weyl 1949, 61f.]

The situation changes, if one turns toward physics and the role of mathematics in physical knowledge. Then more than mere consistency is at stake and mathematics acquires a specific role in a "symbolical representation" of material objectivity, the "transcendent", as Weyl liked to say in counterposition to the "immanent" cognitive reality of the symbols.[2]

4. A dialogue partner for understanding modern physics

In his passage on physical questions in PMN, Weyl referred to Leibniz at different occasions: purely infinitesimal ("near") geometry, orientation of space and time, matter, and the topical complex of causality, law, chance, and freedom.

We saw already that Weyl's turn towards PMN happened after several years of great activities in the mathematical sciences and at the end of a phase of changing perspectives. In 1918 he had invented and proposed his "purely infinitesimal geometry", today

---

[2] More in [Scholz 2005*b*].

one would call it a *scale gauge geometry* which, in contrast to Riemannian geometry, excluded the possibility of a direct comparison of metrical quantities at different points in the manifold. Beginning in 1920 he gave up, step by step, his immediate hope for a geometrically unified field theory of electromagnetism and gravity and a Mie-Hilbert type dynamistic matter theory built upon it. But this did not mean a withdrawal of the conviction that the geometrical invention of a "purely infinitesimal" scale gauge geometry was justified and continued to be valuable. Between 1921 and 1923 he developed a philosophically founded, but mathematically formulated program of a conceptual foundation of a most general metrical infinitesimal geometry, with some Kantian inklings, the "mathematical analysis of the problem of space"[Weyl 1923].[3]

In 1926 Weyl quoted Leibniz as supporter and advisor:

> As the true lawfulness of nature, according to Leibniz's continuity principle, finds its expression in laws of nearby action, connecting only the values of physical quantities at space-time points in the immediate vicinity of one another, so the basic relations of geometry should concern only infinitely closely adjacent points ('near-geometry' as opposed to 'far-geometry'). Only in the infinitely small we may expect to encounter the elementary and uniform laws, hence the world must be comprehended through its behaviour in the infinitely small. [Weyl 1949, 86]

In 1918 he had other, in particular field theoretic reasons to demand such a perception of geometry, but in any case such a view stood in good agreement with a Leibnizian perspective, which could be appealed to in order to foster such a view in geometry.

In his discussion of the relativity of space and time, Weyl briefly mentioned the respective views of Aristotle, Descartes, Galilei, and Leibniz, while he discussed the position of Newton, "the absolutist", at length. To illustrate the relationist position of Leibniz he quoted from the third letter to Clarke:

> Under the assumption that space be something in itself, that it be more than merely the order of bodies among themselves, it is impossible to give a reason why God should have put the bodies (without tampering with their mutual distances and relative positions) just at this particular place and not somewhere else; for instance, why He should not have arranged everything in the opposite order by turning East and West about. If, on the other hand, space is nothing more than just the order and relation of things, if without the bodies it is nothing at all except the possibility of assigning locations to them, then the two states supposed above, the actual one and its transposition, are in no way different from each other. Their apparent difference is solely a consequence of our chimerical assumption of the reality of space itself. ... [Weyl 1949, 97]

Weyl compared this view of Leibniz with Kant's famous argument for the transcendental ideality of space (*Prolegomena* §13 etc.) and sided with Leibniz [Weyl 1949, 80, 97]. Kant illustrated the point of difference by a striking metaphysical thought experiment. He proposed to assume that a "left hand", i.e., a specific chiral spatial object, was

---

[3]Cf. [Scholz 2004*b*]

produced in "the first creative act of God". Kant's position would be that this would have introduced already the character of a "left" object, rather than a "right" one, even without any other object of comparison. This was a result of Kant's realization that, at his time, left- or right-handedness were not definable conceptually, but could be discerned by pure intuition only. Moreover he seemed to assume the transcendental subject of "pure intuition" as timeless as God himself. Weyl did not accept this as a valid argument. According to him, only the comparison of the first object ("left hand") with a chiral object brought about in a "second act of creation" would allow one to make any distinction at all:

> He [God, ES] would have changed the plan of the universe not in the first but in the second act, by bringing forth a hand which was equally rather than oppositely oriented to the first-created one. [Weyl 1949, 97, footnote]

Translated into mathematical terminology: *Pure space* being assumed orientable (at least locally), God's "first creative act" would select an orientation. Only after that, in the "second creative act", does it make sense to ask for a locally defined, oriented object to coincide with or to break the orientation selected in the first step.[4]

In the discussion of the causal structure of relativity theory and its importance for the concept of time order, Weyl made an illuminating excursion to Leibniz again. He started to explain the modern (Einsteinian) causal order:

> Likewise any event happening at O has influence only upon the events at later world points; the past cannot be changed. That is to say, the stratification [of past and future and simultaneity] has a causal meaning; it determines the *causal connection of the world*. [Weyl 1949, 101]

Without any recontextualization he continued:

> This was recognized by Leibniz, who explains in his 'Initia rerum mathematicarum metaphysica' (Math. Schriften VII, p. 18). 'If of two elements which are not simultaneous one comprehends the cause of the other, then the former is considered as *preceding*, the latter as *succeeding*.' (ibid.)

Weyl thus invoked Leibniz just as if he were a contemporaneous (or time-less) dialogue partner who could be asked for advice in questions pertaining to most recent modern physics.

With respect to the concept of matter Weyl was, of course, fond of Leibniz's dynamical (quasi "active") characterization of the latter.

> Leibniz (opposing Descartes) has emphatically stressed the dynamic character of inertia as a tendency to resist deflecting forces; for instance, in a letter to de Volder (Philosophische Schriften, II, p. 170) he writes, 'It is one thing if

---

[4] Indirectly this metaphor may even shed light on the problem of symmetry breaking (also between matter and anti-matter) in the "early universe", discussed in modern elementary particle physics. Philosophically reflected people need not adhere to an interpretation of the "early universe" in the sense of scientific realism, which dominates the imagination of present mainstream physics so strongly, but may take it as what it is: speculative metaphysics in scientific guise. If "generation of the world" is not understood in a quasi natural-historic sense, but as *structural genesis* with only indirect relation to timelike developments in the material world, "God's first creative act" may be read as a rhetorical figure for the first step of (conceptual) structure generation in the formation of our scientific world picture.

something merely retains its state until some event happens to change it –
a circumstance which may occur if the subject is completely indifferent with
respect to either state; it is another thing and signifies much more if the subject
is not indifferent but possesses a power, an inclination as it were, to retain its
state and to resist the cause of change.' [Weyl 1949, 105]

After his own failed attempt to unify forces and the hope for a (Mie-Hilbert type)
dynamist derivation of matter structures, Weyl returned to a more cautious position. In
1926 he spoke of an unreducible duality of matter and (interaction) field, at least for
the moment, and referred to Newton as "entirely dominated by this dualism". Of course
this could not be convincing for him, and he used Leibniz as an early protagonist of the
contrasting position:

The classical philosopher of the dynamical conception of the world is, however, Leibniz. To him, what is real in motion does not lie in the change of
position as such, but in the moving force. 'La substance est un être capable
d'action, une force primitive' transspatial and immaterial. ...
... The ultimate element is the *monad*, an indecomposable unit without extension, from which the force bursts forth as a transcendental power. Only
with regard to the distribution of the monads in space, which itself is merely
a *phaenomenum bene fundatum*, is the body described as an *extended* agent.
Pure activity, however, is all; preestablished harmony takes the place of such
reciprocal effects as we think are carried by the field from particle to particle.
[Weyl 1949, 174]

Weyl then explained the field actions of matter in general relativity and expressed the
hypothesis that mass is established by the "flux of the gravitational field which a particle
sends through an enveloping shell" (p. 175). He even indicated his idea that, maybe, matter
lies "beyond" the singularities of spacetime structure.

Indeed general relativity does not prescribe the topology of the world, and it
may therefore happen that the world has unattainable 'fringes' not only toward the infinite but also inwardly. In line with Leibniz's idea, the material
particle, although embedded in a spatial environment from which its field effects take their start, would itself then be a *monad* existing beyond space and
time. [Weyl 1949, 175]

Weyl continued with a reinterpretation of his own in the context of modern field theory. In
that case, a charge cannot be localized pointwise; it is rather given by a density, such that
integrals over closed surfaces indicate that the latter surround a charge. Comparable to a
neo-Kantian view, the splitting of space and time was for him a question of the subject,
but unlike Kant, no transcendental one, but an empirical one, bound to matter.[5]

He made a surprising move from Leibniz, via his own thoughts about a matter concept compatible with general relativity, to Schelling:

---

[5]Weyl talked about "...the geometrico-physical basis for the splitting of the world into space and time which
takes place within our consciousness, tied as it is to a material body" [Weyl 1949, 176].

Schelling, partially under the influence of Leibniz has expressed ideas which vaguely anticipate this development [toward a general relativist concept of matter in the Weylian sense, ES]. 'Thus there ought to be discernible in experience something', he says on p. 21 of his 'Erster Entwurf der Naturphilosophie' (1799; *Sämtliche Werke*, III, p. 21, Cotta, 1858) 'which without being in space, would be principle of all spatiality.' This 'natural monad' is not itself matter but action, 'for which there is no measure but its own product.' [Weyl 1949, 176].

Weyl continued with a sketch of Schelling's "construction" of continuous spacefilling matter, "a shapeless fluid – which we today would replace by the field." Once one has started to consider most recent structures of mathematical physics from the point of view of this Schellingian scheme, one could also transfer it one stage further, replacing classical fields by quantum fields ("product of activity"). The activity behind the (quantum) field, could just as well be characterized metaphysically by "natural monads", a mathematical representation of which would have to be a quantum agens structure beyond space and time, giving rise to a spacefilling "shapless fluid", expressed on different levels by quantum fields, semiclassical fields or, in the classical limit, by classical fields [Sieroka 2010*b*]. In any case, we can here see a clear impact of Leibnizean thought on Weyl's agency concept of matter [Sieroka 2007, Scholz 2004*a*].

Finally with respect to the challenging relationship between *causality/law* on the material side of the world and *freedom/purposiveness* on the humanistic side of the world, with the grey area of *chance* in between, Weyl developed a clarifying view of his own. Again he did not abstain from allusions to classical positions in philosophy, among others to Leibniz. After a short review of diverse positions, even including premodern world views, Weyl hinted at the contraposition between Hobbes and the latter's "first consistent modern theory of determinism in which natural law appears as the binding force" (rather than God's predetermination, Kismet etc.) and Descartes who "clung to the freedom of will, and (...) had to do so if the self-certainty of thinking guarantees truth as demanded by the principles of his philosophy" [Weyl 1949, 208]. The contradiction between lawful determination in the realm of *res extensa* and self-determination according to clear ideas in the *res cogitans*, arising from the Cartesian approach, remained a philosophical evergreen in philosophy of the modern era.

Weyl explained:

Two quotations from Leibniz may be given here. In an essay on freedom (*Lettres et opuscules inédits de Leibniz*, ed. Foucher de Careil, Paris 1854, p. 178 et seq.) he states 'that there may, or even must, be truths which no analysis can reduce to the identical truths or to the principle of contradiction, which, on the contrary, require an infinite series of reasons for their support; a series which is transparent to God only. And this is the essence of all that one considers free and accidental.' Further in his *Monadology (Philosophische Schriften*, VI, pp. 607–623; Section 79): 'The souls act according to the laws of final causes through appetences, means and ends. The bodies act according to the laws of efficient causes or motion. And these two realms, of final and efficient causes, are in mutual harmony.' [Weyl 1949, 209]

Between these and many more short and striking characterizations of contributions to this topic by philosophers and scientists of the last three centuries, Weyl framed his own view. The recent scientific insights into the unreducible stochastic features of "atomic events", [Weyl 1920], [Weyl 1949, 198], and into the causal structure of relativity, in which the causal future of an event x is not completely determined by its causal past, because the causal past of any future point is strictly wider than that of x [Weyl 1949, 210], undermined or even dissolved, according to Weyl, the "antinomy" of "knowing and being", which had been so acute in the early and classical modern period. Among those who had been struggling with this antinomy was Kant whose argumentation, how freedom of will was reconcilable with classical (Laplacian) determinism, could not achieve a convincing solution.

> Kant, according to the scientific situation of his time, agrees with this view as far as the world of space-time phenomena is concerned [classical determinism, ES], and he tries, by distinguishing between the phenomenal and the intelligible world, to give a transcendental solution of the conflict between natural causality and freedom of will. His solution, however, can hardly be carried through consistently and even remained obscure to himself to such a degree that he was unable to understand the changes in the character of a person. [Weyl 1949, 210]

Weyl did not claim to have a definite solution. As a philosophizing scientist, he looked for answers in better understanding of the scientific base of human nature, although he saw clearly that the scientific knowledge in biology and psychology was not sufficiently far developed to allow anything like a convincing scientific treatment of the question. He was critical of contemporary *vitalist* answers to these questions, expressed, e.g., by Hans Driesch who, by the way, was his coauthor in *Handbuch der Philosophie* with an essay on *Metaphysics of Nature* [Driesch 1927].

> All these questions as to the essence of life and the possibility of spontaneous generation are premature and must rest until the day when the laws of life will be known to us to a much wider extent. [Weyl 1949, 215]

On the other hand future progress of the natural sciences, in particular biology and psychology, might lead to a deeper understanding of how the *open* lawfulness of the natural world, which expressed itself already on the foundational level by the stochastic nature of physical laws, may go hand in hand with organization, life, and even purposiveness of the soul and the intellect. For the moment, the former was perfectly consistent with the latter, but far from being able to "explain" it. Although thus, according to Weyl, the "body-soul problem" still belonged to the class of riddles which were unsolvable at the time, he was optimistic in principle:

> I do not believe that insurmountable difficulties will be encountered in any unprejudiced attempt to subject the entire reality, which undoubtedly is of a psycho-physical nature, to theoretical construction – provided the soul is

interpreted merely as the aggregate of the real psychic acts in an individual. It is an altogether too mechanical conception of causality which view the mutual effects of body and soul as being so paradoxical that one would rather resort, like Descartes, to the occasionalistic intervention of God or, like Leibniz, to a harmony instituted at the beginning of time. (ibid.)

At the end of his PMN, Weyl came to the conclusion that there was a "general agreement regarding the most essential insights of natural philosophy as it is found among all those who approach the problem seriously and with a free and independent mind rather than in the light of traditional schemes", regardless of whether their background was in philosophy, in the sciences, or in mathematics. His own contributions had their "firm foundation in the first, mathematical part". He opened the last paragraph with an appeal:

> Exact natural science, if not the most important, is the most distinctive feature of our culture in comparison to other cultures. Philosophy has the task to understand this feature in its peculiarity and its singularity. [Weyl 1949, 216]

Leibniz had adressed this task, in his time and his way, in a highly elaborate and productive way. Weyl did so at a critical stage of the development of modern science in the early 20th century.

5. In place of a conclusion: Weyl's Leibniz

Philosophical considerations, once suspended, are difficult to resume. Let us, nevertheless, try a short glance back: Like any other of the scientists or philosophers of the 19th or 20th centuries, Weyl adapted Leibniz to his own perspective. We have have seen this effect in topics such as

- *mathesis universalis* realized in modern axiomatic mathematics,
- *characteristica generalis* and *ars combinatoria* realized in Hilbert's foundational approach to mathematics,
- *Ausdehnungslehre* and vector calculus considered as a variant of *analysis situs*,
- the discussion of the *relativity* of space,
- causality and *time order* in modern physics,
- and the *dynamical* character of inertia, including Weyl's own *agens theory of matter*, supported and upgraded by Leibnizian fragments.

This should not be read too critically. Weyl knew clearly what he did; he frankly admitted:

> In conclusion I want to emphasize once more that it has not been my intention to write a history of philosophical thought within the natural sciences. This would require much more comprehensive historical studies ... [reference to Lasswitz and Cassirer, ES] ... Primarily interested in mathematical research, I am wanting, in both time and love, for such work. [Weyl 1949, 216]

Moreover, we found three topics with an indirect or even a direct influence of Leibnizian motifs on Weyl's own work in mathematics or mathematical physics. The latter's view of the *continuum* concept and on *near*, or "purely infinitesimal", *geometry* carried Leibnizian traces; although they were apparently indirectly transmitted by Weyl's reading of Husserl and Fichte after 1916 [Ryckman 2005, Scholz 2000, Scholz 2005a, Sieroka 2007, Sieroka 2010a, Eckes 2011].

Weyl's transition from a field theoretic dynamistic concept of matter to his *agens theory of matter* was triggered by problems inside the mathematical "construction" of empirically adequate matter structures in the frame of classical field theories; but its reflection and its connection to wider philosophical topics seems to be enriched by our protagonist's intense studies of Leibniz during 1926, in additon to that of Fichte already in the years before [Sieroka 2007, Sieroka 2010a, Scholz 2004a].

Finally Weyl's mature understanding of the nature of mathematics and its role in acquiring knowledge of the outside world, which, for lack of a better label, I have called *symbolic realism* elsewhere [Scholz 2005b], was apparently supported by his reading of Leibniz's metaphysics and the role of mathematics in it. In view of all this and in spite of all necessary caution, we may finally conclude that Leibnizian traces are to be found not only in Weyl's PMN. Some of them seem to have left perceivable, although not spectacular, imprints on Weyl's work as a mathematical scientist.

References

[Ashketar e.a. 2003] Ashketar, Abhay; Cohen, Robert S.; Howard, Don; Renn, Jürgen; Sarkar, Sahoptra; Shimony, Abner (eds.). 2003. *Revisiting the Foundations of Relativistic Physics: Festschrift in Honor of John Stachel*. Dordrecht etc.: Kluwer.

[Baeumler/Schroeter 1927] Baeumler, Alfred; Schroeter, Manfred, ed. 1927. *Handbuch der Philosophie. Bd. II. Natur, Geist, Gott*. München: Oldenbourg.

[Driesch 1927] Driesch, Hans. 1927. "Metaphysik der Natur.". In [Baeumler/Schroeter 1927, Bd. II B].

[Eckes 2011] Eckes, Christophe. 2011. *Groupes, invariants et géométrie dans l'oeuvre de Weyl*. Une étude des 'ecrits de Hermann Weyl en mathématiques, physique matématique et philosophie, 1910–1931. Thèse de doctorat en philosophie. Université de Lyon III.
[http://math.univ-lyon1.fr/homes-www/remy/TheseChristopheEckes-26sept2011.pdf]"

[Feferman 2000] Feferman, Solomon. 2000. The significance of Weyl's 'Das Kontinuum'. In [Hendricks e.a. 2000]. pp. 179–194.

[Goenner 2004] Goenner, Hubert. 2004. "On the history of unified field theories." *Living Reviews in Relativity* 2004-2. [http://relativity.livingreviews.org/Articles/lrr-2004-2].

[Goldstein 2003] Goldstein, Catherine; Ritter, Jim. 2003. "The varieties of unity: Sounding unified theories 1920–1930." In [Ashketar e.a. 2003].

[Hawkins 2000] Hawkins, Thomas. 2000. *Emergence of the Theory of Lie Groups. An Essay in the History of Mathematics 1869–1926*. Berlin etc.: Springer.

[Hendricks e.a. 2000] Hendricks, Vincent F.; Pedersen, Stigandur; Jørgensen Klaus F., ed. 2000. *Proof Theory: History and Philosophical Significance*. Dordrecht etc.: Kluwer.

[Hesseling 2003] Hesseling, Dennis. 2003. *Gnomes in the Fog. The Reception of Brouwer's Intuitionism in the 1920s*. Basel: Birkhäuser.

[Leibniz 1890a] Leibniz, Gottfried Wilhelm. 1890a. "Korrespondenz Leibniz – Clarke.". In GP VII.

[Ryckman 2005] Ryckman, Thomas. 2005. *Reign of Relativity. Philosophy in Physics* 1915–1925. Oxford: University Press.

[Scholz 2000] Scholz, Erhard. 2000. Hermann Weyl on the concept of continuum. In [Hendricks e.a. 2000, 195–220].

[Scholz 2004a] Scholz, Erhard. 2004a. The changing concept of matter in H. Weyl's thought, 1918–1930. In *The interaction between Mathematics, Physics and Philosophy from* 1850 *to* 1940, ed. J. Lützen. Dordrecht etc.: Kluwer. [http://arxiv.org/math.HO/0409576].

[Scholz 2004b] Scholz, Erhard. 2004b. "Hermann Weyl's analysis of the "problem of space" and the origin of gauge structures." *Science in Context* 17:165–197.

[Scholz 2005a] Scholz, Erhard. 2005a. "Philosophy as a Cultural Resource and Medium of Reflection for Hermann Weyl." *Revue de Synthèse* 126:331–351. [arxiv.org/math.HO/0409596].

[Scholz 2005b] Scholz, Erhard. 2005b. Practice-related symbolic realism in H. Weyl's mature view of mathematical knowledge. In *The Architecture of Modern Mathematics: Essays in History and Philosophy*, ed. J. Ferreiros, J. Gray. Oxford: UP, 291–309.

[Scholz 2006] Scholz, Erhard. 2006. "Introducing groups into quantum theory (1926–1930)." *Historia Mathematica* 33:440–490. [http://arxiv.org/math.HO/0409571].

[Sieroka 2007] Sieroka, Norman. 2007. "Weyl's 'agens theory of matter' and the Zurich Fichte." *Studies in History and Philosophy of Science* 38:84–107.

[Sieroka 2009] Sieroka, Norman. 2009. "Husserlian and Fichtean leanings: Weyl on logicism, intuitionism, and formalism." *Philosophia Scientiae* 13:85–96.

[Sieroka 2010a] Sieroka, Norman. 2010a. *Umgebungen. Philosophischer Konstruktivismus im Anschluss an Hermann Weyl und Fritz Medicus.* Zürich: Chronos.

[Sieroka 2010b] Sieroka, Norman. 2010b. "Geometrisation versus transcendent matter: A systematic historiography of theories of matter following Weyl." *British Journal for Philosophy of Science* 61:769–802.

[van Dalen 2000] van Dalen, Dirk. 2000. "Brouwer and Weyl on proof theory and philosophy of mathematics.". In [Hendricks e.a. 2000, 117–152].

[Vizgin 1994] Vizgin, Vladimir. 1994. *Unified Field Theories in the First Third of the* 20th *Century.* Translated from the Russian by J.B. Barbour. Basel etc.: Birkhäuser.

[Weyl 1918] Weyl, Hermann. 1918. *Das Kontinuum. Kritische Untersuchungen über die Grundlagen der Analysis.* Leipzig: Veit. ^{2}1932 Berlin: de Gruyter. English [Weyl 1918/1987].

[Weyl 1918/1987] Weyl, Hermann. 1918/1987. *The Continuum.* English translation by S. Pollard and T. Bole. Kirksville, Missouri: Thomas Jefferson University Press. Corrected republication New York Dover 1994.

[Weyl 1920] Weyl, Hermann. 1920. "Das Verhältnis der kausalen zur statistischen Betrachtungsweise in der Physik." *Schweizerische Medizinische Wochenschrift* 50:737–741. GA II, 113–122.

[Weyl 1923] Weyl, Hermann, *Mathematische Analyse des Raumproblems.* Vorlesungen gehalten in Barcelona und Madrid. Berlin etc.: Springer 1923 Reprint Darmstadt: Wissenschaftliche Buchgesellschaft 1963.

[Weyl 1927] Weyl, Hermann, *Philosophie der Mathematik und Naturwissenschaft.* München: Oldenbourg 1927. In [Baeumler/Schroeter 1927, Bd. II A], and separat print. Further editions ^{2}1949, ^{3}1966. English with comments and appendices [Weyl 1949].

[Weyl 1949] Weyl, Hermann, *Philosophy of Mathematics and Natural Science.* 2nd ed. 1950. Princeton: University Press 1949, ^{2}1950.

[Weyl 1954] Weyl, Hermann, "Erkenntnis und Besinnung (Ein Lebensrückblick)." *Studia Philosophica* 1954. In [Weyl 1968, IV, 631–649].

[Weyl 1968] Weyl, Hermann, *Gesammelte Abhandlungen,* 4 *vols.* Ed. K. Chandrasekharan. Berlin etc.: Springer 1968.

Erhard Scholz
Wuppertal University
Department C – Mathematics
D-42907 Wuppertal, Germany
scholz@math.uni-wuppertal.de

Gödel, Leibniz and "Russell's Mathematical Logic" *

Gabriella Crocco

1. Some known facts about Gödel's interest in Leibniz

Kurt Gödel explicitly mentioned Leibniz in only one paper, "Russell's mathematical logic", which appeared in 1944 in the volume of the *Library of Living Philosophers* devoted to Bertrand Russell and edited by A. Schilpp.[1] Nevertheless, the tribute paid by Gödel to Leibniz in this text is so important that this alone suffices in attesting to the role that Leibniz's work played in Gödel's thought.

Besides this text, which will be extensively analysed in the following sections, evidence of Gödel's deep interest in Leibniz can be adduced through Gödel's own more or less public declarations, as reported by his friends or colleagues, and through the unpublished texts of the Gödel archives in the Princeton Firestone Library.

Among the most widely known evidence of the first kind is that reported by Karl Menger. According to him,[2] Gödel began to be interested in Leibniz's work in the early thirties. Giving an account of Gödel's trip to the USA in 1939, Menger remembers that Gödel was particularly worried about the fate of the Leibniz Archives, because of the political situation in Germany at that time. As Hao Wang first reported in his book *Reflections on Kurt Gödel* (1987): "Menger asked Gödel, 'Who could have an interest in destroying Leibniz's writings?' 'Naturally those people who do not want men to become

---

*Since the first draft of this paper, a considerable amount of new transcriptions of Gödel's manuscripts is now available thanks to the project ANR-09-BLA-0313, under my direction. Therefore, the formulations given here could be improved in several places, but the conjectures advanced are confirmed by the new texts.

[1][Gödel 1944], p. 119 and p. 140 (we quote from the edition of [Gödel CWII]). This is the only reference considering exclusively the texts published during Gödel's life. Apart from the letters and the still unpublished material from the Archives there is a further mention of Leibniz in the second preparatory draft of Gödel's paper for Einstein ([Gödel 1949]) recently published in volume III of the collected works. Gödel mentions Leibniz in these terms: "That time and space have no existence independent of and besides the things was asserted already by Leibniz." [Gödel 46] p. 238.

[2][Menger 1994]. Menger says that Gödel had already begun to concentrate on Leibniz in around 1932. A library slip contained in the Archives [Gödel Nachlass] folder 5/54, 050173 attests that Gödel asked for the Gerhard edition of Leibniz's mathematical writings in 1929. See [van Atten Kennedy 2003].

more intelligent' Gödel replied. To Menger's suggestion of Voltaire being a more likely target, Gödel answered 'Who ever became more intelligent by reading Voltaire's writings?'"[3]

If Gödel's level of interest in Leibniz was high in the 1930s, his systematic study of Leibniz's work began in earnest in the 1940s. We know this fact directly from Gödel himself. In 1974 and 1975 Gödel received letters from Burke Grandjean asking him to answer a few questions about his intellectual and educational background. Gödel never answered these letters but replies were written by him and found in the archives.[4]

To the two questions:

"5. When, if at all, did you first study any of the works of the following:
a) Ludwig Boltzmann b) Jan Brouwer c) Paul Finsler d) Immanuel Kant
e) Karl Kraus f) Fritz Mauthner g) Jules Richard h) Ludwig Wittgenstein
6. How much importance, if any, do you attribute to each of the scholars in the above list, in the development of your interests?"

Gödel answered:

"As to 5 I would like to say that only Kant had some infl. on my phil. thinking in gen (& that I got acq. with him about 1922), that I knew Wittgen. very superficially, that I read only two papers by Finsler (in or after 32) finally that the greatest phil. infl. on me came from Leibniz which I studied (about) 1943–46."[5]

There are two facts confirming Gödel's assertion:

a) the incredible amount of notes handwritten by Gödel on primary and secondary literature on Leibniz found in the Gödel Archives attesting to the depth of Gödel's study[6]; and

b) the most significant account of Gödel's unpublished philosophical work, (represented by the so-called "Max Phil" manuscripts written between 1938 and at the earliest 1955), which contain, between 1943 and 1946[7] an increasing amount of references to Leibniz.

---

[3][Wang 1987] p. 103–4. The passage from Wang continues as follows: "Later (perhaps in or after the 1950s) Menger discussed Gödel's ideas about the destruction of Leibniz's writings with O. Morgenstern, who described how Gödel, to supply evidence for his belief, 'took him one day into the Princeton University library and gathered together an abundance of real astonishing material'. The material consisted of books and articles with exact references to published writings of Leibniz on the one hand, and the very series of collections referred to on the other. Yet the cited writings are all missing in one strange manner or another. 'This material was really highly astonishing' said Morgenstern."

[4][Wang 1987] p. 16–7. [Gödel CW IV] p. 441–50.

[5][Gödel CW IV] p. 449–50. This version comes from an undated draft of Gödel's replies.

[6]The notes are contained in [Gödel Nachlass] series 5, boxes 24–38.

[7]The Max-Phil is a series of 15 notebooks containing philosophical remarks which Gödel called "Max 0-XV". One of them, the XIII, was lost by Gödel himself. Up to now there is only a fragmentary transcription of the Max-Phil, made by Cheryl Dawson between 1992 and 1993, which covers around 750 pages out of 1500. This transcription circulates among scholars. It was in my possession recently, thanks to the kindness of the Dawsons. An increasing number of references to Leibniz are made especially in Max-Phil X and XI. Max Phil VII, written between July '42 and Sept. 1942, contains no explicit reference to Leibniz. Max Phil VIII written between the

The end of this intensive study on Leibniz in 1946 did not signify a turning point in Gödel's interests. Different evidence shows just how true are Gödel's replies to Grandjean's questionnaire in 1974 about Leibniz's great philosophical influence on him.

First of all, one of the most significant manuscripts of the "Max Phil", Max Phil XIV, written over a long period between 1946 and 1955, shows how at that time Gödel was trying to develop a really general philosophical system based on Leibniz's monadology where, as in Leibniz's work, questions of physics, logic, mathematics and rational theology interact together in the aim of the construction of a metaphysical system.

Secondly, at least two of Gödel's letters found in his archives show how, till the end of the 1950s and the beginning of the 1960s, Gödel was still trying to improve his knowledge of Leibniz and was searching for such a metaphysical system largely inspired by Leibniz's monadology.

The first is a letter to Walter Pitts, the well-known co-author of *A Logical Calculus of Ideas Immanent in Nervous Activity* [McCulloch and Pitts 1943], one of those mathematicians who were among the fathers of cybernetics. The letter, dated September 23, 1958, is very eloquent on Gödel's lasting conviction about Leibniz's topicality for modern science. Gödel writes to Pitts:

"I was very much interested to hear that you are considering applications of Leibnizian ideas to modern physics [...]. I do think that the possible applications of Leibniz's work to modern science are far from being exhausted. In particular the unpublished manuscripts of Hanover may contain invaluable ideas as to the systematic solution of mathematical, as well as other scientific, problems. In fact this must be so, if Leibniz ever put down on paper what he definitively claimed to have discovered, and if the manuscripts concerned were not lost in the subsequent centuries."[8]

The second is one of the letters written by Gödel to his mother in 1961, about the afterlife. Having explained his philosophical convictions to her, Gödel (mentioning Leibniz) says:

"Of course, today we are far from being able to justify the theological world-view scientifically, but I think already today it may be possible purely rationally (without the support of faith and any sort of religion) to apprehend that the theological world-view is thoroughly compatible with all known facts (including the conditions that prevail on our earth). Two hundred fifty years ago the famous philosopher and mathematician Leibniz already tried to do that, and that is also what I have attempted in my last letter. What I called the theological world-view is the idea that the world and everything in it has meaning

---

15th Sept. 1942 and the 18th of Nov. 1942 contains one explicit reference to Leibniz. Max Phil IX written between the 18th of November 1942 and the 11th of March 1943, contains no explicit reference to Leibniz. Max Phil X, from the 12th of March 1943 and the 27th of January 1944 contains fourteen references to Leibniz. Max Phil XI from the 28th of January 1944 to the November 1944 contains seven references to Leibniz. Max Phil XII, from the 15th November 1944 and the 5th June 1945, is not transcribed. Max Phil XIII, as we said before, was lost by Gödel.
[8][Gödel CW V] p. 159.

and reason to it and in fact a good and indubitable meaning. From that it follows directly that our earthly existence, since it in itself has a very doubtful meaning, can only be a means toward the goals of another existence. The idea that everything in the world has meaning is, after all, precisely analogous to the principle that everything has a cause, on which the whole of science rests"[9]

The third evidence of Gödel's lasting interest in Leibniz's work after 1946 and up until his death is taken from Hao Wang's conversations with Gödel from 1967 to 1972 [Wang 1996].

The image Wang offers of Gödel is that of a mathematician deeply convinced that mathematics describes a non-sensual reality, which exists independently both of the acts and the disposition of the human mind and therefore[10] of a philosopher trying to explore and explain the objective reality of concepts and their relations. According to Wang, as time went by this mathematician and philosopher was more and more engaged in the attempt to cultivate philosophy as a system in the old, classical sense, i.e., as a guide for scientific research and as a means of investigating the meaning of the world.[11] Wang affirms that from 1943 to 1958, under the influence of the Vienna Circle, Gödel approached philosophy by way of its relation to science (logic and mathematics in the first place although from 1947 to 1950 Gödel worked on physics). Wang says that the Leibnizian project of a *Characteristica Universalis*, posing definitions concerning the primitive concepts of all science and developing deductively knowledge in all domains, was alive for Gödel up to the end of the 1950s.[12] Nevertheless, Wang adds, the difficulties he encountered in preparing the Carnap paper,[13] and the impossibility he felt of having a satisfactory account of the content of mathematics and of their concepts pushed him to a change of direction. Wang says that by 1959 Gödel had concluded that philosophy required a new method different from that of science and thought he had found such a method in Husserl's phenomenology. He adds that in around 1972, Gödel affirmed that he had not found what he was looking for.[14]

---

[9][Gödel CW IV] p. 439. The content of these letters on the afterlife is often presented as one of the most clear manifestations of Gödel's fear of death, possibly related to a more deep mental fragility. A brief passage from [Wang 1996] (see below about this text) proves on the contrary a certain "political" or at least ethical ground for such a rational theology, which again relates him deeply to Leibniz. Gödel says to Wang in October 1972: "The rulers find it hard to manipulate the population: so they use materialism to manipulate the intellectuals and use religion to manipulate the workers. Before the communists can conquer the world, they will have some rational religion. The present ideal is not a sufficiently strong motive. Can't reform the world with a wrong philosophy. The founders of science were not atheists or materialists. Materialism began to appear in the second half of the eighteenth century". (p. 146, transcription n. 4.3.15)
[10]The exact explication of this "therefore" depends on the period of Gödel's investigations, but it is a constant feature of his philosophy that mathematical objects depend on concepts.
[11]Cf. [Wang 1996] p. 309 and the letter to his mother mentioned above.
[12][Wang 1987] p. 174.
[13]The one he prepared for the volume in honour of Carnap for the Schilpp's collection and that he never published. There are different drafts of this paper, under the title "Is mathematics syntax of language?" Two of them are published in the third volume of the *Collected Works*, [Gödel 1953].
[14][Wang 1996], Chap. 2, especially pp. 76–81 and p. 88.

What also appears from the transcriptions of Gödel's conversations with Wang is that even though in around 1959 Gödel considered the necessity of changing his method,[15] many logical and metaphysical aspects of Leibniz's thought were still at the centre of Gödel's work after the 1950s. One of these aspects is certainly the search for a general type-free intensional logic of concepts with simple primitives from which all the other concepts can be composed, and the consequent questioning on the difference between concepts and objects. His ideas on this last subject in Wang's account strongly recall Leibniz's discussions about the difference between abstract and concrete, complete and incomplete concepts, Ideas in mente homini and Ideas in mente Dei (see below). The second aspect concerns the importance of monadology especially for physics and biology. It is interesting in this sense to mention four passages reported by Wang, the first two because they are very explicit on the Leibnizian structure of Gödel's own monadology; the last two for their enigmatic, striking assertion of the importance of monads for modern science.

"0.2.1 My theory is a monadology with a central monad [namely God]. It is like the monadology by Leibniz in its general structure.[16]

0.2.2 My philosophy is rationalistic, idealistic, optimistic, and theological.[17]

9.1.8 It is an idea of Leibniz that monads are spiritual in the sense that they have consciousness, experience, and drive on the active side and contain representations (Vorstellungen) on the passive side. Matter is also composed of such monads. We have the emotional idea that we should avoid inflicting pain on living things, but an electron or a piece of rock also has experiences. We experience drives, pains and so on ourselves. The task is to discover universal laws of the interactions of monads, including people, electrons and so forth. For example attraction and repulsion are the drives of electrons and they contain representations of other elementary particles.

9.1.9 Monads [bions] are not another kind of material particles; they are not in fixed parts of space, they are nowhere and therefore not material objects. Matter will be spiritualized when the true theory of physics is found. Monads only act *into* space; they are not *in* space. They have an inner life or consciousness;

---

[15]The conference of 1961 is the one where Gödel affirms explicitly his interest for Husserl and phenomenology [Gödel 1961] in [Gödel CW III] p. 383–4.

[16][Wang 1996] p. 8 (in the sequel we will quote Wang's transcription using only his own numbering.) That Gödel specifies in his conversations that his monadology has a general Leibnizian structure seems to us a clear indication that he was aware of other kinds of monadology, such as that of Mahnke or Renouvier, but that he did not consider them interesting from the point of view of his own philosophical project; (two slips of paper from the archives prove that Gödel asked for *Eine neue Monadologie* by Dietrich Mahnke (1917) and *La nouvelle monadologie* de Charles Renouvier et Pratt (1899). See, nevertheless, [van Atten Kennedy 2003] p. 457 for another evaluation of Gödel's interest in Mahnke's phenomenology.

[17]*Ibid.* Without entering into an analysis which would be out of context here, we can add, on the basis of [Gödel 1951] that Gödel's rationalism can be summarised as the affirmation that for clear questions posed by reason, reason can also find a clear answer; and that idealism and theology have in common that they "see sense, purpose and reason in everything". Idealism can be interpreted in a general way as the doctrine which affirms that space-temporal perceived reality is the result of the relation between subject and objects because in a particular time they have no existence independently from this relation; cf. [Gödel 1946] quoted in note 1 and [Gödel CW IV] p. 527.

222   G. Crocco

in addition to relations to other particles (clear in Newtonian physics, where
we know the relationship between the particles), they also have something inside; in quantum physics the electrons are objectively distributed in space, not
at a fixed place at a fixed moment, but at a ring. Hence it is impossible for
electrons to have different inner states, only different distributions."[18]

2. Open problems and methodological decisions

In spite of the huge amount of evidence for Leibniz's influence on Gödel's philosophical
project, it is hard to evaluate precisely how Gödel, in his work, reconciles this influence
with other ones, which are also clearly attested to (for example Kant, Husserl, German
idealism and especially Hegel whose name is repeatedly mentioned by Gödel to Wang).
The exact reasons for the "turn" in around 1959 (as mentioned by Wang), the precise nature of the difficulties with concepts which prevented Gödel from publishing his Carnap
paper, the way his interest in Husserl can be reconciled with his "Platonism", the relation
between his work in physics and his monadology are at the centre of modern investigations into Gödel's philosophical thinking . Wang's transcriptions of his conversations with
Gödel are precious but not precise enough to develop a clear picture of Gödel's philosophical project and indeed are too general to give an answer to the question which would be
central for us: what is the exact distance from Leibniz that Gödel thought necessary to
take account of modern science within a philosophical system?

What is clearly stated, on the basis of Wang's account, is the metaphysical project
which is strongly inspired by Leibniz: concepts such as "forms" or structures of reality
which cause existence of spatiotemporal objects as well as mathematical objects, the latter
being the limiting case of spatiotemporal objects.[19]

What is not clear enough, from Wang's account, however is the structure of the
system, the method, the difficulties experienced by Gödel on applying it, and his precise
attitude toward the philosophers of the past.

----

[18][Wang 1996] p. 291–2. In the lecture of 1951 mentioned in the previous note, Gödel affirms that the development of the *Zeitgeist* from the Renaissance to modern times has led particularly in physics to the situation
that "the possibility of knowledge of the objectivizable states of affairs is denied, and it is asserted that we must
be content to predict results of observations. This is really the end of all theoretical science in the usual sense
(although the predicting can be completely sufficient for practical purposes such as making television sets or
atomic bombs)." Gödel considered his own contribution to cosmology as an attempt to complete the modern
theory of gravitation, which he considered as deeply unaccomplished, even from the mathematical point of view
(see [Audureau 2004] pp. 145–6). Gödel's image of gravitation in the Leibnizian terms of monadic attraction
and repulsion, as it is suggested in these two passages, is a way of refusing to acknowledge, in very imaginative
terms, this *Zeitgeist*'s diktat. Especially the second passage ("matter will be spiritualized when the true theory
of physics is found") suggests that the difficulties in unifying modern physics do not depend on the question of
finding the right mathematical model, but on the courage to deepen our hypothesis on the reasons for gravitation.
Cf. also [Wang 1996] p. 58. Many passages in Max-Phil X confirm such an interpretation.
[19]Cf. [Wang 1996] 0.02 and also "5.3.11: The beginning of physics was Newton's work of 1687, which needs
only very simple primitives: force, mass law. I look for a similar theory for philosophy or metaphysics. Meta-
physicians believe it possible to find out what the objective reality is; there are only a few primitive entities
causing the existence of other entities. Form (So-Sein) should be distinguished from existence (Da-Sein): the
forms- though not the existence of the objects- were, in the middle age, thought to be within us."

It is a fact that Gödel considered Leibniz's search for the primitive concepts of our knowledge and Husserl's "intuition of essences" as similar in their tasks[20], but it is not clear what exact place Gödel gave to Husserl in his search for an "updating" of Leibniz's project.

Gödel probably thought that Husserl's method could be an alternative to the Leibnizian project of the *Characteristica* intended as a (non-formal) deductive system giving the primitives by definition (through explicit axioms),[21] although he told Wang that he could not arrive at a satisfactory solution. He probably considered and explored both approaches till his death. Some passages of Wang's transcriptions seem to go in this direction, such as the following:

> 5.3.7 Phenomenology is not the only approach. Another approach is to find a list of the main categories (e.g., causation, substance, action) and their interrelation, which however, are to be arrived at phenomenologically. The task must be done in the right manner.

> 9.3.10 Philosophy aims at a theory. Phenomenology does not give a theory. In a theory concepts and axioms must be combined, and the concepts must be precise ones. [...]

Even if this hypothesis is viable, it is still unclear what exactly are these primitive precise concepts, from which philosophy could be derived and what kind of logic he intended to apply to them – in view of arriving at a satisfactory system implementing his ideal of informal rigour. This is not a mere technical problem, as logical questions are fundamental in the construction of Gödel's system being at the heart of physical and metaphysical problems.[22]

---

[20]Cf, [Wang 1996] "5.3.19 Leibniz believed in the idea of seeing the primitive concepts clearly and distinctly. When Husserl affirmed our ability to "intuit essences," he had in mind something like what Leibniz believed. Even Schelling adhered to this ideal, but Hegel moved away from it. True metaphysics is constantly going away. Kant was a skeptic [sic], or at least believed that skepticism [sic] is necessary for the transition to true philosophy."

[21]Gödel spoke about the (non-formal) logic of concepts, which would give a new life to the *Characteristica* project, in the Russell paper (see further) and in the conversation with Wang [Wang 1996] Chap. 8. One of the ways to think of such a system of logic is the system of [Church 1932], [Church 1933], mentioned in [Gödel 1944] and in [Wang 1996] p. 279. The informal aspect of this system is that logical rules cannot be applied on a simple inspection of the symbolic structure of a linguistic expression but in order to apply them we have to take into consideration its meaning.

[22]One of the clearest cases of such a connection between logical and metaphysical problems is the example of possibility, which seems to have an unacceptable deterministic nature in Leibniz's system: "9.4.2 With regard to the structure of the real world, Leibniz did not go nearly as far as Hegel, but merely gave some 'preparatory polemics'; some of the concepts, such as that of possibility are not clear in the work of Leibniz; Leibniz had in mind a build up of the world that has to be so determined as to lead to the best possible world [...]". See also about possibility the interesting 9.4.10 which mentions Hegel's first three categories – being, non-being and becoming: "Independently of Hegel's (particular choice of) primitive terms, the process is not in time, even less an analogy with history. It is right to begin with being, because we have something to talk about. But becoming should not come immediately after being and non-being; this is taking time too seriously and taking it objectively. It is very clear that possibility is the synthesis between being and non-being. It is an essential and natural definition of possibility to take it as the synthesis between being and non-being. Possibility is a 'weakened form of being' " [Wang 1996].

The problem is particularly pressing in relation to a very basic question: in what way are Gödelian concepts tied to Leibnizian ideas and what does Gödel actually mean by concepts? What does Gödel have in mind when, in 1972 during his conversation with Wang, he speaks of concepts in intension, as opposed to concepts in extensions? Answers to these questions are required in order to understand his dissatisfaction with his Carnap paper and his "turn" toward Husserl after 1959. There are very interesting analyses which really assimilate Gödel's concepts in intension with Husserl's notion of intension and that consider the "phenomenological turn" as a definitive abandonment of a primitive naïve Platonism.[23] There are other interpretations which stress how such identification is problematic on the basis of Gödel's famous assertion of the reflexivity of concepts, which is very clearly stated in his conversations with Wang and which seems to be in contradiction with a strict Husserlian frame.[24]

Interpretative conjectures about Gödel's philosophical project can and have been made[25] but they will remain mere conjectures without a serious and complete analysis of the philosophical material of the Archives, which is far from being completed.[26] What is totally missing is an understanding of Gödel's philosophical positions in relation to his scientific work. Contrary to a quite general opinion Gödel continued to be interested in logic, physics and mathematics after the 1940s, as it appears from the Archives and from Wang's account. We do not know how his philosophical work interacted with and guided his reflections on science.

Such a situation invites a methodological choice, which I think is the wisest option considering the present state of our knowledge. The painstaking, almost maniacal attention that Gödel always paid to the editing and publication of his writing commands our respect. We need to reconstruct Gödel's thought-process, taking as a starting point his published work, which is very often misunderstood because of Gödel's style (concise and sometimes enigmatic) but also because of the content of his writings, which are generally contrary to what the *Zeitgeist* considers as standard. We need to go from the published

---

[23]Cf. [van Atten Kennedy 2003], [Tieszen 98]. The idea of a progressive abandonment of Platonism is quite generally accepted (see for example [Parsons 1995]). Many of the commentators quote generally a passage from [Gödel 1933] p. 50, where Gödel, commenting on his own results on the foundations of mathematics, says: "The result of the preceding discussion is that our axioms if interpreted as meaningful statements, necessarily presuppose a kind of Platonism, which cannot satisfy any critical mind". The passage is presented as an early symptom of Gödel turning away from Platonism, which should have then been accomplished completely in the fifties. There is another natural interpretation of this text without any negative weight against Platonism, if we consider that the critical minds are all those people who believe that the Kantian Copernican revolution is a definite progression in philosophy and that, in this sense, accept his critical philosophy (see [Audureau 2004] p. 62). That Platonism concerning ideas is compatible with Leibniz's conception of Ideas *in mente Dei* is a well known result of Leibnizian studies see for example [Mugnai 2001] Chapter 2, or [Nachtomy 2007] Chapter 1. See also below.

[24][Crocco 2006] p. 190 and in particular note 50.

[25]See also [Hintikka 1998], [Buldt 2002].

[26]Most of Gödel's manuscripts and notes in the Archives are written in Gabelsberger. The French ANR "Agence nationale pour l'évaluation de la recherche" financed a four-year project under my direction, to revise and pursue the work of transcription. After one year of work we have a complete transcription of three of the Max Phil (X, XI and XIV) thanks to the work of a group of people including Mark van Atten, Paola Cantù, Eva-Maria Engelen and Robin Rollinger.

works to the unpublished materials and then back again. I would like in the following to apply such a method to the issue of Leibniz's influence on Gödel, trying to use Wang's material and the available transcriptions of the Max-Phil manuscripts in order to clarify and deepen some of Gödel's explicit doctrines presented in the published work.

When we inquire about Leibniz's influence on Gödel it is certainly from the Russell paper that we have to start, i.e., the text of 1944 that, up to his death, Gödel always considered as one of his most important publications in terms of the explanation of his philosophical ideas concerning logic. Making use of passages from the archives and from the conversations with Wang, we will try to solve some of the interpretative difficulties concerning this paper and to reconstruct some central Leibnizian ideas, which Gödel continued to elaborate till the end of his life. Certainly the Russell paper is essentially concerned with questions of logic and the philosophy of mathematics, but Gödel's systematic approach to philosophy allows us, starting with logical questions, to reach some central metaphysical ones. The Russell paper is therefore an essential step on the way to understanding Gödel's philosophical thinking and also a fundamental step in order to understand his critical consideration of the Leibnizian project in the light of modern science.

3. The structure of Gödel's "Russell's mathematical logic": problems of interpretation

Enigmatic, unfair with regard to Russell and disjointed: this is the judgement very often given about this paper; beginning with Russell himself, who, in his reply to contributors, avoided any answer to the questions evoked by Gödel; pursuing with Hermann Weyl, who in his 1946 review of the Schilpp volume says (concerning Gödel's contribution): "it is the work of a pointillist: a delicate pattern of partly disconnected, partly interrelated, critical remarks and suggestions"[27]; ending with the judgement of Charles Parsons, who says in his introduction to the re-publication of the paper in the second volume of Gödel's Collected Works : "the organization of the paper is difficult for the present commentator to analyze". Parsons proposes a division of the paper in eight parts, as follows:

1. Introductory remarks (125–128)
2. Russell's theory of descriptions (128–31)
3. The paradoxes and the vicious circle principle (131–37)
4. Gödel's own realistic view of classes and "concepts" (137–41)
5. Contrast with Russell's "no-classes theory" and the ramified theory of types; limitations of the latter (141–47).
6. The simple theory of types (147–50)
7. The analyticity of the axioms of *Principia* (150–52)
8. Concluding remarks on mathematical logic and Leibniz's project of a universal characteristic (152–53).

This division does not completely fit with that which is suggested by Gödel through very explicit formulations as "I pass now to..." (131), "I now come in somewhat more details

---

[27] [Weyl 1946].

to..." 147, "In conclusion I want to say..." (150). These formulations suggest a division into four parts to which we can add the introduction and the conclusion as follows:

0. the introductory remarks (125–127) containing a definition of mathematical logic as the science underlying all other, a homage to Leibniz, considered as the father of this point of view, and a sketchy comparative analysis of the respective contributions of Peano, Frege and Russell in "put[ting] into effect" the Leibnizian project.

1. the description of Russell's early realistic attitude and, as an example of it, his theory of definite descriptions. (127–131). The end of the introduction and the beginning of this part is marked by the following sentence: "I do not want, however, to go into any more details about either the formalism or the mathematical content of the *Principia*, but want to devote the subsequent portion of this essay to Russell's work concerning the analysis of the concepts and axioms underlying mathematical logic. In this field Russell had produced a great number of interesting ideas some of which are presented more clearly (or are contained only) in his earlier writings. I shall therefore frequently refer to these earlier writings, although their content may partly disagree with Russell's present standpoint." p. 127

2. the discussion of the paradoxes and the critical analysis of Russell's solution of them through the ramified theory of types (the theory of orders as Gödel calls it) and the "constructivistic" vicious circle principle which, according to Gödel, inspired it (pp. 131–147). This part is introduced by the following passage "I pass now to the most important of Russell's investigations in the field of the analysis of the concepts of formal logic, namely those concerning the logical paradoxes and their solutions." (p. 131)

3. the analysis of the simple theory of types (147–150). This part is introduced by the following passage: "I pass now in somewhat more detail to theory of simple types which appears in *Principia* as combined with the theory of orders; the former is however, (as remarked above) quite independent of the latter, since mixed types evidently do not contradict the vicious circle principle in any way." (p. 147).

4. the question of the analyticity of the axioms of mathematics (150–152). This part is introduced by the sentence "In conclusion I want to say a few words about the question of whether (and in which sense) the axioms of *Principia* can be considered to be analytic."

5. the conclusion which comes back to Leibniz and his project. (p. 152–3)

There is no other so explicit transition sentence in the paper.

This question of the organisation of the paper is a not a minor issue but a central one for our analysis. The Russell paper was solicited from Gödel by Schilpp on the 18th of November 1942 and Gödel sent the final version at the end of September 1943. Gödel says that his intensive study of Leibniz began in 1943. The evidence given in Section 1, about Gödel's concern for the fate of the Leibniz Archives in 1939, and the existence in the Gödel Archives of a sheet dated 1941, with a first version of his very Leibnizian proof on the existence of God, are sufficient to exclude that Gödel's work on Leibniz began in 1943. It is just its *intensive* form, which started in 1943, as the increasing amount of direct mentions of Leibniz in Max-Phil X and XI[28] proves. A plausible conjecture is therefore that the Russell paper was the occasion of this intensive study. Having expressed in it what he considered to be the current situation of logical research and the reasons and difficulties associated with the Leibnizian project, Gödel could have then decided to deepen his knowledge of Leibniz's work in order to propose more clearly his own logical and ontological system. The interpretative key of the Russell paper could therefore lie in its structure. It opens and ends with Leibniz and his project of the *Characteristica*. This could mean that the way the arguments follow each other is related to a certain way of interpreting Leibniz at that time when Gödel was convincing himself of the necessity of an intensive study on Leibniz's work. The Russell paper could be interpreted as proposing a sort of programmatic list of the central questions that Gödel intended to explore more deeply, and which, from his point of view, had been treated by Russell in an unsatisfactory way. Before coming to the question of the content of this programmatic list, let us begin with this tribute to Leibniz, which comes in the form of both the introduction and the conclusion to the paper.

In the introduction Gödel defines mathematical logic as a mathematical science, treating specific objects (such as classes, relations and combinations of symbols) and mathematical logic as a science "prior to all others, which contains the ideas and principles underlying all sciences". It was in this second sense, he says, that Logic was considered by Leibniz in his *Characteristica Universalis*, and in this same second sense it was developed by Peano and Frege with the aim of furnishing "a logical calculus really sufficient for the kind of reasoning occurring in the exact sciences". Gödel considers Russell's work as part of the movement initiated by Peano and Frege and he interprets and judges their works as attempts toward the realisation of the Leibnizian project of the *Characteristica*. As a matter of fact, Frege's project runs against the paradoxes and Russell (as Gödel says later) "by analysing the paradoxes to which Cantor's set theory had led, freed them from all mathematical technicalities, thus bringing to light the amazing fact that our logical intuitions (i.e., intuitions concerning such notions as truth, concept being, class) are self-contradictory"(p. 124). The criticism Gödel develops in the paper shows how, in order to accomplish the Leibnizian project and in spite of Russell's work, the fundamental notion of class and concept "need further elucidation" (p. 152). Gödel stresses, in the end, that the accomplishment of this project, if correctly realised, should really improve progress in science and especially in mathematics and reject any kind of interpretation of the paradoxes in terms of the utopian character of the Leibnizian project. He says:

> "It seems reasonable to suspect that it is this incomplete understanding of the foundations [of classes and concepts] which is responsible for the fact that mathematical logic has up to now remained so far behind the high expectations of Peano and others who (in accordance to Leibniz's claims) had hoped that it would facilitate theoretical mathematics to the same extent as the decimal system of numbers has facilitated numerical computations. For how can one expect to solve mathematical problems systematically by mere analysis of the

---

[28] See note 7 before.

concepts occurring if our analysis so far does not suffice to set up the axioms? But there is no need to give up hope. Leibniz did not in his writings about the *Characteristica Universalis* speak of a utopian project [...]."[29]

In the course of his introduction to the Russell paper in the third volume of Gödel's Collected Works, Charles Parsons, considering these lines as "a sort of coda of this intricate paper", expresses his astonishment concerning these Gödelian statements: "His suggestions that the hopes expressed by Leibniz for his Characteristica Universalis might after all be realistic" he says "is one of the most striking and enigmatic utterances". Apparently Parsons judges Gödel's assertions in the light of the development of set theory and considers this theory as having definitively broken up the dream of a *Grande logique*, able to directly give an account by itself (through the notion of extension) of the logical nature of the concept of set. There is clear evidence that Gödel never accepted this verdict and that he was, for all his life, directly committed to the Leibnizian idea of logic as the science of all sciences. Moreover his diagnosis of the difficulties of Frege's and of Russell's attempted solutions suggests the necessity of a return to Leibniz, beyond Frege's and especially Russell's works, compatible with the constraints of a foundation of modern mathematics. The "disconnected" structure of Gödel's analysis can probably disappear through a reconstruction of the Leibnizian background of this paper. Gödel seems to point out, implicitly, that some of what everyone at the time considers as "progress" accomplished by contemporary logic is perhaps nothing of the sort. Beyond the specific solutions or suggestions that Gödel formulates at that time, what is at stake is the rehabilitation of the great Leibnizian project as Gödel understood it.

Let us now come again to the four parts in which Gödel divides his paper.

Charles Parsons recognizes explicitly, beyond the introductory and conclusive remarks on Leibniz, the first, the second, the third (corresponding to 6 in his list) and the fourth (corresponding to 7 in his list) parts that we mentioned before. He divides the second part into three distinct sections (3–5 in his list), which is perfectly justified from an expository point of view. Nevertheless it seems to us that, without recognising the unity of part 2 and without searching for the general relation between the four parts, something of Gödel's intention is lost. What can we conjecture about this intention?

The first part from § 5 to § 9 is introduced, as we saw before, by a sentence (§ 5) where, ending the introduction, Gödel affirms that he wants to devote the rest of the paper to Russell's work concerning the analysis of the concepts and axioms underlying mathematical logic and it is clear from the context that he intended mathematical logic in the sense which he explained it in the introduction (the science "which contains the ideas and principles underlying all sciences"). After some consideration of Russell's early realistic attitude, contrasted with his further constructivism or fictionalism[30], the rest of this part is devoted to an example. Russell's analysis of descriptive phrases is given as an exemplary

---

[29] [Gödel 1944] p. 140.

[30] See in [Gödel 1944] p. 125 the *Author's addition of* 1964 *expanded in* 1972 where Gödel explicitly uses the term "fictionalism". This attitude is so described in the text of 1944: "When he started on a concrete problem, the objects to be analysed (e.g., the classes or propositions) soon for the most part turned into "logical fictions". Though perhaps this needs not necessarily mean (according to the sense in which Russell uses this term) that these things do not exist, but only that we have no direct perception of them".

case of this realistic attitude in the analysis of "fundamental logical concepts" (beginning of § 7, at the end of p. 128). Russell's and Frege's positions are comparatively analysed and criticised but Gödel holds a quite enigmatic position on what he considers the real solution to the problem. What we will suggest in this paper on the basis of some of the Archive texts, is the fact that in the same period Gödel's reflections on the problem of descriptive phrases is abundant and in a constant dialogue with the problem of individual concepts in Leibniz. This material shows how this problem was seen by Gödel in this period as a crossroads of semantic, epistemological and metaphysical questions, which involves in particular the relation between extension and intension, the status of mathematical objects and the nature of concepts, which are the subject matter of the rest of the paper.

The long second section runs from § 10 to § 36. This section is essentially concerned with concepts and classes and the difficulties to which they give rise. The theory of orders is presented as Russell's attempt to clarify their nature and their relationship. A very deep analysis of the vicious circle principle on which it rests is proposed. This part is very coherent in its content and proposes a long, critical analysis of Russell's position in *Principia*.

§ 37 opens the third part of the paper. It begins with an analysis of Frege and Russell's doctrine about concepts and propositional functions as unsaturated or incomplete entities, which is at the basis of the theory of simple types. In a general way Gödel criticises it (§ 37). Nevertheless he considers it as containing the fundamental strategy of the limited ranges of significance, which is clearly, according to him, the most promising one for the solution of the paradoxes through a type-free strategy (§ 38).

In Section 5, we will suggest how the analysis of the second and third parts of the paper, considered with regard to Wang's conversations with Gödel, hints at a very peculiar doctrine of concepts, which in many regards brings to mind Leibniz's analysis. In opposing Frege and Russell, Gödel seems to come back to Leibniz in order to reject both unsaturation and (consequently) distinction of types for concepts.

On the one hand, Gödel's criticism of the ramified theory of types shows, from his point of view, how wrong Russell's treatment of intensions is, through the vicious circle principle and how it obscures the relations between the linguistic terms, their intensions and their extensions, i.e., (according to Gödel) predicates or notions, concepts and classes. From Gödel's point of view, treating correctly the difference between extension and intension, concepts and classes, is a fundamental task which logic has to solve in order to find the correct solution to paradoxes. From Gödel's point of view this is the very Leibnizian problem of the distinction between possible and real, eternal ideas and spatiotemporal (or quasi-spatiotemporal, i.e., mathematical) objects.

On the other hand, the simple theory of types, which is the most coherent realisation of an intensional calculus of concepts, gives Gödel the occasion to radically criticise the Fregean-Rusellian doctrine of the incomplete nature of concepts considered as unsaturated. It seems that, according to Gödel, we could not find the true logic of concept without "coming back" to this fundamental prejudice of contemporary logic, as it derives from an unaccomplished realism of concepts.

In the fourth and last part (§ 41, 42) Gödel distinguishes two senses of analytic, one purely formal (i.e., an axiom is analytic if, by virtue of the definition of the terms occurring in it, it can be defined in such a way that it reduces in a finite number of steps to the law of identity) and a second sense concerning the content of the terms not their definitions. In this second sense, an axiom is analytic if it is "true by virtue of the meaning of the concepts occurring in it". He adds in note 47 that this view also permits a reduction to a special case of $a = a$, if the reduction is effected not by virtue of the definitions of the terms occurring, but by virtue of their meaning, which can never be completely expressed in a set of formal rules. The question of analyticity allows Gödel to show that the Leibnizian idea that truth of reasons are in a certain sense identities is still very much alive. The correct interpretation of the notion of analyticity is essential, from Gödel's point of view, in order to overcome the empiricist-sensualist prejudices of contemporary theory of knowledge, as he argues later in the Carnap paper.

All of the above suggests a kind of hidden interconnection between these four parts, which could be more clearly appreciated if we focused on the Leibnizian background of Gödel's work. Beyond the Fregean-Russellian systematization of modern logic, Gödel seems to be claiming the necessity of coming back to some logical issues re-evaluating Leibniz's ideas: 1) the notion of individual concept and the relations between possibility and existence that it presupposes; 2) the way to look at the opposition between extensions and intensions; and 3) the opposition between concepts, ideas in *mente homini* and ideas in *mente dei*, that is, the opposition between the subjective and objective sides of concepts. On some of these issues, Gödel expresses his disagreement with Russell's solution to some of Frege's difficulties, on others he rejects both Frege's and Russell's analysis of questions on which their agreement was explicit. This hypothesis could be satisfactorily confirmed when the Max Phil has been completely transcribed.[31] Nevertheless with our current state of knowledge, it is possible using some of the texts already transcribed to formulate some important conclusions about this background. We will try in this paper to underline some problems with the four parts of the Russell paper, and to give some hints about the Leibnizian background of Gödel's reflection.

4. Gödel's discussion of descriptive phrases and the Leibnizian metaphysical problem of the individual concepts

If there were only one subject on which Russell's analysis was quite universally celebrated, then it would certainly be that of definite descriptions. Russell's solution for them is still one of the favourite examples of analytic philosophers (perhaps together with the famous adage "to be is to be the value of a variable") to show the possibility of making "progress" in philosophy.[32]

Gödel provides two explicit reasons for considering the issue. The first reason (p. 127) is that it gives a clear example of the fact that, for Russell, questions of logic

---

[31] There are substantial mentions of Frege in Max-Phil VI, VII, VIII, IX, XIV and XV (respectively 1, 7, 6, 8, 2 and 1 mentions) and some on Russell in Max Phil I, IV, V, VII, VIII, IX and XIV (respectively 1, 5, 4, 2, 2, 9 and 3 mentions). But many of them have not yet been transcribed, in particular those from Max-Phil IX, which was written between 18th of November 1942 and the 11th of March 1943. We will use some of the other mentions, in the sequel.

[32] Cf. for example [Quine 1941].

are questions concerning reality, which require definite answers, and are not matters of convention. The second reason (*ibid.* and p. 143) is that it shows how Russell's realistic attitude was stronger in theory than in practice because his logical analysis often obeys a constructivist order of ideas.[33] Nevertheless, this illustrative power of the theory of definite descriptions in Russell's philosophy does not exhaust Gödel's interest. Its treatment by Gödel is quite enigmatic: in no way does Gödel subscribe to Russell's solution,[34] but at the same time he seems hesitant to give a precise diagnosis as to what is wrong with it. After having presented Frege's and Russell's analysis, Gödel affirms: "As to the question in the logical sense, I cannot help the feeling that the problem raised by Frege's puzzling conclusion has only been evaded by Russell's theory of descriptions and that there is something behind it which is not completely understood." The fact is that the logical analysis of the descriptive phrases directly involves semantic and metaphysical problems, which Gödel was working on at that time.

Let us first recall the problem, as Gödel presents it. When we ask: 'what is the meaning of descriptive phrases?' Gödel says that there is an apparently obvious answer that Frege adopted. An expression such as "the author of Waverly" denotes or signifies Walter Scott and in a general way a descriptive phrase denotes the object it describes, if it exists. Nevertheless, this apparently obvious answer, in conjunction with a further apparently obvious axiom stating that the signification of a composite expression, containing constituents which have themselves a signification, depends only on the signification of these constituents (not on the manner in which this signification is expressed) implies "almost inevitably"[35] the unexpected consequence that all true sentences have the same signification, "the True" as it is called by Frege, and that all false sentences signify "the False".[36] The other principles, which lead "almost inevitably" to this consequence are established by Gödel as follows :

1) "$F(a)$" and the "a is the object having the property F and which is identical with a" have the same meaning;
2) Every proposition "speaks about something", i.e., can be brought in the form $\phi(a)$; and
3) For every object a and b, there exists a true proposition of the form $\phi(a, b)$ (as $a \neq b$ or $a = a \wedge b = b$);

Gödel says that from these three principles and from the two implied by the obvious axioms mentioned before, i.e.,

---

[33] [Gödel 1944] p. 143. See also the passage at p. 127 already quoted in note 30.

[34] See especially the last sentence of this part: "There seems to be only one purely formal respect in which one may give preference to Russell's theory of descriptions. By defining the meaning of sentences involving descriptions in the above manner, he avoids in his logical system any axiom about the particle "the" [...] Closer examination, however shows that this advantage of Russell's theory over Frege's subsists only as one interprets definitions as mere typographical abbreviations, not as introducing names for objects described by the definitions, a feature which is common to Frege and Russell." [Gödel 1944] p. 130–1.

[35] *Ibid*, p. 129.

[36] It's important to stress that Gödel speaks about the "almost metaphysical sense" in which Frege intended this doctrine of the truth. Beyond the question of the correctness of Frege's assertion lies the fact that for Gödel such a problem does have definitively metaphysical consequences.

4) The possibility to substitute co-denotative expressions, independently of the way they are denoted and

5) The fact that descriptive phrases denote the objects they describe,

then we can derive Frege's conclusion.

Let us do that explicitly, just to fix terms, following Alonzo Church's suggestion[37].

Church restricts Gödel's argument to sentences containing descriptive phrases and shows how passing from the sentence "Sir Walter Scott is the author of Waverley" to the sentence "The number of counties in Utah is 29" through the equivalent sentences: "29 is the number of novels on Waverley that Walter Scott wrote altogether" and "29 is the number of counties in Utah". We will use this suggestion to generalize Church's argument to any sentence whatever, using principle 1 above.[38]

Suppose that "$F(a)$" and "$G(b)$" are two true sentences which "speak about" two different objects a and b, through two different properties F and G. Using our five principles we can derive the second from the first, the only semantic aspect common to them being their truth.

i) $F(a)$	By hypothesis
ii) $a = (\imath x)(F(x) \wedge x = a)$	From i), by principles (1), (5), using Peano's definition of 'the' (reversed iota).
iii) $a = (\imath x)(F(x) \wedge x = a \wedge b = b)$	From ii) by principles (3), (4), supposing that the two predicates "$= (\imath x)(F(x) \wedge x = a)$" and "$= (\imath x)(F(x) \wedge x = a \wedge b = b)$" are co-denotative because they have the same extension.
iv) $b = (\imath y)(y = b \wedge a = a)$	From iii) by the principles (2) and (3)

[37] The argument in its informal structure was presented by Alonzo Church in his *Introduction to mathematical logic* (1956) p. 24–25, but it was clearly anticipated by Gödel on pages 128–129 of his paper on Russell. There is a very well-known modern version of this argument given by Davidson ("Truth and Meaning" 1967) which refers in a note to Church (1956), without mentioning Gödel. Nevertheless Davidson's formulation hides the question of extension and intension using the general extra-hypothesis that logically equivalent terms have the same reference, which never was a Gödelian thesis. Moreover the structure of Davidson's argument is more closely related to [Church 1943], which, as Parsons also remarks in his introduction to the Russell paper, depends "on somewhat different assumptions" ([Gödel CW II] p. 104). Actually [Church 1943] is addressed to Carnap's *Introduction to semantics*, and is therefore different from [Church 1956], which on the contrary focuses on Frege's analysis. For a general presentation of the problem see "The philosophical significance of Gödel's slingshot" [Neal 1995].

[38] "[...] the sentence "Sir Walter Scott is the author of Waverley" must have the same denotation as the sentence "Sir Walter Scott is the man who wrote twenty-nine Waverley novels altogether", since the name "the author of Waverley" is replaced by another name of the same person; the latter sentence, it is plausible to suppose, if it is not synonymous with "The number such that Sir Walter Scott is the man who wrote that many Waverley novels altogether is twenty nine", is at least so nearly so as to ensure its having the same denotation; and from this last sentence in turn, replacing the complete subject by another name of the same number, we obtain, as still having the same denotation, the sentence "The number of counties in Utah is twenty-nine"" [Church 1956] p. 24–25.

| v) $b = (\imath y)(y = b \wedge G(y))$ | From iv) by principles (3), (4), supposing again that the two predicates, having the same extension, are co-denotative. |
| vi) $G(b)$ | From v) by principles (1) and (5) |

Therefore, from a true sentence we can derive a true one, whatever its content is. And from this, it follows that, if sentences have a denotation at all, this can be only their truth-value, which (according to the doctrine of saturated and unsaturated expressions) are objects.

In contrast with this Fregean doctrine of the denotation of sentences, there is the Russellian one, apparently more obvious from Gödel's point of view. According to the Russellian doctrine the things which correspond to true sentences, in the real world, are facts, whilst false sentences correspond to nothing. Nevertheless, because of the "almost inevitable" conclusion from the fact that descriptive phrases indicate objects to the fact that sentences indicate truth-values, Russell gives a different, less obvious interpretation of descriptive phrases. A descriptive phrase, according to Russell, denotes nothing, but has a meaning only in the context of a sentence. To be short and using the famous Russellian example, "The author of Waverley" says nothing about Scott, but is only a "roundabout way of asserting something about the concepts occurring in the descriptive phrase." (p. 130).

Gödel adds that Russell adduces two arguments in favour of this view, namely:

"that a descriptive phrase can be meaningfully employed even if the object described does not exist (e.g., in the sentence "The present king of France does not exist"); that one may very well understand a sentence containing a descriptive phrase without being acquainted with the object described, whereas it seems impossible to understand a sentence without being acquainted with the objects about which something is being asserted."[39]

This means that from Russell's point of view there are semantic and epistemological reasons, which justify his deflationist analysis of descriptive phrases. Does Gödel accept such arguments?

A short text found in the Nachlass with item number 040269, and designated by the editor of Gödel's Collected Works as Reprint E,[40] concerns some of Gödel's commentaries on Bernays' review, in 1946, of Gödel's Russell paper. To Bernays's general

---

[39] In note 7 of his text Gödel explicitly mentions Frege's doctrine of sense and contrasts it with Russell's analysis in these terms "From the indication (*Bedeutung*) of a sentence has to be distinguished what Frege called its meaning (*Sinn*) which is the conceptual correlate of the objective existing fact (or "the True"). This one should expect to be in Russell's theory a possible fact (or rather the possibility of a fact, which would exist also in the case of a false proposition. But Russell, as he says, could never believe that such "curious shadowy" things really exist."

[40] [Gödel CW II] p. 321–2.

remark of a sort of a lack of clarity in Gödel's paper about the distinction between Sense and Denotation[41] Gödel answers with two interesting remarks, which are the following:

4) Das Probl. der Beschreibung ist durch "Sinn" und "Bedeutung" in befriedi-
gender Weise gelöst.

5) Das Extents. axiom gilt nicht für Begriffe.[42]

On the basis of this text and the content of the paper let us give a first interpretation of Gödel's analysis in the following terms. Frege's conclusion is unacceptable from Gödel's point of view. Truth-values are not objects denoted by sentences. But how is it possible to avoid such a conclusion? It suffices to look at the theses, which lead to Frege's puzzling conclusion. It seems plausible that principles (1), (2) and (3) are quite deep, rational requirements from Gödel's point of view, just consequences of the meaning of identity and of predication. This leaves (4), and (5) which can be doubted.

Principle (4) can be doubted at least if we assume that predicates denote concepts[43] and that the principle of extensionality does not hold for concepts. This principle is evoked, in a very hypothetical form, some pages further on in the paper. Gödel says in § 22, speaking of concepts: "It may even be that the axiom of extensionality or at least something near to it holds for concepts". But the pages about Bernays's review definitively reject this hypothesis.[44] What then does it mean to reject principle (4) ?

Considering Church's example, we cannot substitute the fact that "29 is the number of the counties in Utah" for "29 is the number of novels on Waverley that Walter Scott wrote altogether", because "to be the number of the counties in Utah" and "to be the number of the novels on Waverly written by Walter Scott" do not denote the same concepts, even if these concepts have the same extensions. In the same way we cannot pass from "to be equal to the object which has the property F and is equal to a" to "to be equal to the object which has the property F and is equal to a and $b = b$" because these predicates denote different concepts even if they have the same extensions. Now, if we reject the principle that two concepts which have the same extension are co-denotative, the derivation from $F(a)$ to $G(b)$ is stopped, and we can assume that sentences indicate facts, but descriptive phrases, if they have *Bedeutungen* at all, indicate individuals.

This interpretation seems quite natural. Nevertheless, Gödel closes the analysis with the enigmatic assertion we quoted before: "As to the question in the logical sense, I cannot help the feeling that the problem raised by Frege's puzzling conclusion has only been evaded by Russell's theory of descriptions and that there is something behind it which is not completely understood." Is that because he doubts (5) in place of or together with (4)? What really are the consequences of the semantic interpretation of descriptive phrases from a logical-metaphysical point of view? What are the questions raised by Frege's puzzling conclusion that Russell's analysis evaded? What does Gödel not clearly understand about definite description? Is it just his hesitation concerning the extensional principle for

---

[41] [Bernays 1946]. See also [Crocco 2006] Section 3.

[42] 4) The problem of description is solved in a satisfactory way by "sense" and "denotation". 5) The axiom of extensionality does not hold for concepts.

[43] I argued in favour of such a thesis in [Crocco 2006] giving as the most important evidence a passage from [Gödel 1951] in [Gödel CW III] p. 320.

[44] See [Wang 1987], p. 307.

concepts that bothers him? In what precise sense is the Fregean Sinn-Bedeutung distinction interpreted by Gödel? How did Gödel think he could solve the problem of descriptive phrases without *Bedeutungen*? Only a complete analysis of the material in the Archives and especially of the Max-Phil can give us a definite answer to these questions.

Nevertheless with regard to the accessible content of the Max-Phil there are some important facts which strongly corroborate the interpretative hypothesis stated in Section 3, and they are as follows:

1) The problem of descriptive phrases is discussed by Gödel in several pages of the Max-Phil before 18th November 1942, the date of Schilpp's first letter to Gödel, asking for the Russell paper. The context of Gödel's discussion of this subject matter, before November 1942, is the analysis of the paradoxes and in particular of the problem of the existence of the paradoxical classes which are mere pluralities. The philosophical discussion concerns more generally the kind of unity which is presupposed by an object and in particular by a mathematical object.

In Max-Phil IV[45] Gödel says on p. 195 in a *Bemerkung Grundlagen*:

"Das Wesentliche der Verstandestätigkeit scheint also zu sein, Mehrheiten als neue Einheiten zu betrachten und das Nichts als Einheit zu betrachten. Die Sinngebung von anfangs Sinnlosem ($\imath x$) geht immer weiter, aber kann prinzipiell nie vollständig zu Ende geführt werden (auch die im Lebesguesschen Sinn messbaren Mengen können noch erweitert werden) oder es kann durchgeführt werden bis (im wesentlichen) auf einen einzigen Fall (vgl. Division)."

Understanding is essentially the activity of unifying the manifold and creating from nothing new unities. The attribution of a Sinn (Sinngebung) to expressions of unities initially deprived of any interpretation (as some descriptive phrases) can always be improved, but this extension is essentially indefinite and incomplete. Exceptions and limiting cases, where the expression becomes "sinnlos", will always subsist, as far as incompleteness is the essential feature of understanding. The allusion to the operation of dividing by zero as a limiting case of division, and the "singular points" or limiting points" of significance, is reminiscent of the discussion of the strategy of the limited ranges of significance in the Russell paper where Gödel, who is very sympathetic with the diagnosis of the paradoxes guiding such a strategy, says that through it, "paradoxes will appear as something analogous as dividing by zero" ([Gödel 1944] p. 150).[46]

Peano's analysis of descriptive phrases and of the reversed iota operator is analysed in detail (Max Phil IV, p 191-3 and Max Phil VII p. 509) but also comparatively in respect with Russell's and Frege's solutions (Max-Phil VII p. 528-29)[47].

2) More striking, there is a passage in Max-Phil VIII, (p. 563)[48] which lists the possible solutions of the problem and indicates what for Gödel is the 'right one'. This latter has a clear connection with the Leibnizian notion of individual concepts conceived in the logical space, that is, for Leibniz in God's understanding, although Leibniz is not here

---

[45] Max-Phil IV was written between the first of May 1941 and the 30th of April 1942.
[46] See later Section 5.2 about this point.
[47] Max-Phil VII was written between the 15th of July and the 10th of September 1942.
[48] Max Phil VIII was written between the 15th of September 1942 and the 18th of November 1942.

explicitly mentioned. This *Bemerkung* constitutes a grammatical remark such as those that the Max-Phil usually opposes to the philosophical (or foundational or mathematical or physical or theological ones). What is at stake in these grammatical remarks is the analysis of what is said or affirmed through language, i.e., what is the most obvious and natural content of our sentences in respect to and in agreement with the metaphysical and logical structure of reality.

> "Bemerkung (Grammatik): Die Bedeutung einer Zeichenverwendung ist der Gegenstand über den im Satz gesprochen wird (dessen Vorstellung im Zuhörer erweckt werden will und im Redner vorschwebt). Die wahre Bedeutung der Zeichenverwendungen der Form $(\imath x)\phi(x)$ <ist>:
>
> 1. Nach Frege der betreffende Gegenstand (widerlegt durch Scotts Argument).
>
> 2. Nach Russell nichts. (Es ist eine zweckmäßige "façon de parler", aber was zweckmäßig <ist> und was grammatisch wie ein Eigenname behandelt wird, bedeutet wahrscheinlich auch etwas. Auch der fruchtbaren Aufgabe diese zu finden wird durch Russellsche "Ausflucht" gerade ausgedichtet.)
>
> 3. Es könnte den Begriff ϕ bedeuten. Aber was ist dann der Bedeutungsunterschied von "Sohn des A" und "der Sohn des A"? Außerdem hat man das Gefühl, dass es etwas "wirklich bedeutet. (Das ist der Fregesche Sinn).
>
> 4. Es bedeutet eine Seite (oder einen Teil) des Gegenstandes (Scott ist nicht nur <der> Autor des Wav<erly>* (*Nap.<oleon>, der die Gefahr für seine Flanke erkannte, ist gewissermaßen ein anderer <Teil>). Es bedeutet Scott, insofern er der Aut.<or> des Wav.<erly> ist.
>
> 5. (Richtige Lösung): Es bedeutet: "einen Ort im logischen Raum* (*oder besser "das was sich an diesem Ort befindet". Das ist aber nicht der ganze Scott, sondern ein Teil), d.h. etwas, was sich zur abstrakten Begriffswelt so verhält wie der Raum (und die Punkte) zur Welt der Sinnbegriffe. Die Dinge überdecken jedes einen bestimmten Teil dieses logischen Raums. Eine nicht eindeutige Kennzeichnung bedeutet einen endlichen Teil des logischen Raums. Diese Gegenstände (Mathematik und logischen Raum und ihre Punkte) bilden eine dritte Art von Entitäten zwischen Dingen und Begriffen. Der logische Raum sieht verschieden aus je nachdem, was für einen "Möglichkeitsbegriff" man zugrunde legt. Der Sinn verhält sich zur Bedeutung in diesem Sinn wie die Def<inition> eines Punktes zum Punkt. Diese Bedeutung liegt also, was "Vielfältigkeit" oder "Grad der Identifizierung" oder "Nähe zum Ding und Ferne von der Erscheinung" betreffen in der Mitte zwischen Sinn und Bedeutung."[49]

What Russell considers (wrongly) as an incomplete symbol, indicates according to Gödel a very precise *Bedeutung*, which is what is presupposed in the understanding of the linguistic exchange. Nevertheless, Gödel says, different solutions can be proposed for the

---

[49] The passage continues some pages after (p. 592) as follows "Fort<setzung von S. 563> 6. Eine Kennzeichnung bedeutet nicht einen bestimmten Gegenstand, sondern eine Funktion, welche jeder 'Welt' einen bestimmten Gegenstand zuordnet. (Daher bedeuten Scott und Verfasser des Wav<erly> nicht dasselbe."

interpretation of this *Bedeutung*. The first is Frege's interpretation, but the Scott argument shows the implausibility of its consequences. The second is Russell's solution, which evades (*Ausflucht*) the important problem to find a (metaphysically) fecund solution to this semantic problem.

The third and fourth solutions were considered by Gödel on p. 529 of Max-Phil VII. They agree with the principle of compositionality and with the hypothesis that the *Bedeutungen* of sentences are states of affairs (or propositions) and not truth-values. The "Scott argument" does not apply to them because they give a purely intensional interpretation of language (nouns, descriptive phrases, and concepts). Both these solutions are therefore quite Russellian except for the fact that they consider concepts as real entities. The third assimilates the *Bedeutung* of a descriptive phrase with a concept, so eliminating the opposition between concepts and objects and therefore between what is independent from space and time and what is not so. The fourth is a generalisation of the third, inasmuch as it considers that the concept, *Bedeutung* of the descriptive phrase, is a complex one, composed of concepts each of which represents an aspect, a part of it (the *Bedeutung* in quantum such and such).

Finally Gödel presents the right solution, which saves the extensional (Fregean) and intensional (Russellian) interpretation, by way of a very Leibnizian doctrine about the relationship between possible and real. A place in the logical space (in the "region des vérités éternelles" as Leibniz calls it, evoking Augustine in the 11th chapter of the 4th book of the *Nouveaux essais*) is a well-determined portion of it, representing a third kind of entities between things and concepts. Things are space-temporal. Concepts are independent from space and time, but also incomplete, in the Leibnizian sense that they can be applied to different subjects. Well-determined portions of the logical space are individual concepts, which are neither things nor concepts, but are related to both of them.

Can this third kind of entities be assimilated to Senses (Sinn)? In a way yes, because they are a perspective on a possible *Bedeutungen*. Are they *Bedeutungen*? In another way yes, because they are what the descriptive phrase refers to. Therefore they are intermediate ("*in der Mitte zwischen Sinn und Bedeutung*").

3) In Max Phil X, partially written during the redaction of the Russell paper (March 1943–July 1944), Gödel comes back to the question on several pages (11, 17, 21, 26, 30, 32, 33, 37, 51, 63, 66–7, 70–1). The context of the discussion here mentions Frege's distinction between *Sinn und Bedeutung* and his analysis of concepts (pp. 9, 44, 92); Leibniz's analysis of existence (in space and time) and logical possibility (p. 57), predication (pp. 56 and 65) and the relationship between what is simple and what is complex (pp. 74–5).

There is also a *Bemerkung Philosophie* (p. 66–7) confirming the idea expressed in Max-Phil VIII about the right solution for the analysis of descriptive phrases:

Bemerkung (Philosophie): Eine leere Beschreibung (der König von Frankreich) ist etwas Ähnliches wie ein Raumpunkt, an dem sich kein Körper befindet (nämlich im logischen Raum). Der Raumpunkt ist sogar ein Spezialfall davon, da er bedeutet: derjenige Körper, welcher in so und so einer Lage zu gewissen gegebenen Körpern sich befindet. Es gibt wahrscheinlich außer dem körperlichen Raum (welcher das Schema der räumlichen Relationen ist, d.h. einen Überblick über die hinsichtlich der räumlichen Relationen bestehenden Möglichkeiten gewährt) auch Raum für andere Relationsgenera (sobald dieser Raum bekannt ist, <ist> es wahrscheinlich ebenso naheliegend Theorien aufzustellen wie die Newtonsche Physik). Ein Beispiel <ist> der Verwandtschaftsraum: Der Existenz von Isomorphien im körperlichen Raum entspricht im allgemeinen wahrscheinlich die durchwegige Erfüllbarkeit der "philosophischen Proportion" $x : a = b : c$. Das gibt auch die Möglichkeit, dass, obwohl z.b. die Allklasse nicht existiert, es sehr fruchtbar sein könnte, mit ihr zu operieren, denn es existieren: 1.) der Begriff der Allklasse (intens<ional>) 2.) die entsprechenden Punkte des logischen Raums. Die Raumpunkte sind gewissermaßen ein Mittelding zwischen Nichts und Etwas.

4) In Max Phil XIV, written from 1946 onwards, Gödel explicitly comes back to the connection between existence and possibility invoked by descriptive phrases and insists on the metaphysical dimension of the logical problem. Leibniz is this time mentioned explicitly.

Bem<erkung> (Gram<matik>): Der objektive Gehalt der Behauptungen "Der König ist der kleinste Mann" und "Der kleinste Mann ist der König" ist derselbe, und sie unterscheiden sich nur dadurch, dass die "Mitteilung" sich auf einen anderen Teil des ganzen Sachverhalts bezieht (und ein anderer Teil vorgesetzt wird). Im ersten Fall wird auf die Existenz der Verbindung der betreffenden Person mit "kleinstem Mann", das andere Mal mit"König von Frankreich" aufmerksam gemacht. *Das heißt, im ersten Fall wird (impliciter) mitbehauptet, dass der König nicht 170 groß ist etc., im zweiten Fall, dass der kleinste Mann nicht Schuster ist.* Ganz ebenso verhalten sich Darstellungen durch Aktiv und Passiv, obwohl aber in diesen Fällen bloß verschiedene Aspekte derselben Sache dargestellt werden. So gibt es doch wieder "natürliche" Aspekte, nämlich wo das Wichtige behauptet (d.h. *betont*) wird und wo die Existenz der Gegenstände der Beschreibung dem Hörer bekannt ist. Bei einem natürlichen Aufbau der Gesamtwissenschaft sollen also offenbar prädikative Aussagen (d.h. Allsätze) mit Existenzsätzen (aufgrund der vorhergehenden Allsätze) abwechseln. Zwischen ∈ und = besteht kein so wesentlicher Unterschied wie zwischen ∃ und diesen beiden. Eigentlich sollten daher zwei Arten von Urteilen unterschieden werden: solche, die etwas zuschreiben (d.h. die Existenz von Verbindungen behaupten), und solche, welche <eine> andere Existenz behaupten. Aber es besteht die Tendenz in der Sprache, auch das auf die Existenz einer Verbindung zurückzuführen, nämlich Verbindung mit der Welt (es ist "in der Welt", nicht"abgeschieden" oder mit Gott), d.h. "Verbindung schlechthin", (= est) soviel wie Existenz (oder heißt es: Verbindung mit etwas). Verbindung sollte vielleicht lieber als ein "Innewohnen" (Teilsein) aufgefasst werden (Leibniz) oder als "Einsein". "Etwas kennen" bedeutet nicht bloß: ein Bild (das es von anderen zu unterscheiden gestattet) in sich haben, sondern ein das Wesen ausdrückendes Bild. Zum Beispiel: Man kann die Verbindung zwischen *A* und *B* nur dann <kennen>, wenn man sie als "Verbindung zwischen *a* und *b*" kennt (oder insofern <man sie als "Verbindung zwischen *a* und *b*" kennt>).[50]

The beginning is very Fregean in the "mood" of the *Begriffschrift* by the reference to the difference of the *Betonen* to the aspect of a fact, and the choice between active and passive in the presentation of the same content. The second part is more Leibnizian, as it concerns the relation between predication and existence.

The two first examples (considering how they are rephrased by Gödel : *Im ersten Fall wird auf die Existenz der Verbindung der betreffenden Person mit "kleinstem Mann", das andere Mal mit "König von Frankreich" aufmerksam gemacht*) seem to concern the current King of France, i.e., a non-existent individual, where existence here means existence in space and time.[51] What these two sentences express is that the *betreffenden Person*, presented by the descriptive phrase in the grammatical position of subject, that is the person who should be known by the hearer according to the way we use descriptive phrases, is the same as the person described by the descriptive phrase in the position of the nominal predicate. How can it be so if the person in question does not exist? How can it be so if we are talking about the present King of France? There are two possibilities. The Fregean interpretation is that in this case the sentence has a content, a sense, but lacks a denotation, and therefore has no truth-value. The second one, the Leibnizian way of considering such a problem, is that the true denotation of the descriptive phrases are individual concepts in the Leibnizian sense, that is complete concepts of individuals. Some of these complete concepts correspond to individual substances. Others correspond to nothing in the spatiotemporal word. They nevertheless exist in a logical sense.

Gödel stresses that the content affirmed in *Der König ist der kleinste Mann* and *Der kleinste Mann ist der König* is the same. The *Sachverhalt* referred to is the same, only the message is different. The sentences only differ in their accent, in what is (negatively) implicated through the assertion, in what is excluded by it: *Das heißt, im ersten Fall wird (impliciter) mitbehauptet, dass der König nicht* 170 *groß ist etc., im zweiten Fall, dass der kleinste Mann nicht Schuster ist.* Again in a very Fregean way of speaking, Gödel draws the parallel between these two phrases and the ones which differ from their verb mood, active or passive. Considering the Fregean example of the *Begriffschrift*, the sentence "The Greeks defeated the Persians at the battle of Platea" and the phrase "The Persians were defeated by the Greeks at the battle of Platea" have the same content, express the same fact, even if with differences in emphasis, in the way the speaker wants to capture the attention of the hearer. While in the first case the question concerns aspects of the (possible) denotation, which are identified (being the King of France, being a little

---

[50] [Gödel Nachlass] Series III, Box 6b, Folder 72, Max Phil XIV, pp 036–038.

[51] It could be possible that *König von Frankreich"* here does not mean the present king of France, and that the phrases have to be interpreted as talking of, for example, Pepin the Short. It seems to us just a little disturbing that such a "famous" example, as that of "the King of France", echoing such a multitude of logical and metaphysical discussions, could have been mentioned in such a "naïve" way by Gödel. Anyway the substance of our analysis does not change if we consider that the descriptive phrases refer "in extension" to a real individual.

man, not-being a shoemaker), on the active/passive case the question concerns only the insistence, the emphasis we put on some component of the fact.

What seems interesting here for Gödel is to remark how there are natural semantic and pragmatic aspects of both of these phrases implied by their use (*So gibt es doch wieder "natürliche" Aspekte, nämlich wo das Wichtige behauptet (d.h. betont) wird und wo die Existenz der Gegenstände der Beschreibung dem Hörer bekannt ist*). When we use a descriptive phrase we normally presuppose the existence of the person described, nevertheless the assertion of this existence need not be considered as part of the objective content expressed by the phrase. In a similar way the accent, which is given through the active or passive form of a phrase, is not a part of its content, of its denotation. The criticism of Russell's analysis seems here quite explicit. This passage clarifies Gödel's remark about his Russell paper, found in the *Nachlass* before. The distinction between *Sinn und Bedeutung* does indeed solve in a satisfactory way the problem of the descriptive phrases. But what is existence and how should we interpret a descriptive phrase without *Bedeutung*?

The second part of the passage answers these questions in a very "Leibnizian mood."

After having affirmed that when we (humans) try a general natural construction (*Aufbau*) of our knowledge, we have to switch from universal to existential assertions because our knowledge is primarily a knowledge of what is real and actually exists in space and time, Gödel makes a distinction between the assertion of the existence of a link between concepts (predication or identity) and the assertion of the existence of an object. The latter is also the assertion of a link but in the sense that an object is a part of the actual space-temporal world. A sentence where a descriptive phrase does not refer to an existing individual (such as the ones concerning the King of France) asserts the link between two aspects, two parts of an individual concept, two conceptual components of it and it is therefore perfectly understandable, without considering or being acquainted with an actual individual. The transition from concepts to objects is expressed through the differentiation of two kinds of link. The first link (the one expressed by predication through $\in$ for *esti* or through identity by $=$) is a link between concepts in intension, a link independent of the world or "with God" as Gödel says; the second one (the one expressed by $\exists$) is a link of the objects, existing in space and time, within the real world. "A is B" should express two kinds of judgements, says Gödel, one stressing a conceptual link between A and B through predication (A being or not being an individual concept and predication expressing a purely logical link) and the other expressing simple existence. Nevertheless it is possible to interpret the latter as expressing the actual link between a property B and an actual object A and therefore the link between A and the world to which it belongs (of which it is a part) because, as in Leibniz, the real world is just the maximal consistent set of concepts permitting the existence of the maximum number of individuals possible. In this sense existence can be considered as expressing a link with something that exists (within the real world) instead of a purely logical link.

What is important is that, in spite of this distinction of judgement, both of these links, since they express a connection between parts, would be better expressed through the Leibnizian *inesse*, (*innewohnen*). Existence and predication are therefore different logical concepts, although they can be considered as two aspects of the same logical relation of *inesse*.

A first conclusion can be drawn here, in the perspective of the Russell paper. A definite description as "the author of Waverley" is definitively about Walter Scott. Nevertheless it could be about him in two different ways. It could be about the individual complete concept of Walter Scott or about the spatiotemporally existing individual object, which corresponds extensionally to such a concept. In both cases, the concepts used to identify the reference (the complete concept or the real individual) express aspects of him and therefore cannot be substituted with other concepts only extensionally equivalent to them. From a logical point of view the concept of existence (expressed by the existential quantifier) can be used with the usual logical rules, when we talk of objects which are supposed to exist in space and time or closed to space and time (such as mathematical objects). When we talk of individual concepts or more generally of nonexistent objects, we should perhaps use another kind of rules, which a logic of concepts should provide.

Therefore, all these remarks confirm that in order to avoid Frege's puzzling conclusion, Gödel considers the best solution to be the partial rejection of both principles (4) and (5).

Moreover, the remarks also confirm the fact that the treatment of the problem of descriptive phrases is not at all episodic in the Russell paper, but rather it runs through the entire paper, as it requires a better understanding of the opposition between extensions and intensions of concepts.

5. Concepts, classes and Gödel's return to Leibniz

There is a deep link between the second and third parts of the Russell paper. Both have as a background the same questions. What is the correct definition of concepts and classes, which frees them from paradoxes? What is the correct characterisation of propositional functions[52] in intension (concepts as Gödel called them, underlining that concepts are different from the combination of symbols expressing them) and of propositional functions in extension (classes)?

On one side, the second part of the Russell paper explicitly considers Russell's diagnosis of paradoxes (§ 10, p. 131). According to Gödel, who mentions Russell's paper of 1906, Russell came to the conclusion that "the erroneous axiom [generating the paradoxes] consists in assuming that for every propositional function there exists the class of objects satisfying it, or that every propositional function exists *as a separate entity*".

On the basis of such a diagnosis there were two possible paths of analysis open to Russell:

The first, rejecting the existence of classes and concepts in general, aimed at determining under what further hypothesis (concerning the propositional functions) these entities (i.e., concepts and classes) do exist (§ 11, p. 132).

Gödel mentions the two directions explored by Russell, according to this first possibility. The Zig-Zag theory (which he calls the intensional solution) makes the existence of a class or a concept depend on some constraints of simplicity of the content or of the

---

[52]Gödel seems to consider a propositional function as what is simply abstracted from a proposition.

meaning of the propositional function.[53] The limitation of size (which he calls the extensional solution) makes the existence of a class or of a concept depend on some constraints of the extension of the propositional function (requiring that it is not too big).[54]

The second possibility, rejecting altogether the existence of classes and concepts, was based on the more radical "no-class theory", according to which classes and concepts never exist as real entities. (14, p. 133).

According to Gödel, Russell's work after 1906 was largely based on the second possibility, the no-class theory, although from *Principia* on Russell presents it under the auspices of the logical principle that justifies it, that is, the Vicious Circle Principle. The rest of the second part of the paper is devoted to the analysis of this principle, of its presuppositions and implications.

On the other side, the third part of the Russell paper is devoted to a second possible diagnosis of the paradoxes that Russell never formulated as such and that Gödel affirms to be specially suited for their intensional form: this is the limited ranges of significance theory. "It consists in blaming the paradoxes not on the axiom that every propositional function defines a concept or class but on the assumption that every concept gives a meaningful proposition, if asserted of any arbitrary object or objects as arguments" (§ 38, p. 148).

Before going into more detail about these two parts of the Russell paper, it is important to underline a tenacious source of misunderstanding about Gödel, when speaking of concepts and intensions. The notion of intension used here by Gödel seems to conform to that used at the time by Russell, following a tradition coming from Leibniz and Peano.[55]

To a 'term' (a sign of a function or a predicate), there corresponds two aspects, which cannot be dissociated from one another (like two sides of the same coin): the intension which was traditionally considered as the logical product of the simple primitive operations or concepts composing the function or the concept to which the predicate refers, and the extension which was the class of the individuals falling under the intension. One of the classical places where Leibniz refers to this double interpretation (whose root is in Porphyry's work) is in the *Nouveaux Essais, Livre IV De la connaissance*, Chap. XVII, § 8, where Leibniz opposes *"la manière vulgaire [qui] regarde plutôt les individus, [à] celle d'Aristote [qui] a plus d'égard aux idées ou aux universaux"*.

---

[53]This is not the only place (p. 132) where Gödel seems to distinguish the content from the meaning of an expression (see for example p. 139 in this same paper: "it is true that such property ϕ (or such proposition $\phi(a)$) will have to contain themselves as constituents of their content (or of their meaning)." (the stress is mine). We will come back to this problem in Section 6. One possible interpretation of this distinction is the following: a propositional function (considered as a linguistic element) expresses a content (a *Sinn* in Frege's terms) and has a meaning (a *Bedeutung* in Frege's terms), which can be considered as a class (extensionally) or a concept (intensionally). (cf. below our discussion of the notion of intension in this same section).

[54]"The second one would make the existence of a class or concept depend on the extension of the propositional function (requiring that it be not too big), the first one on its content or meaning (requiring a certain kind of "simplicity" the precise formulation of which would be the problem)" Ibid. p. 132. Axiomatic set theory in Zermelo's or Von-Neumann's version are given as examples of implementing the extensional solution, Quine's stratification as implementing the intensional one.

[55]See [Crocco 2006] Section 4, for more details.

This tradition has to be opposed to the one coming very plausibly from Kant for whom concepts are ultimately rules of production of objects, our knowledge being essentially knowledge of objects and not of universals. According to this tradition, which was spread through the (wrong) interpretation of Frege given by Carnap, and which is common to Carnap, Weyl and Husserl, concepts in intension are assimilated to a way of grasping, forming or unifying a plurality of things, whereas concepts in extensions are those pluralities. This second sense of intension can be likened to Frege's *Sinn* of a predicate.

Can intensions in the first sense (intension $_1$) be likened to intensions in this second sense (intension $_2$)? The answer seems to be in the negative, inasmuch as the notion of a logical product of simple concepts does not necessarily have anything to do with the way we form or unify a plurality. Moreover nothing prevents an intension $_1$ to be grasped or given to us in a particular way (*in quantum* such-and-such) or through a given perspective (intension $_2$), i.e., through a specific sense, which has to be distinguished from the concept itself (intension $_1$). The first notion is ontological, the second mainly epistemological.[56]

Therefore it seems that, at this time, Gödel distinguishes from the level of language the Fregean level of *Sinn* (which he generally calls "content", and that we have identified with the notion of intension $_2$) and the double level of reference (meaning, as he called it, or in the Fregean terms *Bedeutung*) which nevertheless can be considered from the double point of view of intension $_1$ and extension.

Gödel repeatedly uses the expression "classes and concepts" to refer to the entities which are referred to as propositional functions (intended as fragments of propositions). The words "classes and concepts" are used in a synonymous way with the words "intensions and extensions", and the deepest question at stake for Gödel is clearly to determine the correct nature of such entities.

5.1. The theory of orders

The second part of "Russell's mathematical logic" focuses on the theory of order, which is, according to Gödel, based on the Vicious Circle Principle (VCP). A double confusion seems to affect Russell's diagnosis of paradoxes through the VCP. The first confusion concerns language and what it expresses, which is in particular (according to Gödel) the confusion between predicates and concepts made by Russell through the notion of propositional functions. The second one concerns the two sides of what is the reference of the linguistic expressions, that is the intensional and the extensional sides.

To clarify these confusions is an important step in the realisation of Leibniz's project, which is after all Gödel's ultimate aim.

The first move made by Gödel to this effect is to distinguish three forms of the VCP in the paper. The strongest states that: 'no totality can contain members definable only in terms of this totality'. The second weaker form of the VCP is obtained by replacing the term 'definable only in terms of' with 'involving'. The third form is obtained from the second one by replacing the term 'involving' with the term 'presupposing'. When the notion of 'presupposing' means 'presupposing for the existence', the third form, the

---

[56]This intension $_2$ can be what Gödel calls the content of a propositional function, distinguishing it from its meaning, cf. note 53 before.

weakest, becomes a rational exigency that every theory should satisfy, but the first and second forms can be denied, at least from a realistic point of view on the entities of logic and mathematics.

The second move made by Gödel is to carefully distinguish the intensional and extensional versions of these principles. For the intensional version, concerning propositional functions, in order to prevent intensional paradoxes, we have to add the principle that a propositional function presupposes the totality of its arguments and of its values (Gödel 1944, p. 126). It seems clear from Gödel's claims that we have, as a consequence, three corresponding forms of VCP for propositional functions. The first states that 'nothing defined in terms of a propositional function can be a possible argument of this function'. The second form is obtained by replacing 'defined in terms of' with 'involving' and the third with the term 'presupposing'.[57]

The first form of the VCP (in both extensional and intensional versions) makes impredicative definitions impossible. The ramified theory of types conforms to it. Gödel gives two arguments in order to explain that this first form is implausible. The first argument concerns the interpretation of the totality that it implies. Actually it implies the assimilation of "all" (expressing the totality) to an infinite conjunction, i.e., that "reference to a totality implies reference to all single elements of it" (p. 136). There is no reason to accept such an interpretation and Gödel mentions Carnap and Langford's analysis of "all" in terms of analyticity. The second argument concerns the interpretation of the entities (concepts in the intensional version or classes in the extensional one) involved in it: "even if "all" means an infinite conjunction it seems that the vicious circle principle applies only if the entities involved are constructed by ourselves" (ibid.). He adds that there is nothing in the least absurd in the existence of totalities containing members that can be described (i.e., uniquely characterized) only by reference to this totality. Only the constructivist prejudice against concepts and classes as real entities can therefore allow the first form of the VCP. On the other hand, Gödel says that concepts are objective entities, i.e., properties and relations between objects (Gödel 1944, p. 128). As real entities, they have to be distinguished from predicates (which are linguistic entities) and from notions, where the term notion means a symbol of a predicate together with a rule of translation of sentences containing this symbol into sentences not containing it. Two different definitions determine different notions. On the contrary, says Gödel, the same concept can be expressed in different ways,

Concerning the second form of the VCP, Zermelo-Fraenkel set theory (which does not satisfy the first form of the VCP, admitting impredicative definitions) conforms indeed to the second one in its extensional version. Actually, it forbids that a set could belong to itself, that is, that a totality (a set) can contain (involve) itself as an element.

Finally the theory of simple types, if it is interpreted intensionally (with quantifier ranging over properties and relations considered as real entities), and if the universal quantifier is interpreted as an infinite conjunction (thus involving the properties over which it

---

[57]Gödel specifies (p. 134) that in the second edition of the *Principia* this intensional version of the VCP is dropped because Russell adopts the principle according to which functions can occur in propositions only through their values, i.e., extensionally. In that case, he said, the paradoxes are avoided by the theory of simple types, interpreted extensionally.

ranges), violates the second form of the VCP in its version for propositional functions. This fact motivated Gödel's assertion on p. 127, according to which the theory of simple types 'has nothing to do with the vicious circle principle', which means that the theory of simple types contradicts the first form of the VCP and is independent from the second one.[58]

By the analysis of the last two examples, Gödel in some way contrasts concepts with classes (as pluralities of objects). The difference between concepts and classes is grounded in self-reflexivity. It is perfectly coherent to think that concepts are able to support self-reflexivity, and that therefore the VCP in its second form does not also apply to concepts. Of concepts (as opposed to notions), it is possible to refer to their totality, to claim that some of them can be described only by reference to all of them (or at least all of a given type) and to say that a property can contain itself as a constituent of its meaning or content (p. 139, cf. note 57 above). Gödel adds that an approximation of this kind of self-reflexivity is in fact given in his theorems of incompleteness, where a proposition contains, as part of its meaning, the assertion of its own demonstrability, and where the demonstrability of a proposition (in the case where the axioms and the rules of inference are correct) implies the proposition in question. Concerning classes, on the contrary, we can consider the fact that the VCP applies in the second and third form as a plausible assumption, sufficient for the development of all contemporary mathematics. Impredicative definitions are allowed for classes, but it is impossible to say that $x \in y$ when x is not less than y. Therefore one is then led to something like the Zermelo set theory. Zermelo's theory is based on an iterative notion of set, where sets are split up into levels, obtained by the relation 'set of' on a specific level, and where mixture and transfinite types are allowed. Gödel clearly considers Zermelo's solution sufficient for the needs of mathematics. The last thesis clearly proposes that we search for different solutions for paradoxes of concepts and those of classes by opposing the needs of logic to the more restricted needs of mathematics, that is, by opposing the notion of structure in the abstract sense and the iterative notion of set. Gödel stresses that abstract structures can contain elements without presupposing them (ontologically), just as a sentence or a phrase belongs to a language, without having ontological primacy over it. On the contrary, sentences presuppose a language, even if they are involved in (belong to) it. In the same way, concepts (which are abstract structures) can contain concepts (can have other concepts as parts of them) without ontologically presupposing them.

Recognizing that the same logical relation of predication can be interpreted both in intensional and extensional terms, giving an explicit preference to the intensional form, opposing the needs of logic to the more peculiar needs of mathematics are all common aspects of a very profound return to Leibniz beyond Frege's analysis. Russell, contrary to Frege, had been very sympathetic towards the Leibnizian interpretation of intension and

---

[58] Actually, the theory of simple types can conform to the second weaker form (when considered extensionally as in Zermelo-Fraenkel set theory, allowing mixed types) or not conform to it (when considered intensionally with universal quantification taken as an infinite conjunction and with or without mixed types) and so independently from the fact of allowing mixed types.

extension of terms[59] but the way he treats propositional functions in intension through the ramified theory of types is unacceptable for Gödel. As suggested in Wang 5.3.17, 8.6.18, the relation of predication consists, in intensional terms, in the application of a concept to an object or to another concept. On the contrary, extensionally, it can be interpreted in terms of an object belonging to a class of objects. Classes are paradoxical objects indeed, but we can describe an iterative process of engendering non-paradoxical multiplicities of objects, i.e., sets representing extensionally a concept. The confusion or the reduction between these two aspects of predication (the intensional and the extensional one) has been constant in modern logic since Frege and Russell's works. Therefore, Gödel stresses explicitly the difference between them, asserting that the extensional predication cannot be reflexive and the intensional one is in general reflexive. This fact explains, according to him, the difference between set theory and logic. As Gödel puts it in his conversation with Wang:

> 8.6.15 For a long time there has been a confusion between logic and mathe-matics. Once we make and use a sharp distinction between sets and concepts, we have made several advances. We have a reasonably convincing founda-tion for ordinary mathematics according to the iterative concept of set. Going beyond sets becomes understandable and, in fact, a necessary step for a com-prehensive conception of logic. We come back to the program of developing *a grand logic* [sic] except that we are no longer troubled by the consequences of the confusion between sets and concepts.

> 8.6.2 Mathematicians are primarily interested in extensions and we have a systematic study of extensions in set theory, which remains a mathematical subject except in its foundations. Mathematicians form and use concepts, but they do not investigate generally how concepts are formed, as is to be done in logic. We do not have an equally well-developed theory of concepts compara-ble with set theory.

> 6.1.11 Because of the unsolved intensional paradoxes for concepts like *con-cept, proposition, proof*, and so on, in their most general sense, no proof us-ing the self-reflexivity of these concepts can be regarded as conclusive in the present stage of development of logic, although, after a satisfactory solution of these paradoxes, such an argument may turn out to be conclusive.

---

[59]The most explicit text on intensions is in [Russell Whitehead 1910] Chapter III, p. 72 of the Introduction, where Russell presents classes as incomplete symbols, mere symbolic or linguistic conveniences, 'not genuine objects as their members are if they are individuals'. Then, he goes on to claim: "It is an old dispute whether formal logic should concern itself mainly with intensions or with extensions. In general logicians whose training was mainly philosophical have decided for intensions, while those whose training was mainly mathematical have decided for extensions. The fact seems to be that, while mathematical logic requires extensions, philosophical logic refuses to supply anything except intensions. Our theory of classes recognises and reconciles these two apparently opposite facts, by showing that an extension (which is the same as a class) is an incomplete symbol, whose use always acquires its meaning through a reference to intension." These intensions, for Russell, are nothing but the propositional functions, which are the constituent parts of propositions and presuppose them ontologically.

Nevertheless, Gödel stresses that recognising such a distinction between logic and mathematics in no way precludes the possibility of a "calculus of conversion" between the intensional and the extensional interpretation of concepts. It just means that in the actual state of our knowledge we have not found such a "calculus of conversion" and therefore, that the theory of sets is the only satisfying frame which we can use for the foundations of mathematics . Gödel seems to think that even if to some concept there would correspond no set, i.e., no object that could be obtained through the iterative idealised repetition of the "set of" operation, there is still the possibility to conjoin to each set its "defining concept":

> 8.6.4 "It is not in the ideas (of *set* and *concept*) themselves that every set is the extension of a concept. Sets might exist which correspond to no concepts. The proposition "for every set, there is a [defining] concept" requires a proof. But I conjecture that it is true. If so, everything in logic and mathematics is a concept: a set, if extensional; and a concept (only) otherwise."

Wang says that the opposition between concepts and objects (sets) increased in Gödel's thoughts after the Russell paper. Nevertheless besides the logical property that we mentioned above (concepts can be applied to themselves, objects cannot belong to themselves) in the conversations with Wang, objects are contrasted with concepts also through a metaphysical characterisation, which closely resembles Leibniz's analysis:

> 9.1.27 Monads are objects. Sets (of objects) are objects. A set is a unity (or whole) of which the elements are constituents. Objects are in space or closed to space. Sets are the limiting case of spatiotemporal objects and also of wholes. Among objects, there are physical objects and spatiotemporal objects. Pure sets are sets that do not involve non-set objects – so that the only Urelement in the universe of pure sets is the empty set. Pure sets are the mathematical objects and make up the world of mathematics.

> 8.2.4 Sets are the limiting case of spatiotemporal objects- either as an analogue of constructing a whole physical body as determined entirely by its parts (so that the interconnections of the parts play no role) or as an analogue of synthesizing various aspects to get one object, with the difference that the interconnections of the aspects are disregarded. Sets are quasi-spatial.[60]

What is the exact relation between spatiotemporal objects, monads and mathematical objects? Only a complete transcription of the Max-Phil can answer this question.

5.2. The theory of simple types

The second and the third parts of "Russell's mathematical logic" contain a main thesis, which intimately correlates them. This main thesis states that the theory of simple types is independent from the theory of orders (p. 127, 147) and it is essential from Gödel's point of view to show the fundamental difference between them, in order to present the new principle, which the theory of simple types brings in (i.e., the principle of the limited ranges of significance) as totally independent from the constructive theory of orders.

---

[60] See Section 6 below about these assertions.

The third part specifies two points of view from which the theory of simple types can be justified. The first one involves the principle of unsaturation (or typical ambiguity)[61] for propositional functions and is, according to Gödel, explicitly adduced by Russell: "The reason adduced (in addition to its "consonance with common sense") is very similar to Frege's who, in his system had already assumed the theory of simple types for functions, but failed to avoid the paradoxes, because he operated with classes, (or, rather, functions in extension) without any restrictions. This reason is that (owing to the variable it contains) a propositional function is something ambiguous or, as Frege says something unsaturated, wanting supplementation) and therefore can occur in a meaningful proposition only in such a way that this ambiguity is eliminated (e.g., by substituting a constant for the variable or applying quantification to it). The consequences are that a function cannot replace an individual in a proposition, because the latter has no ambiguity to be removed, and that functions with different kinds of arguments (i.e., different ambiguities) cannot replace each other; which is the essence of the theory of simple types." ([Gödel 1944] p. 147–8). Propositional functions are in this sense "fragments" of propositions, "which have no meaning in themselves, but only insofar as one can use them for forming propositions by combining several of them, which is possible only if they "fit together", i.e., they are of the appropriate types."

The second point of view (called realistic by Gödel at p. 149 § 2) is independent from this "fictionalist standpoint" because it considers (extensionally) classes as essentially formed by an infinite iteration of the process of forming classes out of infinitely many individuals. The same process seems applicable to form propositional functions in the intensional sense, if we consider predication of a concept to an argument x as expressing a proposition of infinite lengths of the form

$$x = a_1 \lor x = a_2 \lor \cdots \lor x = a_n$$

where a_i are individuals or classes obtained on a (logically) precedent level of iteration.

Generally speaking the theory of simple types is considered by Gödel as "a stepping stone" (end p. 148) towards a more satisfactory system. On the one hand it has the clear advantage of bringing in a new idea for the solution of paradoxes: "[i]t consists in blaming the paradoxes not on the axiom that every propositional function defines a concept or a class, but on the assumption that every concept gives a meaningful proposition, if asserted of every object or objects as argument". On the other hand it is not completely plausible from a realistic point of view "since what one would expect to be a concept (such as, e.g., "transitivity" or the number two) would seem to be something behind its various "realizations" on the different levels and therefore does not exist according to this theory of types" (p. 148 end).

This example is particularly important to understand Gödel's criticism of the theory of simple types: transitivity is a property of relations. A relation is called transitive if and only if when applied to arguments a, b, c, if it relates a to b and b to c it relates also a to c (where a, b, and c can be objects, classes or concepts). Since, in the theory of

[61] As a matter of fact Russell calls ambiguity what Frege called unsaturation, or at least something very close to it. See [Russell & Whitehead 1910] Section IV of the introduction of the first edition. This notion of ambiguity has to be distinguished from the systematic ambiguity of truth and falsehood of Section III.

simple types, types are mutually exclusive, it follows that transitivity should be defined for relations of types 1, 2, and so on. According to the theory of simple types, a concept (property or relation) exists insofar as it has a homogeneous range of arguments from the point of view of types. Therefore, the concept of transitivity which applies to relations of each type does not exist, and speaking of it is a sort of *abus de langage*.

Considering the conversations with Wang, we see that this diagnosis of the theory of simple types relies on a deeper rejection of Frege's and Russell's analysis. Rejecting Frege's and Russell's idea of the priority of propositions over concepts and the identification of concepts with propositional functions is a way to come back to Leibniz on a very fundamental issue of modern logic.[62] The theory of unsaturation, which commands this priority in Frege's work, seems to be, according to Gödel, completely false. Propositional functions are, through the doctrine of unsaturation, considered as "fragments" of propositions, which have no meaning in themselves, but only in so far as one can use them to form propositions. On the contrary, Leibniz's ideas do not come from propositions but from simpler concepts. In the same sense, from Gödel's point of view, unsaturation has nothing to do with concepts which are entities composed by primitive concepts:

> 9.1.26 *Concepts*. A concept is a whole – a conceptual whole– composed out
> of primitive concepts such as negation, existence, conjunction, universality,
> object, (the concept of) concept, whole, meaning and so on. We have no clear
> idea of the totality of all concepts. A concept is a whole in a stronger sense
> than sets; it is a more organic whole, as a human body is an organic whole of
> its parts.[63]

The rejection of the notion of unsaturation coupled with the criticism about the lack of plausibility of the theory of simple types from a realistic point of view (see above) opens the problem of the notion of predication. The doctrine of unsaturation allows Frege and Russell to avoid an explicit relation of predication, that is: a relation connecting terms to form a proposition. There is, nevertheless, a price to pay for such an elimination of an *explicit* relation of predication: considering a (propositional) function as a fragment of a proposition implies distinguishing different types of concepts according to the different arguments they accept. Such a classification gives rise to an infinite *implicit*[64] hierarchy of logical relations of predication each one differing fundamentally from the others.[65] It is a well-known fact, remarked upon by Gödel, that in spite of this differentiation of types – a consequence of the distinction of an infinite[66] hierarchy of logical relations of predication– Frege failed to avoid paradoxes because he operated with classes (functions in extensions) without restrictions. Types for concepts and constraints of types for classes are fundamental for Russell too, both in his first sketchy attempt with simple types, and in his second one with ramified types. As an infinite hierarchy of types means an infinite

---

[62] [Crocco 2006]

[63] Cf. also the Carnap paper [Gödel 1953] p. 360.

[64] Implicit means here that the logical relation of predication is completely determined by the nature of the concepts to which it applies and of the arguments that they receive. See also [Angelelli 1967] pp. 92–101.

[65] Cf. the end of "Über Funktion und Begriff" [Frege 1891] and "Begriff und Gegenstand" [Frege 1892] § 17.

[66] Actually Frege just needs three different levels because of its function ∩, which allows him to pass to classes, which are not typified.

hierarchy of relations of predication, Gödel rejects types openly from 1944. For Gödel there is only one fundamental relation which is meaningful predication (in the double form of application or belonging)[67] and which warrants the unity of logic:

> 5.3.17 The basis of every thing is meaningful predication, such as *Px*, *x* belongs to *A*, *xRy*, and so on. Husserl had this. Hegel did not have this; that is why his philosophy lacks clarity. [...]

> 8.6.18 [...] Logic studies only what a concept applies to. Application is the only primitive concept apart from the familiar concepts of predicate logic with which we define other concepts.

> 9.1.16 The significance of mathematical logic for philosophy lies in its power to make explicit by illustrating and providing a frame for axiomatic method. Mathematical logic makes explicit the central place of predication in the philosophical foundation of rational thought.

The search for a type-free logic, allowing self application of concepts, except for limiting cases is one of the main tasks of logic, necessary to implement Leibniz's project.

The existence of one unifying logical relation between terms is a central thesis in Leibniz's logical work. To reduce the connection between terms to the unique relation of "*inesse*" gives him the opportunity to unify the different kinds of syllogism. Modern logic rejects Leibniz's analysis on this issue. Frege and Russell both criticised what they considered a lack of distinction (between predication, subsumption and existence) resulting from this unification. Is there any evidence that Gödel was willing to rehabilitate Leibniz's notion of predication? And in this case how did he consider the relations between predication, subsumption and existence? Some hints have been given in Section 4 in the passage from Max-Phil XIV talking about the Leibnizian *inesse*, but again, only a complete analysis of the Max-Phil could answer these questions.

6. Analyticity and truths of reason

Russell pleaded for the logical nature of mathematics at least since the *Principles of mathematics* in 1903, mentioning Leibniz (through Couturat's work of 1901) as the one who strongly advocated that "all mathematics is deduction by logical principles from logical principles" (Chap. 1, Section 5). He added: "owing partly to a faulty logic, partly to belief in the logical necessity of Euclidian geometry, he [Leibniz] was left into hopeless errors, in the endeavour to carry out in detail a view which, in its general outline, is known to be correct" (ibid.).

Gödel does not mention Russell's (and Frege's) notion of analyticity (deduction by logical principles from logical principles) but refers directly to Leibniz's notion of truth of reason (truths which can be reduced in a finite number of steps to identities) and asks the question whether the axioms of *Principia* can be said to be analytic in this sense. He then distinguishes two means for accomplishing such a reduction: by virtue of explicit definitions (or by rules for eliminating them from sentences containing them); and by

---

[67] See below in this section, for the reasons of this distinction.

virtue of the meaning of the concept occurring in them, where this meaning may perhaps be indefinable.

In the first sense, his incompleteness theorem shows that even the theory of integers cannot be said to be analytic. But this is not the case of the second sense, and Gödel quite explicitly suggests that this is a way to revert to the very Leibnizian notion of truths of reason. Actually, in this second sense "every mathematical proposition could perhaps be reduced to a special case of $a = a$, namely if the reduction is effected not by virtue of the definitions of the terms occurring, but by virtue of their meaning, which can never be completely expressed in a formal set of formal rules" (ibid. note 47).

Analyticity (in the second sense) can be considered as expressing some kind of identity of meaning. Gödel said that all the axioms of *Principia* (except the axiom of infinity) can be considered as analytic in this sense for certain interpretations of the primitive terms, namely if the term "predicative function" is replaced either by "class" (in the extensional sense) or (leaving out the axiom of choice) by "concepts" since nothing can express better the meaning of the term "class" than the axiom of classes [Zermelo's *Aussonderungsaxiom*] and the axiom of choice and since, on the other hand, the meaning of the term "concept" seems to imply that every propositional function defines a concept.

The same idea is stated at the end of the 1951 paper ([Gödel CW III] pp 320 and 321). In 1951 Gödel gives an account of analytic truths (truths of reason) in terms of the logical relations between concepts and specifies how far this idea of analyticity is from that of the Vienna circle: mathematical truths are analytical but even if they are true whatever the situation of the world is, this fact does not imply that they are "void of content" because their content consists exactly in the relations between concepts or between concepts and abstract objects, which both exist in a strong sense independently of our sensations of the world of the things.

> " [...] it is correct that a mathematical proposition says nothing about the physical or psychical reality existing in space and time, because it is true already owing to the meaning of the terms occurring in it, irrespectively of the world of real things. What is wrong, however, is that the meaning of the terms (that is the concepts they denote) is asserted to be something man-made and consisting merely in semantical conventions. The truth, I believe, is that these concepts form an objective reality of their own, which we cannot create or change, but only perceive and describe." (p. 320)

In the second passage the two adjectives "incomplete" and "indistinct" again remind us of Leibniz very closely.

> "I wish to repeat that "analytic" here does not mean "true owing to our definitions" but rather "true owing to the nature of the concepts occurring [therein]" in contradistinction to the properties and behaviour of things". This concept of analytic is so far from meaning "void of content" that it is perfectly possible that an analytic proposition might be undecidable (or decidable only with a [certain] probability). For our knowledge of the world of concepts may be as limited or as incomplete as that of the world of things. It is certainly un-

deniable that this knowledge, in certain cases, not only is incomplete but even indistinct." (p. 321)

There are also two interesting passages in the Carnap paper, where the Fregean distinction between *Sinn* and *Bedeutung* is evoked as useful in order to clarify this notion of analyticity. Gödel, after reiterating the assertion that "analytic" cannot mean void of content affirms: "The neglect of the conceptual content (i.e., the "sense" according to Frege) as something objective (i.e., non-psychological) is also responsible for the wrong view that the conclusion in logical inference, objectively, contains no information beyond that contained in the premises [...] for the conclusion represents the empirical (or more generally the extra-logical) content of the premises, or part of it, in a conceptually different form and that the conclusion is implied by the premises is itself an objective fact concerning the primitive terms of logic occurring [in] and specific for these terms."[68] That means that we have to distinguish the conceptual content of a sentence from the factual one, and that this conceptual content plays a specific role in explaining the nature of mathematical and logical sentences, which are true irrespectively of the world of real things.

As it is explicitly stated in version V of the Carnap paper, this conceptual content is composed in the case of concepts by "primitive ones" (Gödel CW III end of p. 360) and in the case of propositions by the very concepts occurring in a sentence: "one may very well say that the proposition mentioned above [it will rain or it will not rain tomorrow] although it says nothing about rain, does express a property of "not" and "or"" ([Gödel CW III] p. 362)

An open problem has to be pointed out here, connected with the interpretation of the oppositions intension/extension, content/meaning mentioned in Section 5. Frege's distinction between *Sinn* and *Bedeutung* concerning predicates, prescribes that the *Bedeutung* of a predicate is the concept itself and its *Sinn* is the way of presenting it. Frege himself never pronounces on what these concepts such as *Bedeutungen* are supposed to be. In particular, although he considers that concepts are "composed" by their "characteristic marks" he never pronounced on the nature of this composition from characteristic marks. He never says explicitly if there are primitive concepts (such as *Bedeutungen*) through which all the others can be composed and what they are. Nevertheless the principle of extensionality, in Frege's system, guarantees an identity criterion. Two concepts (such as *Bedeutungen*) are identical if and only if so are their extensions. This last principle cannot fit with Gödel's concepts, as we saw in Section 4, and so we are sent back to the problem of understanding in what precise sense is the Fregean *Sinn-Bedeutung* distinction interpreted by Gödel? Does Gödel accept the idea that we have to distinguish the way we apprehend a concept from the concept itself?

Many passages in Max-Phil refer to Frege's distinction between *Sinn* and *Bedeutung*. A definitive answer to our question can be only given when the transcriptions have been completed.

We conjectured (cf. Section 5 above) that Gödel distinguishes the three levels: language, sense (*Sinn*, content, intension $_2$) and meaning (*Bedeutung* as intension $_1$ or extensions).

---

[68][Gödel 1953] version III page 350 note 40

In the present state of our knowledge, we can only say that there is a Leibnizian distinction, which could have inspired Gödel here. Leibniz draws a distinction between concepts and ideas *in mente homini,* which is paralleled by that between thoughts and truths.[69] Leibniz considered concepts as depending on acts of the mind, which consist in grasping ideas, (where the latter exist independently from this act and are a sort of disposition which is activated at the occasion of experience). It is not impossible that Gödel thought, at that time, that Frege's distinction between *Sinn* and *Bedeutung* is an actualisation of the Leibnizian one, considering that the different ways to affirm a *Bedeutung* do not depend on a psychological act, but are deeply connected with their expressions in language. It is also possible that these same concepts (considered as senses of the ideas *in mente homini*) are what would be studied through phenomenology, as Gödel says in his paper of 1961.

It remains that these concepts "in *mente homini*" are in a relation of "expression" with the ideas "in *mente dei*", that is with the structures of the possible and of the real, which can be at most recognised by intuition or perception and never constructed by us. Gödel speaks with Wang about *ideas,*[70] supposed to be different from concepts inasmuch as they are independent from languages. He gives as examples of ideas: *concept, absolute truth, absolute definability* ([Wang 1996], 6.1.11–16) and says that if the idea of absolute proof can be clarified, then we can solve the intensional paradoxes, and therefore prove the superiority of minds over computers. For Gödel, as for Frege, before 1959,[71] senses are probably those double-faced entities, looking both toward language and toward reference (the latter being human concepts in intension $_1$ or in extension). On the contrary Gödel, after 1959, probably did not consider language as being central for grasping concepts (intended as actualisation of ideas *in mente homini* expressing ideas *in mente dei,* that is structures of reality). He thinks that while the axiomatic method (as opposed to explicit definitions) is the essential way to clearly and precisely define objects and concepts, it is not the only method to grasp them. Gödel probably took it to be necessary to provide a definition of concepts in a way independent from language. The concept of calculability is the most striking example of a definition, which is independent of language. Gödel hoped that in the future the same kind of characterisation would be found in general for other concepts. To arrive at such a characterisation of concepts we should try to grasp simple and primitive concepts and clarify what their knowledge consists of. Husserl's phenomenology appeared to him to be a possible tool for clarifying the content of a concept, its sense, in other words for approaching as nearly as possible the right perspective on it. Following this interpretation, what Gödel seeks in Husserl is a method with which to grasp concepts, that is, a way to clarify senses *independently from language.* Nevertheless, the sense of a concept, the acts we have to perform to grasp it, need not necessarily

---

[69] A VI 4B, p. 1369, VI, 6, p. 106. See also [Mugnai 2001], Chapter 1. As is well known, the difference is stated by Leibniz in order to contrast the Lockian anti-innatism and his "static"conception of knowledge.

[70] [Wang 1996] p. 268, and especially 8.4.20–22. Wang refers to Kantian ideas, opposed to concept, but 8.4.23 specifies that "Kant's distinction between ideas and concepts is not clear. But it is helpful in trying to define precise concepts.

[71] Cf. the passage in [Gödel 1953] quoted in Section 4, and its reference in note 38.

be identified with the concept itself (neither with the idea *in mente homini*, nor with the idea *in mente Dei*, using Leibniz's way of speaking).

This interpretation is compatible with the letter of the conference of 1961 [Gödel CW III] page 382. Gödel speaks here of *Sinn*:

Nun gibt es ja heute den Beginn einer Wissenschaft, welche behauptet, eine systematische Methode für eine solche Sinnklärung zu haben, und das ist die von Husserl begründete Phänomenologie. Die Sinnklärung besteht hier darin, dass man die betreffenden Begriffe schärfer ins Auge fasst, indem man die Aufmerksamkeit in einer bestimmten Weise dirigiert, nämlich auf unsere eigene Akte bei der Verwendungen dieser Begriffe, auf unsere Mächte bei der Vollführung unsere Akte, etc.

It is also compatible with the general Leibnizian inspiration of Gödel's analysis, which is the lasting background to his philosophical research.

Acknowledgment

Different parts of this paper were presented at the IHPST in Paris, in September 2003, at the seminar of the *Philosophische Fachbereich* of Konstanz, in June 2004, at the international conference organised by Pierre Cassous-Noguès at the university of Lille, in May 2006 and at a workshop organised by myself for the Ceperc of the university of Aix-Marseille 1, in December 2006. The rest of this research was sustained by the ANR "Gödel" project n. ANR-09-BLA-0313.

I would like to thank all the people who helped me with their remarks and in particular: Philippe de Rouilhan, Jean Mosconi, Mark van Atten, Göran Sundholm and Eric Audureau. Thanks also to the three transcribers of Gödel's manuscripts, Cheryl Dawson, Eva-Maria Engelen and Robin Rollinger. A special thanks to Eva-Maria Engelen and Paola Cantù, members of the ANR group. Without both of them the pursuit of the "Gödel" project would have been impossible.

References

[Angelelli 1967] Angelelli, I., *Studies on Gottlob Frege and traditional philosophy*. Reidel, Dortrecht, 1967.

[Audureau 2004] Audureau, E., *Kurt Gödel, critique de la théorie de la relativité générale*. Doctoral thesis of the University of Aix-Marseille 1. November 2004

[Bernays 1946] Bernays, P., 'Review of Gödel's "Russell's mathematical logic"', *The Journal of Symbolic logic* **11** (1946), 75–79.

[Buldt] Buldt, B. *et al.* (eds), *Kurt Gödel. Wahrheit und Beweisbarkeit*. öbv&hpt, Wien 2002.

[Church 1932] Church, A., "A set of postulates for the foundations of logic" *Annals of mathematics*, **2**, n. 33, 1932, pp. 346–366.

[Church 1933] Church, A., "A set of postulates for the foundations of logic (second paper)" *Annals of mathematics*, **2**, n. 34, 1933, pp. 839–864.

[Church 1943] Church, A., "Review to Carnap's Introduction to semantics", *Philosophical Review*, vol 52, n. 3, May 1943, pp. 298–304.

[Church 1956] Church, A., *Introduction to mathematical logic*, Princeton University Press, 1956.

[Crocco 2003] Crocco, G., "Gödel, Carnap and the Fregean Heritage", *Synthese*, **137**, n° 1–2, 2003, pp. 21–41.

[Crocco 2006] Crocco, G., "Gödel on concepts", *History and Philosophy of Logic*, **27** Mai 2006, pp. 171–191

[Frege 1891] Frege, G., "Funktion und Begriff" in *Kleine Schriften* Angelleli I. ed. Hildesheim, Georg Olms, (1967) pp. 125–142. First edition 1891.

[Frege 1892] Frege, G., "Über Begriff und Gegenstand" in *Kleine Schriften* Angelleli I. ed. Hildesheim Georg Olms, (1967) pp. 167–178. First edition 1892.

[Davidson 1984] Davidson, D., *Inquiries into Truth and Interpretation*, Oxford University Press 1984.

[Gödel 1933] Gödel, K., "The present situation in the foundations of mathematics", in [Gödel CW III] pp. 45–53.

[Gödel 1944] Gödel, K., "Russell's mathematical logic" in *The philosophy of Bertrand Russell*, P.A. Schilpp ed. *Library of living philosophers*, Open Court La Salle 1944, pp. 123–153. Reprinted in [Gödel CW II] pp. 119–141

[Gödel 1946] Gödel, K., "Some observations about the relationship between theory of relativity and Kantian philosophy" in [Gödel CW III] pp; 230–246.

[Gödel 1949] Gödel, K., "A remark about the relationship between relativity theory and idealistic philosophy" in *Albert Einstein, philosopher – scientist* P.A. Schilpp ed. *Library of living philosophers* Open Court La Salle 1949, pp. 555–562; reprinted in [Gödel CW II] pp. 202–207,

[Gödel 1951] Gödel, K., "Some basic theorems on the foundations of mathematics and their philosophical implications" in [Gödel CW III] pp. 304–323.

[Gödel 1953] Gödel, K., "Is mathematics syntax of language ?" version III, in [Gödel CW III] pp. 334–356, version V, in [Gödel CW III] pp. 356–362.

[Gödel 1961] Gödel, K., "The modern development of the foundations of mathematics in the light of philosophy" in [Gödel CW III] pp. 374–387.

[Gödel CW II] Gödel, K., *Collected Works vol. II. Publications* 1938–1974. Feferman S. *et al.* eds. New York and Oxford. Oxford University Press 1990

[Gödel CW III] Gödel, K., *Collected Works vol. III. Unpublished essays and lectures.* Feferman S. *et al.* eds. New York and Oxford. Oxford University Press 1995

[Gödel CW IV] Gödel, K., *Collected Works vol. IV. Correspondence A–G.* Feferman S. *et al.* eds. New York and Oxford. Oxford University Press 2003

[Gödel CW V] Gödel, K., *Collected Works vol. V. Correspondence H–Z.* Feferman S. *et al.* eds. New York and Oxford. Oxford University Press 2003

[Gödel Nachlass] Gödel, K., *Gödel Nachlass*, Firestone Library Princeton.

[Hintikka 1998] Hintikka, J., "On Gödel's philosophical assumptions" *Synthese* 114, 1998, pp. 13–23

[McCulloch and Pitts 1943] McCulloch, W., Pitts W., "A logical calculus of the ideas immanent in nervous activity", *Bulletin of mathematical biophysics* **5**, pp. 115–133. 1943

[Menger 1994] Menger, K., "Memories of Kurt Gödel". In McGuinness, B. (ed.) *Reminiscences of the Vienna Circle and the Mathematical Colloquium.* Dordrecht: Kluwer (1994)

[Mugnai 1990] Mugnai, M., "Leibniz nominalism and the reality of ideas in the Mind of God" in Heinekamp, Lenzen and Schneider (eds.) *Mathesis rationis*. Münster, Nodus, 1990 pp. 153–67.

[Mugnai 2001] Mugnai, M., *Introduzione alla filosofia di Leibniz*. Torino, Einaudi 2001.

[Nachtomy 2007] Nachtomy, O., *Possibility, agency, and individuality in Leibniz's metaphysics.* Dordrecht, Springer 2007.

[Neale 1995] Neale, S. "The philosophical significance of Gödel's slingshot" Mind, vol. 104, 761–825.

[Parsons 1995] Parsons, Ch., "Platonism and mathematical intuition in Kurt Gödel's thought" *Bulletin of Symbolic logic*, 1, n. 1 (1995) pp. 44–74.

[Quine 1941] Quine, W.V.O., "Whitehead and the rise of modern logic" in *The philosophy of Alfred North Whitehead*, P.A. Schilpp ed. Open Court La Salle 1941.

[Russell Whitehead 1910] Russell, B., Whitehead, A.N., *Principia mathematica*, Cambridge, Cambridge university Press 1963. First edition 1910.

[Tieszen 1990] Tieszen, R., "Kurt Gödel's path from the incompleteness theorem (1931) to phenomenology (1961)", *The Bulletin of Symbolic logic*, **4** n. 2 1998, pp. 181–203.

[van Atten Kennedy 2003] Van Atten, M., Kennedy, J., "On the philosophical development of Kurt Gödel", *The Bulletin of Symbolic Logic* **9**, n. 4, 2003, pp. 425–476.

[Wang 1974] Wang, Hao, *From Mathematics to Philosophy*, London, Routledge & Kegan Paul 1974.

[Wang 1987] Wang, Hao, *Reflections on Kurt Gödel* Cambridge: The MIT Press (1987)

[Wang 1996] Wang, Hao, *A logical journey from Gödel to philosophy* Cambridge: The MIT Press (1996)

[Weyl 1946] Weyl, Hermann, "Review of Schilpp 1944", *American Mathematical Monthly*, 53, 208–214 (1946) Reprinted in Weyl's *Gesammelte Abhandlungen*, vol. 4, p. 599–605.

Gabriella Crocco
Université de Provence
Centre d'Epistémologie et d'Ergologie Comparatives
CNRS – CEPERC – UMR 6059
29 avenue Robert Schuman
F-13621 AIX EN PROVENCE CEDEX 1, France
gabriella.crocco@wanadoo.fr

Chaitin, Leibniz and Complexity

Herbert Breger

Let me begin with a general remark on Leibniz's reputation. If one wants to set down clearly an outcome of the priority dispute between Newton und Leibniz, independent of national differences, then one has presumably to concede that Leibniz was the loser, at least inasmuch as his adherents had to defend him constantly in the 18th und 19th centuries. Even in Germany the reputed historian of mathematics Moritz Cantor assumed that Leibniz had tried to forge the date of one of his documents in infinitesimal calculus [Cantor 1901, 182–183].

In the 20th century the books by Couturat, Russell and Cassirer promoted the interest in Leibniz's philosophy, and in the wake of the relativity theory Reichenbach's hypothesis that Leibniz's theory of space and time should be assessed as a precursor may have reinforced this trend. But Child renewed the accusation of plagiarism, and it was only in 1949 that Hofmann's book (perhaps only the English translation of 1974) brought this discussion to an end. Nonetheless Leibniz remained a somewhat dubious figure in the English-speaking world and the development of the non-standard analysis by Schmieden and Laugwitz on the one hand and Robinson on the other did not really succeed in changing the situation. But during the past two or three decades Leibniz has suddenly risen considerably in general estimation, even in Great Britain and the USA. If Leibniz has been unfairly treated for three hundred years, then we can safely say that this is hardly the case today, whereas Newton, at least in Djerassi's well-known play about the priority dispute, is definitely not being dealt with fairly. This amazing development is presumably connected with a more general change in the intellectual atmosphere; I am referring to the triumphant career of the personal computer. Although the binary numerical system is of lesser significance as a mathematical accomplishment, it is being mentioned and explained everywhere; moreover there are also Leibniz's logical calculi, his project of a characteristica universalis and his observations on artificial intelligence.

The general interest in or even the enthusiasm about the computer has changed the whole climate of thought considerably. The renowned physicist John Archibald Wheeler has formulated a number of "really big questions", as he called them; one of them is "It from Bit?" Can the concept of information become a central concept of physics? Subsequent to Wheeler's contribution, physicists have been discussing this question using examples from numerous and diverse areas of physics. Baeyer's book *Information. The New*

Language of Science reports about these efforts that are still underway [cf. also Chaitin 2005, 61–62]. The well-known ecologist Jeremy Rifkin [1999, 211–212] speaks of the fact that in biology a new view of nature is emerging in which the computer is playing an outstanding role. Stephen Wolfram even proposes *A New Kind of Science*; this is meant to be a kind of natural science that explains complicated natural processes as a result of simple computer programs. I do not know whether Wolfram's book is really a serious contribution to sciences, but I mention him because Chaitin mentions his book as an important contribution to the discussion [Chaitin 2004, 282, 283; Chaitin 2005, 15, 157; Chaitin 2005, 59–60, 200]. Finally, there are also philosophers who imagine a new philosophical approach based on computers: The Italian philosopher Ugo Pagallo wrote an *Introduzione alla filosofia digitale*, and the subtitle of his book reads *Da Leibniz a Chaitin*, – from Leibniz to Chaitin. Chaitin too speaks on occasion of a digital philosophy – an idea that is suggested by the computer [Chaitin 2005, XIII] and in the final analysis can be traced back to Leibniz [Chaitin 2005, 117, 142].

So I have now arrived at my topic. In the first section I would like to consider Chaitin's mathematical achievements and his philosophical considerations, in the second section I will deal with his reference to Leibniz. In the third section I will look into a topic in which Chaitin does not refer to Leibniz, although he might well have done so.

In the 1960s and 70s Gregory Chaitin (born 1947) and Andrej Kolmogorov developed the mathematical theory of automata; I restrict my observations here to Chaitin. Chaitin begins by defining the complexity of a binary string. The complexity is defined as the size (measured in bits) of the shortest computer program that is necessary to produce this string as the output of a computer. Naturally I am skipping various specifications here that are necessary for this definition to meet the demands of mathematical rigour. Because the number π can be calculated by a simple infinite series, the binary notation of π is of a very small degree of complexity. The example also shows that the time needed for the calculation (π does have an infinite binary notation) is irrelevant for this definition. If on the other hand one produces a binary string by means of ten thousand throws of a coin, then the complexity of this string lies somewhere in the order of 10 000, because hardly any kinds of simplification will be possible: the computer program will more or less have to consist of the explicit enumeration of the noughts and ones that resulted from the coin-tossing. From this one derives a fairly obvious definition of chance: a binary string of the length n is called random, if its complexity lies in the order of n.

It is likewise an obvious step to apply this to physical theories. One thinks of theory as an important part of a computer program that has the observation data as its output. If the complexity of the physical theory is not much less than the length of the observation data as a binary string, then it is evidently a bad theory. One expects of a good theory that it has far less complexity, i.e., that it is capable of compressing the observation data. The compression of images is after all for all of us something we experience every day: for example if in a picture the sky has a uniform blue colour, then we need less information to describe this area; the picture can by compressed.

Chaitin also applies his considerations more specifically to mathematical theories. The axioms of a mathematical theory, including the rules for logical conclusions, can be seen as the compression of theorems that this theory can derive. This at least is how we are used to see the relationship of axioms and theorems. One always implicitly assumes that the axioms are simple. For this conventional idea Chaitin likes to refer to Hilbert, occasionally also to Leibniz: a manageable area of axioms should be sufficient to derive all mathematical knowledge [Chaitin 2005, 118; Chaitin 1998, 2, 5].

Now Chaitin is interested in precisely those areas of mathematics in which the axioms cannot be regarded as a compression of the theorems. Chaitin calls a mathematical statement irreducible, if it can only be derived from an equally complex set of axioms, rather than from a simple set of axioms, in other words more or less only from itself. In such a case deriving a statement from axioms is simply futile. For mathematicians this is of course a rather worrying idea; it means that there are mathematical statements that in the sense of the definition mentioned above are "random". These are mathematical facts that are "true for no reason" [Chaitin 2005, 129]. Chaitin, who not only possesses a considerable skill for enthusing about something, but also remarkable rhetorical skills, finds catchy formulations for this such as "The Limits of Mathematics", as one of his books is called [Chaitin 1998], or even "The Limits of Reason", as the title of one of his articles in the *Scientific American* [Chaitin 2006] reads.

A first example of such an irreducible statement is the value of the real number Ω that Chaitin introduced and that one could also call Chaitin's number (Chaitin 2006; Chaitin 2005, 129–141). This number is the probability that a randomly selected Turing machine will halt. For simplicity's sake in what follows the word "program" will be used in connection with a Turing machine.

In order that the number Ω be well defined, one has to decide on a certain programming language and comply with a few further technical details that I will not look into here. In binary notation Ω is defined as an infinite sum, whereby every program of the length of N bits that halts, contributes exactly $\left(\frac{1}{2}\right)^{N}$ to this sum. In other words: every program of the length of N bits that halts, contributes exactly 1 to the binary notation of Ω at the Nth place. This is an indisputable definition, but it is impossible to calculate Chaitin's number. If one were able to calculate the number, then one could also solve Turing's halting problem, but we know that this problem is not capable of being resolved. It is perfectly possible to calculate several digits of the binary expansion of Chaitin's number, but the first N digits cannot be calculated with a program that is markedly shorter than N bits. If a particular program for calculating the first N digits in the binary notation of Chaitin's number is given, then an infinite number of digits will always remain that the program cannot calculate.

So Chaitin's number is irreducible. From this one can immediately conclude that in a formal system one can only calculate, with the exception of a constant, as many places in Chaitin's number as the formal system's complexity states (if this were not the case, then Chaitin's number would be reducible). If one wants to calculate more places, then one must feed more information into the system of axioms. So we have here the same situation as in Gödel's incompleteness theorem; Chaitin talks of an "extremely strong incompleteness result" [Chaitin 2005, 132].

Until now one might still regard Chaitin's number and equally Gödel's incompleteness theorem as an oddity in some border areas of mathematics. But Chaitin has placed his number in relation to Diophantine equations, and that does lie in the field of everyday mathematics. Hilbert's tenth problem was: is there a method of deciding whether a given Diophantine equation is solvable? Matijasevič showed in 1970 that Hilbert's 10th problem is equivalent to the Turing halting problem. Assuming a Turing machine is given, then one can construct a Diophantine equation which has a solution if this Turing machine halts, and only then. Vice versa, if a Diophantine equation is given, then one can construct a Turing machine that halts if the equation is solvable, and halts only then [Chaitin 2002, 84]. Chaitin carried on from this result and explicitly specified a Diophantine equation that has 17 000 positive integer variables and is 200 pages long when printed out. One of the variables is a parameter, so that one can also look at the equation as a system of an infinite number of equations. This system of equations differs from a Diophantine equation in the conventional sense in that the variables can also occur in the exponents. This system of equations thus relates to Chaitin's number in the following manner: if the Nth place of Ω is a 0, then the equation belonging to the parameter N has a finite number of solutions; if the Nth place is 1, then this equation has an infinite number of solutions [Chaitin 2002, 122–123; Chaitin 2005, 135–139].

So due to Chaitin's results the problems of undecidability und incompleteness have drawn very much closer to normal mathematics. Added to this is the fact that in past decades mathematics (likewise under the influence of the computer) has been devoting itself to problems of such complexity that dealing with them would previously have been regarded as a hopeless undertaking. This is true of engineering science and climatology; but it also holds true for the proof of the four-colour theorem and questions of encoding with very large prime numbers. Chaitin therefore proposes that mathematicians should change the way they see themselves; it is simply not the case that relatively few and relatively simple axioms suffice to derive all mathematical knowledge. Our traditional concept of mathematics is not determined by the objective structure of mathematical knowledge, but simply by the tools that mathematicians have hitherto used. In the words of Chaitin: "And the map of our mathematical knowledge resembles a highway running through the desert or a dangerous jungle; if you stray off the road, you'll be hopelessly lost and die!" [Chaitin 2005, 24–25].

Because according to Chaitin the problems of incompleteness, first discovered by Gödel, have drawn very close to normal mathematics, it is only logical that Chaitin should plead for Lakatos's quasi-empiricism [Chaitin 1998, 26, 82–83; Chaitin 2002, 69–70; Chaitin 2005, 141; Chaitin 2006, 60]. Axioms are not necessarily self-evident truths; mathematicians should sooner proceed like physicists, i.e., introduce new axioms every now and then. The Riemann hypothesis is, he believes, a suitable candidate for a new axiom [Chaitin 1998, 23, 86; Chaitin 2002, 77].

Another candidate for a new axiom would be the statement (phrased here rather colloquially) that the class of the quickly solvable problems and the class of problems with quickly verifiable solutions differ from one another [Chaitin 2006, 61]. Encoding with very large prime numbers is based precisely on the assumption, as yet unproven, that one can quickly determine whether one of two given natural numbers is the divisor of the other, but that one can by no means quickly draw up a list of the prime factors of a given number. Naturally one must, Chaitin concedes, also reckon with the possibility in such a quasi-empirical procedure that one must later withdraw an axiom.

I would like now to turn to Chaitin's references to Leibniz. The first thing that a visitor of Chaitin's home page sees (http://www.umcs.maine.edu/˜chaitin/) is not the name Chaitin, it is a portrait of Leibniz and a quotation from Leibniz's Discours de metaphysique. Similarly an article by Chaitin in the *Scientific American* on his number Ω awakens the impression when skimming over the contents that one might find all this in Leibniz's works [Chaitin 2006, 54, 57, 59].

Chaitin reports – with a twinkle in his eye – that Leibniz beat him to it: "In fact, you can only really appreciate Leibniz if you are at *his* level. You can only realize that Leibniz has anticipated you *after* you've invented a new field by yourself" [Chaitin 2005, 58]. He had been invited to Bonn for the German Congress of Philosophy (2002) and remembered while preparing himself for his talk that he had read an interesting Leibniz quote in Hermann Weyl [Weyl 1926, 146–147; Weyl 1932, 40–42; Weyl 1949, 191; Weyl 1976, 243] that he had not hitherto looked into any further. Leibniz had made a "key observation about complexity and randomness" [Chaitin 2005, 58; cf. also Chaitin 2004, 179, 278; Chaitin 2006, 57] and his own reflections in the 1960s had begun with precisely this observation.

In paragraph 6 of the Discours de metaphysique Leibniz explains that one can draw a curve through dots distributed randomly on a piece of paper following a uniform rule and he continues: "quand une règle est fort composée, ce que luy est conforme, passe pour irregulier" [A VI, 4, 1537–1538]. From this it may be concluded: however God had made the world, it would always have been regular and followed a certain order. But God has created the world in the most perfect manner and hence with the simplest laws and the greatest wealth of phenomena. Likewise Leibniz called it a fundamental rule of his philosophical system of thought [A VI, 6, 490] that the variety of manifestations should be derived from just a few principles. What interested Hermann Weyl and Chaitin about this place in the Discours de metaphysique is clearly the insight that what is important is not the existence of laws or a uniform rule, it is the law or rule's simplicity. If the law or rule is of similar complexity to that of the phenomena produced, then the law or rule is uninteresting. This certainly relates to an important idea of Chaitin, but the general atmosphere of the computer age presumably contributes to attributing a fundamental aspect to Leibniz's comment.

I would now like to point out briefly a few mistakes of slight significance. As far as I can yet say in view of the unpublished estate, Leibniz did not prove Wilson's theorem, as Chaitin maintains [Chaitin 2005, 57; Dantzig 1947, 51]; he merely assumed the theorem inductively and then due to a miscalculation he considered the correct supposition to be wrong and changed it into an incorrect supposition [Mahnke 1912, 42, footnote]. In addition, Chaitin sees a relationship between human limitations, divine transcendence and Leibniz's use of the word "transcendent" [Chaitin 2005, 95]. However, Leibniz uses the mathematical term "transcendent" simply in the sense that every degree of an algebraic equation is transcended. An equation with the square root of 2 in the exponent of the variable is called by Leibniz accordingly "interscendent" [Breger 1986]; this terminology admittedly has not been able to assert itself. Finally, I don't know whether the Leibniz quotation that Chaitin places at the beginning of his book *Meta Math!* really is from Leibniz, at least I know of no proof. But Leibniz definitely did say something similar, namely "Les Mathematiciens ont autant besoin d'estre philosophes que les philosophes d'estre Mathematiciens" [GP I, 356].

Chaitin's enthusiasm for Leibniz even led him to call into question the recognised text printed in the Academy edition of Leibniz's Discours de metaphysique. In paragraph 6 of the Discours de metaphysique we read "Je dis qu'il est possible de trouver une ligne geometrique dont la notion soit constante et uniforme suivant une certain regle;" [A VI, 4, 1538]. According to Chaitin this is obviously a printing error [Chaitin 2004, 279]; it ought to read "Je dis qu'il est possible de trouver une ligne geometrique dont la motion soit constante et uniforme suivant une certaine regle". The meaning of "motion" in the sense of spatial movement is little known; but it did exist in the French language [Littré 1971, 480]. When you first look at it, Chaitin's correction appears meaningful, but five lines later Leibniz is already talking again about "notion", and this time "motion" cannot be meant. To be sure, a few lines later on Leibniz talks of "mouvement". But in fact this is his normal usage; he uses the word "mouvement" to denote movement in space in French. Heinrich Schepers kindly searched for me in his data base of Leibniz texts; the outcome is that in his correspondence with Clarke, Leibniz did indeed use "motion" in the sense of movement in space [GP VII, 366, 377]; but then this is surely inspired by the use of the English word "motion" by his English correspondent. However, the decisive argument is to be found by consulting the four manuscripts in Leibniz's estate that contain the place in question. In all four manuscripts one finds "notion". Three of the manuscripts were written by a scribe, but Leibniz entered corrections in his own handwriting in these manuscripts and did not correct the word "notion" [LH IV, III, 7 fol. 2 v°; LH I, III, 1 fol. 14 v°; LH I, III, 1 fol. 23 v°]. The fourth manuscript is in Leibniz's own handwriting (LH I, III, 1 fol. 2 v°). So we can continue to use the present text of the Discours de metaphysique.

Finally, I turn to Chaitin's remarks on real numbers. When talking on this topic Chaitin does not mention Leibniz; but he could have done, as I would now like to demonstrate.

In the course of his life Chaitin arrived at a certain doubt as to whether natural numbers of any size do really exist or in what sense it might be the case [Chaitin 1998, 85]. "I don't really believe in real numbers anymore and I don't even think I believe in positive integers anymore." [Chaitin 2002, 47]. No physicist has ever measured a number with more than twenty places behind the decimal point [Chaitin 2002, 48, 129]. Chaitin's critical reflections on the concept of the real number go back among other things to a treatise by Emile Borel [Chaitin 2005, 105–116, 121–125; Chaitin 2002, 128–129]. This affinity is naturally no matter of chance: Borel's manner of reasoning, which rests on concrete examples and questions of calculation or on explicit stateability, fits in particularly well with Chaitin's way of thinking, which is influenced by the computer and is hence "constructivist". A type of thinking influenced by the computer is naturally not particularly platonic.

Borel showed that if one really believes in the concept of the real number as an infinite sequence of numbers then one might put the sum of human knowledge into one single real number. One might for example digitalise the *Encyclopaedia Britannica* (something that happens anyway with every text before it is printed) and interpret this as a binary notation of a real number. That is nothing to get excited about, for this is only a finite amount of information and the real number that thus arose would even be rational. But Borel went further in 1927 and looked at an infinite amount of information. One can count all sorts of texts in French, first all texts with only one character, then all texts with two characters etc. Then one can define Borel's number as follows: let the nth digit in the decimal expansion of this number be 0, if the nth text is meaningless; let it be 1, if the nth text is not a yes-or-no question; let it be 2, if it is such a question, but this question cannot be answered; let it be 3 or 4, if the text is a yes-or-no question and the answer is yes or no.

Borel posed the question whether there is a reason why we should regard the existence of such a real number as possible, and his answer was "no". Turing showed in 1936 that the amount of calculable real numbers is countable, since the amount of possible programs on a Turing machine is countable. Since there is an uncountable number of real numbers, thus most of the real numbers are not calculable. To put it more plainly: a randomly chosen real number is not calculable with the probability of 1 [Chaitin 2005, 109]. By the same argument it can be shown that a randomly chosen real number cannot be named with a probability of 1; there is no way, whether constructively or non-constructively, of characterising this particular real number in such a way that it can be distinguished from the other real numbers [Chaitin 2005, 113]. "In summary: Why should I believe in a real number if I can't calculate it, if I can't prove what its bits are, and if I can't even refer to it? And each of these things happens with probability one!" [Chaitin 2005, 115].

From these remarks one begins to piece together an intuitive picture of real numbers: they form a sticky mass, from which only a countable amount of individual numbers can be extracted. Precisely this fact is reflected in Leibniz's theory of the continuum, which for the most part transfers the Aristotelian theory of the physical continuum to mathematics. According to Leibniz the continuum is not given as a number of individual points, it is given as a whole, in which only in the course of a mathematician's reflections do individual points emerge in the continuum through division [Breger 1992]. In doing so, clearly, at the most a countable number of divisions can be carried out and hence at the most a countable number of points emerges. Naturally Leibniz did not have the terms "countable" and "uncountable" at his disposal, but he was aware that the divisions could never all be conducted at the same time, in other words, that the continuum can never be given as a number of individual points.

In his notes De libertate, contingentia et serie causarum, providentia, Leibniz reports that he had succeeded in solving the problem of freedom by an analogy with the continuum. I do not propose to expand on the problem of freedom; I am only interested here in what Leibniz says in these notes about infinite decimals. An infinite decimal is for Leibniz in-finite in the literal sense, i.e., nothing positive, rather something for which there is no end. Even God cannot therefore successively progress through the individual places of an infinite decimal fraction expansion. A mathematical proof that were based on such a successive progression through the whole decimal fraction expansion is impossible for God too [A VI, 4, 1658]. Quite in contrast to the assertion repeatedly found in the secondary literature that Leibniz used an actual infinity in mathematics, it must be plainly stated that

Leibniz was quite serious about a mathematical infinite as something which fundamentally cannot be concluded. In Leibniz's metaphysics this results in a small modification that I find important to point out here. Because in reality, according to Leibniz, everything is individual, and every individual is characterised by an infinity of determinations [A VI, 6, 230–231, 289–290, 57], for God the mathematician things are also basically too complex. Determining the best of all possible worlds cannot happen mathematically in the strict sense, it can only happen "infallibili visione" [A VI, 4, 1658]. This observation from the year 1689 is in strong contrast to the well-known remark made in 1677 "Cum Deus calculat et cogitationem exercet, fit mundus." [GP VII, 191]. So if Chaitin speaks of the limits of mathematics [Chaitin 1998; Chaitin 2002, 102], one can safely say that for Leibniz too mathematics comes up against limits that are caused by complexity.

Acknowledgement

I would like to thank James G. O'Hara, who drew my attention to Chaitin.

References

[Baeyer 2005] Baeyer, Hans Christian von, *Information. The New Language of Science*, London 2003.

[Breger 1986] Breger, Herbert, Leibniz' Einführung des Transzendenten. In: 300 *Jahre 'Nova Methodus' von Leibniz* (1684–1984) (= *Studia Leibnitiana*, Sonderheft 14), ed.: A. Heinekamp, Stuttgart, pp. 119–132.

[Breger 1992] Breger, Herbert, Le continu chez Leibniz. In: *Le labyrinthe du continu*, eds.: Salanskis/Sinaceur, Paris 1992, pp. 76–84.

[Cantor 1901] Cantor, Moritz, *Vorlesungen zur Geschichte der Mathematik*, vol. 3, second edition, Leipzig 1901.

[Chaitin 1986a] Chaitin, Gregory, Information-Theoretic Computational Complexity, in: Tymoczko (ed.): *New Directions in the Philosophy of Mathematics*, Boston, Basel, Stuttgart 1986, pp. 287–299.

[Chaitin 1986b] Chaitin, Gregory, Gödel's Theorem and Information, in: Tymoczko (ed.): *New Directions in the Philosophy of Mathematics*, Boston, Basel, Stuttgart 1986, pp. 300–311.

[Chaitin 1998] Chaitin, Gregory, *The Limits of Mathematics*, Singapore 1998.

[Chaitin 2002] Chaitin, Gregory, *Conversations with a Mathematician*, London 2002.

[Chaitin 2004] Chaitin, Gregory, Leibniz, Information, Math and Physics, in: *Knowledge and Belief. Proceedings of the 26th International Wittgenstein Symposium. Wissen und Glauben. Akten des 26. Internationalen Wittgenstein-Symposiums*, eds: Löffler/Weingartner, Vienna 2004, pp. 277–286.

[Chaitin 2005] Chaitin, Gregory, *Meta Math! The Quest for Omega*, New York 2005.

[Chaitin 2006] Chaitin, Gregory, The Limits of Reason, *Scientific American*, March 2006, pp. 54–61.

[Dantzig 1947] Dantzig, Tobias, *Number. The Language of Science. A Critical Survey Written for the Cultured Non-Mathematician*, third edition, London 1947 (first edition 1930).

[Littré 1971] Littré, Emile, *Dictionnaire de la langue française*, Paris, volume 5.

[Mahnke 1912] Mahnke, Dietrich, Leibniz auf der Suche nach einer allgemeinen Primzahl-gleichung, *Bibliotheca Mathematica. Zeitschrift für Geschichte der mathematischen Wissenschaften*, 3rd series, vol. XIII, 1912, pp. 29–61.

[Pagallo 2005] Pagallo, Ugo, *Introduzione alla filosofia digitale. Da Leibniz a Chaitin*, Torino 2005.

[Rifkin 1999] Rifkin, Jeremy, *The Biotech Century. How Genetic Commerce Will Change the World*. London 1999.

[Weyl 1926] Weyl, Hermann, Philosophie der Mathematik und Naturwissenschaft, in: *Handbuch der Philosophie*, eds: A. Baeumler/M. Schröter, Abteilung II (Natur, Geist, Gott), Munich and Berlin 1926. pp. 1–162.

[Weyl 1949] Weyl, Hermann, *Philosophy of Mathematics and Science*, Princeton 1949.

[Weyl 1976] Weyl, Hermann, *Philosophie der Mathematik und Naturwissenschaft*. Darmstadt 1976 (4th edition).

[Weyl 1932] Weyl, Hermann, *The Open World*, New Haven 1932.

[Wolfram 2002] Wolfram, Stephen, *A New Kind of Science*, Champaign (Illinois) 2002.

Herbert Breger
Eichstr. 7
D-30161 Hannover, Germany
Herbert.Breger@gmx.de

Abbreviations

A   *Gottfried Wilhelm Leibniz: Sämtliche Schriften und Briefe*, Deutsche Akademie der Wissenschaften (Darmstadt and Berlin: Akademie-Verlag, 1923–), cited by series, volume, and page number.

C   *Opuscules et fragments inédits de Leibniz*, ed. L. Couturat (Paris: Alcan, 1903).

Dutens   Leibniz: *Opera omnia nunc primum collecta in Classes distributa praefationibus & indicibus exornata*, studio Ludovici Dutens, 6 vols., Fratres de Tournes: Geneva 1768.

E   *Leibnitii opera philosophica quae extant Latina Gallica Germanica omnia*, 2 vols., ed. J.E. Erdmann, Berlin 1839/40.

GM   *Leibnizens Mathematische Schriften*, ed. C.I. Gerhardt (Berlin: Asher and Schmidt, 1849–63), 7 vols.

GP   *Die Philosophischen Schriften von Leibniz*, ed. C.I. Gerhardt (Berlin: Weidman, 1875–90), 7 vols.

Grua   *G.W. Leibniz: Textes inédits*, ed. by G. Grua (Paris: Presses universitaires de France, 1948), 2 vols.

LH   Niedersächsische Landesbibliothek Hannover: Leibniz-Handschriften

LBr   Niedersächsische Landesbibliothek Hannover: Leibniz-Briefwechsel

Author Index

Subject Index